An Introduction to Quiver Representations

GRADUATE STUDIES IN MATHEMATICS 184

An Introduction to Quiver Representations

Harm Derksen
Jerzy Weyman

American Mathematical Society
Providence, Rhode Island

2010 *Mathematics Subject Classification.* Primary 16G20, 16G10, 16G70, 14L24, 13A50.

For additional information and updates on this book, visit
www.ams.org/bookpages/gsm-184

Library of Congress Cataloging-in-Publication Data

Derksen, Harm, 1970- | Weyman, Jerzy, 1955-
Title: An introduction to quiver representations / Harm Derksen, Jerzy Weyman.
Description: Providence, Rhode Island : American Mathematical Society, [2017] | Series: Graduate studies in mathematics ; volume 184 | Includes bibliographical references and index.
Identifiers: LCCN 2017016426 | ISBN 9781470425562 (alk. paper)
Subjects: LCSH: Directed graphs. | Representations of graphs. | Vector spaces. | AMS: Associative rings and algebras – Representation theory of rings and algebras – Representations of Artinian rings. msc | Associative rings and algebras – Representation theory of rings and algebras – Auslander-Reiten sequences (almost split sequences) and Auslander-Reiten quivers. msc | Algebraic geometry – Algebraic groups – Geometric invariant theory. msc | Commutative algebra – General commutative ring theory – Actions of groups on commutative rings; invariant theory. msc
Classification: LCC QA166.15 .D47 2017 | DDC 512/.46–dc23
LC record available at https://lccn.loc.gov/2017016426

∞ The paper used in this book is acid-free and falls within the guidelines
established to ensure permanence and durability.
Visit the AMS home page at `http://www.ams.org/`

10 9 8 7 6 5 4 3 2 1 22 21 20 19 18 17

Contents

Preface

The idea of writing this book came from the collaboration of the authors. We started working on quiver representation when the first author visited Northeastern University as a Research Scholar in the academic year 1997/1998. We both came from other fields of algebra so we had to learn the subject from scratch. At the same time we were interested in connections of quiver representations with invariant theory, representations of algebraic groups, and algebraic geometry.

This experience made us realize that (except for Crawley-Boevey's notes [**18**] available online), there is no introductory text which would allow a person without any knowledge of Artin algebras to learn the subject quickly.

Later both authors taught courses on quiver representations at University of Michigan and Northeastern University, respectively. The second author also taught a two month course on quiver representations at Tor Vergata University in Rome in the spring of 2007.

We are grateful to Calin Chindris, Jiarui Fei, Ryan Kinser, Andrea Appel, Andrew Carroll, Sachin Gautam, Daniel Labardini, Kavita Sutar, Salvatore Stella, Riccardo Aragona, Ada Boralevi, Cristina DiTrapano, Luca Moci, and other students who took part in these courses.

The book is addressed to non-specialists, who want to learn the subject without going through the extended preparation in algebra, just starting from basic linear algebra. It turned out to be impossible to be completely elementary, so in some places we use some basic algebraic geometry (mainly the dimension counts).

We do not prove the results in full generality, working over the field $\mathbb{C}$ of complex numbers. We also work mostly with acyclic quivers. We only cover Auslander-Reiten duality in the case of hereditary algebras.

The book reflects our point of view, so the semi-invariants are covered in detail and we stress their role in our approach. Some of the results could be proved just using stability conditions without mentioning semi-invariants, but we find the combinatorics of the rings of semi-invariants quite fascinating.

Still many important topics are left out, for example, Ringel-Hall algebras and Nakajima quiver varieties.

In recent years the field developed very quickly. New concepts and connections emerged. We wanted this development to be reflected in the book, hence there are chapters on orthogonal categories, exceptional sequences and cluster categories.

We stress the connections of quiver representations with representations of algebraic groups and moduli problems.

The authors thank the National Science Foundations for its support during the writing of this book.

We discussed the subject of the book with many mathematicians, with some of whom we coauthored our papers. We benefited from discussions with Prakash Belkale, Calin Chindris, Bill Crawley-Boevey, Jose Antonio de la Pena, Jiarui Fei, Lutz Hille, Kiyoshi Igusa, Ryan Kinser, Mark Kleiner, Visu Makam, Kent Orr, Charles Paquette, Idun Reiten, Claus Ringel, Ralf Schiffler, Aidan Schofield, Gordana Todorov, Michel Van den Bergh, and Andrei Zelevinsky.

The second author also wants to thank Piotr Dowbor, Daniel Simson and Andrzej Skowroński whose questions about the invariant theory of quivers aroused his interest in the subject.

Chapter 1

Introduction

1.1. Basic Definitions and Examples

A **quiver** is a directed graph where multiple edges and loops are allowed. Formally, we define the following.

Definition 1.1.1. A quiver Q is a quadruple $Q = (Q_0, Q_1, h, t)$ where Q_0 is a finite set of **vertices**, Q_1 is a finite set of **arrows**, and h and t are functions $Q_1 \to Q_0$. For an arrow $a \in Q_1$, $h(a)$ and $t(a)$ are called the **head** and the **tail** of a.

In examples, we will often use $1, 2, 3, \ldots$ or $x, y, \ldots$ for vertices, and $a, b, c, \ldots$ for arrows. We will write ha and ta instead of $h(a)$ and $t(a)$. This eases the notation.

Example 1.1.2. For the graph

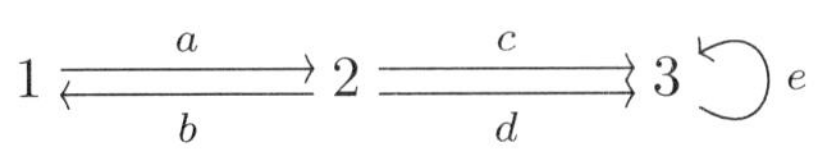

we have $Q_0 = \{1, 2, 3\}$, $Q_1 = \{a, b, c, d, e\}$, $ta = hb = 1$, $tb = ha = tc = td = 2$ and $hc = hd = he = te = 3$.

We call $Q' = (Q'_0, Q'_1, h', t')$ a **subquiver** of $Q = (Q_0, Q_1, h, t)$ if $Q'_0 \subseteq Q'_0$, $Q'_1 \subseteq Q_1$ and h and t restrict to h' and t' respectively.

Throughout this book, with the exception of Section 8.4 of Chapter 8, we will work over the field $\mathbb{C}$ of complex numbers. The representation theory of quivers can be done over an arbitrary field, but we will stick to $\mathbb{C}$ for simplicity.

Definition 1.1.3. We get a **representation** V of $Q = (Q_0, Q_1, h, t)$ if we attach to every vertex $x \in Q_0$ a finite dimensional $\mathbb{C}$-vector space $V(x)$ and to every arrow $a \in Q_1$ a $\mathbb{C}$-linear map $V(a) : V(ta) \to V(ha)$.

Definition 1.1.4. Suppose that V and W are representations of $Q = (Q_0, Q_1, h, t)$. We get a **morphism** $\phi : V \to W$ if we attach to every vertex $x \in Q_0$ a $\mathbb{C}$-linear map $\phi(x) : V(x) \to W(x)$ such that for every $a \in Q_1$ the diagram

$$\begin{array}{ccc} V(ta) & \xrightarrow{V(a)} & V(ha) \\ {\scriptstyle \phi(ta)}\downarrow & & \downarrow{\scriptstyle \phi(ha)} \\ W(ta) & \xrightarrow[W(a)]{} & W(ha) \end{array}$$

commutes, i.e., $\phi(ha)V(a) = W(a)\phi(ta)$.

The set of all morphisms from V to W is denoted by $\operatorname{Hom}_Q(V, W)$. A morphism $\phi = (\phi(x), x \in Q_0) \in \operatorname{Hom}_Q(V, W)$ can be thought of as an element in $\bigoplus_{x \in Q_0} \operatorname{Hom}(V(x), W(x))$. Consider the $\mathbb{C}$-linear map

$$d_W^V : \bigoplus_{x \in Q_0} \operatorname{Hom}(V(x), W(x)) \longrightarrow \bigoplus_{a \in Q_1} \operatorname{Hom}(V(ta), W(ha)) \tag{1.1}$$

defined by

$$d_W^V(\phi) = \big(\phi(ha)V(a) - W(a)\phi(ta), a \in Q_1\big),$$

where $\phi = (\phi(x), x \in Q_0)$. By definition, the kernel of d_W^V is $\operatorname{Hom}_Q(V, W)$. This shows that $\operatorname{Hom}_Q(V, W)$ is a finite dimensional $\mathbb{C}$-vector space. We will see later that the *cokernel* of d_W^V has an interesting interpretation as well.

Example 1.1.5. Consider the quiver

$$Q : 1 \xrightarrow{a} 2 .$$

A representation of Q is a linear map $V(a) : V(1) \to V(2)$ where $V(1)$ and $V(2)$ are finite dimensional $\mathbb{C}$-vector spaces. A morphism $\phi : V \to W$ is a pair $(\phi(1), \phi(2))$ such that $\phi(1) : V(1) \to W(1)$ and $\phi(2) : V(2) \to W(2)$ are linear, and $\phi(2)V(a) = W(a)\phi(1)$. The last condition is equivalent to the diagram

$$\begin{array}{ccc} V(1) & \xrightarrow{V(a)} & V(2) \\ {\scriptstyle \phi(1)}\downarrow & & \downarrow{\scriptstyle \phi(2)} \\ W(1) & \xrightarrow[W(a)]{} & W(2) \end{array}$$

being commutative.

Example 1.1.6. Consider the loop quiver

$$Q: 1 \circlearrowleft a .$$

A representation of Q is a finite dimensional $\mathbb{C}$-vector space $V(1)$ together with an endomorphism $V(a) : V(1) \to V(1)$. An element $\phi(1) \in \mathrm{Hom}(V(1), V(1))$ is a morphism if and only if it commutes with $V(a)$.

The **dimension vector** $\underline{\dim}(V)$ of a representation V is the function $Q_0 \to \mathbb{N}$ defined by $\underline{\dim}(V)(x) = \dim V(x)$ for all $x \in Q_0$. Here $\mathbb{N} = \{0, 1, \dots\}$ is the set of nonnegative integers. The set of dimension vectors is $\mathbb{N}^{Q_0}$, the set of all $\mathbb{N}$-valued functions on Q_0. For a dimension vector α, we define $|\alpha| = \sum_{x \in Q_0} \alpha(x)$.

1.2. The Category of Quiver Representations

We define a category $\mathrm{Rep}(Q)$ where the objects are representations of Q and the morphisms are as defined in the previous section. In this section (and the exercises) we will study this category and investigate the interpretations of monomorphisms, epimorphisms, isomorphisms, kernels, cokernels, products and coproducts in this category.

Suppose that $\phi : V \to W$ is a morphism. In Category Theory, ϕ is an **epimorphism** if for every object Z and every two morphisms $\psi_1, \psi_2 : Z \to W$ with $\psi_1\phi = \psi_2\phi$ we have $\psi_1 = \psi_2$. The morphism ϕ is called a **monomorphism** if for every object Z and every two morphisms $\psi_1, \psi_2 : W \to Z$ with $\phi\psi_1 = \phi\psi_2$ we have $\psi_1 = \psi_2$. We call ϕ an **isomorphism** if there exists a morphism $\psi : W \to V$ such that $\psi\phi$ and $\phi\psi$ are equal to the identity morphisms 1_V and 1_W respectively.

A morphism $\phi : V \to W$ is called injective (respectively surjective) if $\phi(x)$ is injective (respectively surjective) for all $x \in Q_0$. We call ϕ bijective if it is injective and surjective, i.e., if $\phi(x)$ is bijective for all $x \in Q_0$.

Lemma 1.2.1. *Suppose that $\phi : V \to W$ is a morphism of quiver representations.*

(a) *ϕ is injective if and only if ϕ is a monomorphism;*

(b) *ϕ is surjective if and only if ϕ is an epimorphism;*

(c) *ϕ is bijective if and only if ϕ is an isomorphism.*

Proof. The proof is Exercise 1.2.1. □

Definition 1.2.2. A representation W is called a **subrepresentation** of a representation V if $W(x)$ is a subspace of $V(x)$ for all $x \in Q_0$, and $V(a) : V(ta) \to V(ha)$ restricts to $W(a) : W(ta) \to W(ha)$ for all $a \in Q_1$.

Suppose that W is a subrepresentation of V. For every vertex $x \in Q_0$, let $\iota(x) : W(x) \to V(x)$ be the inclusion map. Then $\iota = (\iota(x), x \in Q_0)$ is an injective morphism.

Suppose that $\phi : V \to W$ is a morphism of representations of $Q = (Q_0, Q_1, h, t)$. For every $x \in Q_0$, define $K(x)$ and $C(x)$ to be the kernel and cokernel, respectively, of $\phi(x) : V(x) \to W(x)$. For every $x \in Q_0$, let $\iota(x) : K(x) \to V(x)$ be the inclusion and let $\pi(x) : W(x) \to C(x)$ be the projection. For every arrow $a \in Q_1$, we have

$$(\phi(ha) \circ V(a))(K(ta)) = (W(a) \circ \phi(ta))(K(ta)) = 0,$$

so $V(a)(K(ta)) \subseteq K(ha)$. This means that we can restrict $V(a)$ to a linear map $K(a) : K(ta) \to K(ha)$. So K is a subrepresentation of V. We call the subrepresentation K the **kernel** of ϕ. The inclusion morphism $\iota : K \to V$ is the kernel of $\phi : V \to W$ in the categorical sense (see Exercise 1.2.2).

In a similar fashion, there is a unique way to define $C(a)$ for all $a \in Q_1$ such that the projection $\pi : W \to C$ is a morphism. The quotient representation C is called the **cokernel** of ϕ. The quotient map $\pi : W \to C$ is a cokernel in the categorical sense (see Exercise 1.2.3).

Given two representations V and W, we can form the **direct sum** $V \oplus W$ as follows. For $x \in Q_0$ we define $(V \oplus W)(x) := V(x) \oplus W(x)$, and for every arrow $a \in Q_1$ the map

$$(V \oplus W)(a) : V(ta) \oplus W(ta) \to V(ha) \oplus W(ha)$$

is given by the matrix

$$\begin{pmatrix} V(a) & 0 \\ 0 & W(a) \end{pmatrix}.$$

There are natural inclusions $\iota_1 : V \hookrightarrow V \oplus W$ and $\iota_2 : W \hookrightarrow V \oplus W$ defined by

$$\iota_1(x) = \begin{pmatrix} 1_{V(x)} \\ 0 \end{pmatrix}, \iota_2(x) = \begin{pmatrix} 0 \\ 1_{W(x)} \end{pmatrix}.$$

With these inclusions, $V \oplus W$ is the **coproduct** in the categorical sense (see Exercise 1.2.5). There are also natural projections $\pi_1 : V \oplus W \to V$ and $\pi_2 : V \oplus W \to W$ defined by

$$\pi_1(x) = \begin{pmatrix} 1_{V(x)} & 0 \end{pmatrix}, \pi_2(x) = \begin{pmatrix} 0 & 1_{W(x)} \end{pmatrix}.$$

With these projections, $V \oplus W$ is the **product** in the categorical sense (see Exercise 1.2.4).

One can verify that $\operatorname{Rep}(Q)$ is an abelian category (see Exercise 1.2.6).

Exercises.

Exercise 1.2.1. *Prove Lemma* 1.2.1.

Exercise 1.2.2. *A morphism $\iota : K \to V$ is called the* **kernel** *of a morphism $\phi : V \to W$ if it has the following properties:*

(1) $\phi\iota = 0$;

(2) *if Z is any object and $\psi : Z \to V$ is any morphism with $\phi\psi = 0$, then there exists a unique morphism $\gamma : Z \to K$ such that $\psi = \iota\gamma$.*

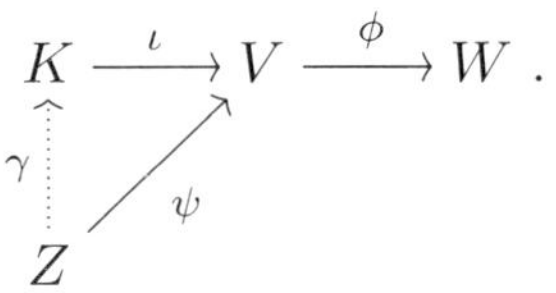

For a morphism $\phi : V \to W$ of quiver representations, let K be the kernel of ϕ, and let $\iota : K \hookrightarrow V$ be the inclusion. Show that $\iota : K \to V$ is a kernel of ϕ in the categorical sense. Also, show that every monomorphism between representations is the kernel of some morphism.

Exercise 1.2.3. *A morphism $\pi : V \to C$ is called the* **cokernel** *of a morphism $\phi : V \to W$ if it has the following properties:*

(1) $\pi\phi = 0$;

(2) *if Z is any object and $\psi : W \to Z$ is any morphism with $\psi\phi = 0$, then there exists a unique morphism $\gamma : C \to Z$ such that $\psi = \gamma\pi$.*

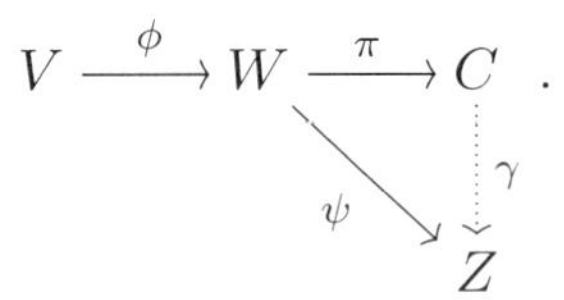

For a morphism $\phi : V \to W$ of quiver representations, let $C(x)$ be the cokernel of $\phi(x)$ for all $x \in Q_0$, and define $\pi(x)$ as the projection $W(x) \to C(x)$. Show that there is a unique way to define $C(a)$ for all $a \in Q_1$, such that π is a morphism. Show that $\pi : W \to V$ is a cokernel in the categorical sense. Also, show that every epimorphism between quiver representations is the cokernel of a morphism.

Exercise 1.2.4. *An object P with two morphisms $\pi_1 : P \to V$ and $\pi_2 : P \to W$ is called a product (in the categorical sense) of V and W if it has the following property. For every object Z and every two morphisms $\psi_1 : Z \to V$ and $\psi_2 : Z \to W$ there exist a unique morphism $\gamma : Z \to P$ such that $\psi_1 = \pi_1\gamma$ and $\psi_2 = \pi_2\gamma$. If V and W are two quiver representations, show that $V \oplus W$ together with the projections $\pi_1 : V \oplus W \to V$ and $\pi_2 : V \oplus W \to W$ form a product in the categorical sense.*

Exercise 1.2.5. *An object C with two morphisms $\iota_1 : C \to V$ and $\iota_2 \to W$ is called a coproduct (in the categorical sense) of V and W if it has the following properties: For every object Z and every two morphisms $\psi_1 : Z \to V$ and $\psi_2 : Z \to W$, there exists a unique morphism $\gamma : C \to Z$ such that $\psi_1 = \gamma\iota_1$ and $\psi_2 = \gamma\iota_2$. Show that the $V \oplus W$ together with the inclusions $\iota_1 : V \to V \oplus W$ and $\iota_2 : V \to V \oplus W$ form a coproduct.*

Exercise 1.2.6. *Verify that* $\operatorname{Rep}(Q)$ *is an abelian category.*

1.3. Representation Spaces

In this section we will see that the isomorphism classes of representations of a fixed dimension α have a nice geometric structure. They correspond to orbits in a certain representation $\operatorname{Rep}_\alpha(Q)$ of a certain algebraic group GL_α. This structure allows us to study the classification of quiver representations using (Geometric) Invariant Theory in later chapters. This geometric approach shows that for certain quivers, the classification of quiver representations of a given dimension corresponds to classical problems in linear algebra, such as studying matrices under left and right multiplication by invertible matrices, and studying matrices up to conjugation.

Suppose that $Q = (Q_0, Q_1, h, t)$ and α is a dimension vector. A **framed representation** of dimension α is a representation V for which $V(x) = \mathbb{C}^{\alpha(x)}$ for all $x \in Q_0$. For every $a \in Q_1$, $V(a)$ is an element of $\operatorname{Mat}_{\alpha(ha),\alpha(ta)}$, where $\operatorname{Mat}_{a,b}$ denotes the set of $a \times b$ matrices with complex entries. Define **representation space** of α-dimensional representations by

$$\operatorname{Rep}_\alpha(Q) = \prod_{a \in Q_1} \operatorname{Mat}_{\alpha(ha),\alpha(ta)} .$$

So $\operatorname{Rep}_\alpha(Q)$ can be viewed as the set of all framed representations of dimension α. If V is any representation of dimension α, Then we can choose an isomorphism $\phi(x) : V(x) \to \mathbb{C}^{\alpha(x)}$ for all $x \in Q_0$. If we define

$$W(a) = \phi(ha)V(a)\phi(ta)^{-1} \in \operatorname{Mat}_{\alpha(ha),\alpha(ta)},$$

then $W \in \operatorname{Rep}_\alpha(Q)$, and $\phi : V \to W$ is a morphism, because $W(a)\phi(ta) = \phi(ha)V(a)$ for all a. Since $\phi(x)$ is invertible for all $x \in Q_0$, ϕ is an isomorphism. So every representation is isomorphic to a framed representation.

Let GL_a denote the set of all invertible $a \times a$ matrices with complex entries. Define

$$\mathrm{GL}_\alpha = \prod_{x \in Q_0} \mathrm{GL}_{\alpha(x)} .$$

Suppose that $\phi = (\phi(x), x \in Q_0) \in \mathrm{GL}_\alpha$. The group GL_α acts on $\operatorname{Rep}_\alpha(Q)$ by

$$(\phi \cdot V)(a) := \phi(ha)V(a)\phi(ta)^{-1}, \quad a \in Q_1. \tag{1.2}$$

If $\phi : V \to W$ is an isomorphism, and $V, W \in \mathrm{Rep}_\alpha(Q)$, then

$$\phi(ha)V(a) = W(a)\phi(ta)$$

and

$$W(a) = \phi(ha)V(a)\phi(ta)^{-1} = (\phi \cdot V)(a)$$

for all $a \in Q_1$. So we have $W = \phi \cdot V$. Conversely, if $W = \phi \cdot V$, it is easy to see that $\phi : V \to W$ is an isomorphism. It follows that two representations $V, W \in \mathrm{Rep}_\alpha(Q)$ are isomorphic if and only if they lie in the same GL_α-orbit.

Corollary 1.3.1. *There is a bijection between the isomorphism classes of α-dimensional representations, and GL_α-orbits in $\mathrm{Rep}_\alpha(Q)$.*

Example 1.3.2. Consider the quiver

$$Q : 1 \xrightarrow{a} 2 .$$

If $\alpha = (m, n)$, then $\mathrm{Rep}_\alpha(Q) = \mathrm{Mat}_{n,m}$, and $\mathrm{GL}_\alpha = \mathrm{GL}_n \times \mathrm{GL}_m$. The group $\mathrm{GL}_n \times \mathrm{GL}_m$ acts on $\mathrm{Mat}_{n,m}$ by

$$(\phi_1, \phi_2) \cdot A = \phi_1 A \phi_2^{-1}.$$

Example 1.3.3. Consider the quiver

$$Q : 1 \circlearrowright a .$$

If $\alpha = (n)$, then $\mathrm{Rep}_\alpha(Q) = \mathrm{Mat}_{n,n}$, and $\mathrm{GL}_\alpha = \mathrm{GL}_n$. The action of GL_n on $\mathrm{Mat}_{n,n}$ is given by conjugation:

$$\phi \cdot A = \phi A \phi^{-1}, \quad \phi \in \mathrm{GL}_n, A \in \mathrm{Mat}_{n,n} .$$

Remark 1.3.4. Define $H_\alpha \subseteq \mathrm{GL}_\alpha$ as the set of all

$$(\lambda 1_{\alpha(x)}, x \in Q_0)$$

with $\lambda \in \mathbb{C}^\star = \mathbb{C} \setminus \{0\}$. Then H_α acts trivially on $\mathrm{Rep}_\alpha(Q)$.

Exercises.

Exercise 1.3.1. *Consider the quiver*

$$Q : a \circlearrowleft 1 \circlearrowright b$$

and let $\alpha = (n)$. What are $\mathrm{Rep}_\alpha(Q)$ and GL_α? How does GL_α act on $\mathrm{Rep}_\alpha(Q)$?

Exercise 1.3.2. *Consider the quiver Q with vertices $\{0, 1, \ldots, m\}$ and arrows $a_1, \ldots, a_m$, where $a_i : i \to 0$. Suppose that $V_1, V_2, \ldots, V_m$ are subspaces of $\mathbb{C}^n$. We can define a representation V of Q by $V(0) = \mathbb{C}^n$, $V(i) = V_i$ for all i, and $V(a_i) : V(i) \to \mathbb{C}^n$ is the inclusion map. Similarly, given subspaces $W_1, \ldots, W_m$ of $\mathbb{C}^n$ we can construct a representation W. Show that V and W are isomorphic if and only if there exists an invertible matrix $A \in \mathrm{GL}_n$ such that $A(V_i) = W_i$ for all i.*

1.4. Indecomposable Representations

Indecomposable representations are the building blocks in the representation theory of quivers. Every representation of a quiver is a direct sum of indecomposable representations. In Section 1.7 we will see that such a decomposition is essentially unique. For the problem of classifying matrices up to conjugation, indecomposable representations correspond to Jordan blocks in the Jordan normal form.

A representation V of $Q = (Q_0, Q_1, h, t)$ is called **trivial** if $V(x) = \{0\}$ for all $x \in Q_0$.

Definition 1.4.1. A nontrivial representation V is called **decomposable** if V is isomorphic to $V_1 \oplus V_2$ for some nontrivial representations V_1, V_2 of Q. A nontrivial representation that is not decomposable is called **indecomposable**. A nontrivial representation V is called **irreducible** or **simple** if for every subrepresentation W of V is trivial or equal to V.

For every $x \in Q_0$, we can define a simple representation S_x by $S_x(x) = \mathbb{C}$ and $S_x(y) = 0$ for $x \neq y$. It is clear that every simple representation of Q is indecomposable. Example 1.4.2 below shows that the converse is not true.

Example 1.4.2. As in Example 1.3.2, consider the quiver

$$Q : 1 \xrightarrow{a} 2 .$$

Suppose that V is an indecomposable representation of Q. If $V(1) = 0$, then $V(2) \neq 0$. If $\dim V(2) > 1$, then we can write $V(2) = W_1 \oplus W_2$ for two nonzero subspaces W_1, W_2 of $V(2)$. Therefore, V is isomorphic to

$$(0 \xrightarrow{0} W_1) \oplus (0 \xrightarrow{0} W_2),$$

so V is decomposable. Contradiction, so we have $\dim V(2) = 1$, and V is isomorphic to

$$0 \longrightarrow \mathbb{C} .$$

Similarly, if $V(2) = 0$, then V is isomorphic to

$$\mathbb{C} \longrightarrow 0 .$$

Suppose that $V(1)$ and $V(2)$ are both nonzero. Suppose that $V(a)$ is not injective. Let K be the kernel of $V(a)$ and choose a complement L to K in $V(1)$. Then V is isomorphic to the direct sum

$$(K \longrightarrow 0) \oplus (L \longrightarrow V(2)),$$

where the map $L \to V(2)$ is the restriction of $V(a)$. Contradiction, so $V(a)$ is injective. A similar argument shows that $V(a)$ is surjective. So $V(a)$ is an isomorphism. If $\dim V(1) > 1$, then we can write $V(1) = W_1 \oplus W_2$, where

W_1, W_2 are nonzero subspaces of $V(1)$. If we define $Z_i = V(a)(W_i)$, then $V(2) = Z_1 \oplus Z_2$ and V is isomorphic to

$$(W_1 \longrightarrow Z_1) \oplus (W_2 \longrightarrow Z_2),$$

where the maps $W_1 \to Z_1$ and $W_2 \to Z_2$ are restrictions of $V(a)$. Contradiction, so $\dim V(1) = \dim V(2) = 1$, and V is isomorphic to

$$\mathbb{C} \xrightarrow{\ 1\ } \mathbb{C}\,.$$

Up to isomorphism, the only indecomposable representations of Q are

$$\mathbb{C} \longrightarrow 0\,,\ 0 \longrightarrow \mathbb{C}\,,\ \mathbb{C} \longrightarrow \mathbb{C}\,,$$

and these representations are clearly indecomposable. The first two representations are simple, but the last one is not. If we define $W(1) = 0$ and $W(2) = \mathbb{C}$, then W is a proper subrepresentation, because obviously $V(a)(W(1)) = \{0\} \subseteq W(2)$.

Let $Z_{r,s,t}$ be the representation which is a direct summand of r copies of $\mathbb{C} \to \mathbb{C}$, s copies of $\mathbb{C} \to 0$ and t copies of $0 \to \mathbb{C}$. Then $Z_{r,s,t}(a)$ is given by the matrix

$$\begin{pmatrix} 1_r & 0_{r\times s} \\ 0_{t\times r} & 0_{t\times s} \end{pmatrix}, \tag{1.3}$$

where $0_{a\times b}$ denotes the $a \times b$ with zero entries. Our classification of the indecomposable representation implies that every representation V of Q is isomorphic to $Z_{r,s,t}$ for some r, s, t. The integers r, s, t are uniquely determined by V. (This also follows from the Krull-Remak-Schmidt theorem discussed in Section 1.7.)

Of course, this is a well-known result in linear algebra: Any linear transformation L between finite dimensional vector spaces has a matrix of the form (1.3) with respect to some choice of bases. Here r is the rank of the linear map L.

Example 1.4.3. As in Example 1.3.3, consider the quiver

$$Q:\ 1 \circlearrowright a\,.$$

Suppose that V is a representation of Q. For some choice of basis of $V(1)$, the matrix of $V(a)$ has the block form

$$\begin{pmatrix} J_1 & 0 & \cdots & 0 \\ 0 & J_2 & \cdots & 0 \\ \vdots & \vdots & \ddots & 0 \\ 0 & 0 & \cdots & J_r \end{pmatrix},$$

where J_i is a $d_i \times d_i$ **Jordan block** for $i = 1, 2, \ldots, r$. This is the **Jordan normal form** of $V(a)$. Let V_i be the representation

$$\mathbb{C}^{d_i} \xrightarrow{J_i} \mathbb{C}^{d_i} .$$

Then V is isomorphic to $V_1 \oplus V_2 \oplus \cdots \oplus V_r$. So if the Jordan normal form of $V(a)$ has more than 1 Jordan block, then V is decomposable. Conversely, if the V is decomposable, then the Jordan normal form has more than 1 Jordan block (see Exercise 1.4.1). So a representation V is indecomposable if and only if the Jordan normal form of $V(a)$ consists of a single Jordan block. Moreover, the uniqueness of the Jordan normal form shows that distinct Jordan blocks correspond to nonisomorphic representations. So we have a bijection between isomorphism classes of indecomposable representations and Jordan blocks.

Exercises.

Exercise 1.4.1. *Show that if V is a decomposable representation of the quiver*

$$1 \circlearrowright a ,$$

then the Jordan normal form of $V(a)$ has more than one Jordan block.

Exercise 1.4.2. *Show that every indecomposable representation of Q is simple if and only if Q has no arrows.*

Exercise 1.4.3. *Find two $n \times n$ matrices A and B such that the $\mathbb{C}$-algebra generated by A and B is $\mathrm{Mat}_{n,n}$. Define a representation of the quiver*

$$b \circlearrowleft 1 \circlearrowright a$$

by $V(1) = \mathbb{C}^n$, $V(a) = A$ and $V(b) = B$. Show that V is simple.

Exercise 1.4.4. *Consider the quiver*

$$Q : \quad \begin{array}{ccc} 1 & \xleftarrow{a} & 2 \\ {\scriptstyle d}\downarrow & & \uparrow{\scriptstyle b} \\ 4 & \xrightarrow[c]{} & 3 \end{array} .$$

For $\lambda \in \mathbb{C}^\star = \mathbb{C} \setminus \{0\}$, define a representation V_λ by $V_\lambda(1) = V_\lambda(2) = V_\lambda(3) = V_\lambda(4) = \mathbb{C}$, $V_\lambda(a) = V_\lambda(b) = V_\lambda(c) = 1_{\mathbb{C}}$ and $V_\lambda(d) = \lambda \cdot 1_{\mathbb{C}}$. Show that the $V_\lambda, \lambda \in \mathbb{C}^\star$ are simple and pairwise nonisomorphic.

Show that the only simple representations of Q are S_1, S_2, S_3, S_4 and V_λ, $\lambda \in \mathbb{C}$.

Exercise 1.4.5. *Consider the quiver*

$$1 \longrightarrow 2 \longrightarrow 3 .$$

Show that the indecomposable representations are

$$\mathbb{C} \longrightarrow 0 \longrightarrow 0\,,\ 0 \longrightarrow \mathbb{C} \longrightarrow 0\,,\ 0 \longrightarrow 0 \longrightarrow \mathbb{C}\,,$$

$$\mathbb{C} \longrightarrow \mathbb{C} \longrightarrow 0\,,\ 0 \longrightarrow \mathbb{C} \longrightarrow \mathbb{C}\,,\ \mathbb{C} \longrightarrow \mathbb{C} \longrightarrow \mathbb{C}\,.$$

1.5. The Path Algebra

In this section we will define an associative $\mathbb{C}$-algebra $\mathbb{C}Q$ for every quiver Q. We will show that the category $\operatorname{Rep}(Q)$ of quiver representations is equivalent to the category $\mathbb{C}Q$–mod of finite dimensional left $\mathbb{C}Q$-modules. This allows us to apply techniques for the theory of modules to quiver representations. In Chapter 3 we will study finite dimensional algebras and their modules. As we will see later, it suffices to study finite dimensional algebras that are of the form $\mathbb{C}Q/J$ where J is a certain two-sided ideal in $\mathbb{C}Q$. The role of path algebras in the theory of finite dimensional associative algebras is similar to the role of polynomial rings for commutative algebras.

Definition 1.5.1. A **path** p in a quiver $Q = (Q_0, Q_1, h, t)$ of length $\ell \geq 1$ is a sequence $p = a_\ell a_{\ell-1} \cdots a_1$ of arrows in Q_1 such that $ta_{i+1} = ha_i$ for $i = 1, 2, \ldots, \ell - 1$. We define $h(p) = hp = ha_\ell$ and $t(p) = tp = ta_1$. For every $x \in Q_0$, we introduce a **trivial path** e_x of length 0. We define $he_x = te_x = x$.

We think of paths as "going from right to left". A path $p = a_\ell a_{\ell-1} \cdots a_1$ corresponds to the concatenation

$$\circ \xleftarrow{a_\ell} \circ \xleftarrow{a_{\ell-1}} \cdots \xleftarrow{a_2} \circ \xleftarrow{a_1} \circ\ .$$

Suppose that V is a representation of Q. For a path $p = a_\ell a_{\ell-1} \cdots a_1$, we define

$$V(p) = V(a_\ell)V(a_{\ell-1}) \cdots V(a_1).$$

For $x \in Q_0$, we define $V(e_x) := 1_{V(x)} : V(x) \to V(x)$. For all paths p, q we have

$$V(p)V(q) = V(pq) \tag{1.4}$$

whenever $tp = hq$.

Definition 1.5.2. The **path algebra** $\mathbb{C}Q$ is a $\mathbb{C}$-algebra with a basis labeled by all paths in Q. We denote by $\langle p \rangle$ the element of $\mathbb{C}Q$ corresponding to the path p in Q. The multiplication in $\mathbb{C}Q$ is given by

$$\langle p \rangle \cdot \langle q \rangle = \begin{cases} \langle pq \rangle & \text{if } tp = hq, \\ 0 & \text{if } tp \neq hq. \end{cases}$$

Here pq denotes the concatenation of paths, and we use the conventions $pe_x = p$ if $tp = x$, and $e_x p = p$ if $hp = x$.

One easily verifies that the multiplication is associative, because the concatenation of paths is associative. We will often write p instead of $\langle p \rangle$ if there is no danger of confusion.

We have $e_x^2 = e_x$ for all $x \in Q_0$ and $e_x e_y = 0$ if $x \neq y$. So $e_x, x \in Q_0$ form an orthogonal set of idempotents. One easily verifies that $1_{\mathbb{C}Q} := \sum_{x \in Q_0} e_x$ is a 1-element in the associative algebra.

An **oriented cycle** is a nontrivial path p with $hp = tp$. A quiver without oriented cycles is called an **acyclic quiver**.

Lemma 1.5.3. *A quiver Q is acyclic if and only if $\mathbb{C}Q$ is a finite dimensional algebra.*

Proof. If p is an oriented cycle, then $p, p^2, p^3, \dots$ are distinct paths. So $\mathbb{C}Q$ is infinite dimensional. If Q is acyclic, then $\mathbb{C}Q$ is finite dimensional by Exercise 1.5.2. □

For an associative algebra A, we denote the category of finite dimensional left A-modules by A–mod, and the category of finite dimensional right A-modules by mod–A.

Proposition 1.5.4. *The categories* $\operatorname{Rep}(Q)$ *and* $\mathbb{C}Q$–mod *are equivalent.*

Proof. First we define a function $\mathcal{F} : \mathbb{C}Q\text{–mod} \to \operatorname{Rep}(Q)$. For a $\mathbb{C}Q$-module M we define $\mathcal{F}(M)$ as follows. We define $\mathcal{F}(M)(x) = e_x M$ for all $x \in Q_0$. Let $L_a : M \to M$ be the left-multiplication by a. Then we have

$$L_a(e_{ta}M) = ae_{ta}M = aM = e_{ha}aM \subseteq e_{ha}M.$$

We define $\mathcal{F}(M)(a) : e_{ta}M \to e_{ha}M$ as the restriction of L_a. If $\phi : M \to N$ is a homomorphism of $\mathbb{C}Q$-modules, then we define $\mathcal{F}(\phi) : \mathcal{F}(M) \to \mathcal{F}(N)$ as follows. For every $x \in Q_0$, we have

$$\phi(e_x M) = e_x \phi(M) \subseteq e_x N,$$

so we define $\mathcal{F}(\phi(x)) : e_x M \to e_x N$ as the restriction of ϕ. It is clear that $\mathcal{F}$ is a functor.

Conversely, we define a functor $\mathcal{G} : \operatorname{Rep}(Q) \to$ mod $\mathbb{C}Q$. For a representation V of Q, we define $\mathcal{G}(V) = \bigoplus_{x \in Q_0} V(x)$. If p is a path, and $v \in V(x)$, then we define

$$p \cdot v = \begin{cases} V(p)(v) & \text{if } tp = x, \\ 0 & \text{otherwise.} \end{cases}$$

From (1.4) follows that this gives $\mathcal{G}(V)$ the structure of a $\mathbb{C}Q$-module. If $\phi : V \to W$ is a morphism of representations, then we define $\mathcal{G}(\phi) : \bigoplus_{x \in Q_0} V(x) \to \bigoplus_{x \in Q_0} W(x)$ as the direct sum of all $\phi(x), x \in Q_0$. One can easily verify that $\mathcal{G}$ is a functor.

The compositions $\mathcal{F}\circ\mathcal{G}$ and $\mathcal{G}\circ\mathcal{F}$ are naturally equivalent to the identity. $\square$

We will show that certain tensor algebras are path algebras. This will be useful later in Section 3.3. Suppose that R is an algebra and M is an R-bimodule. We define a **tensor algebra**

$$T_R(M) = \bigoplus_{n=0}^{\infty} M^{\otimes n},$$

where $M^{\otimes 0} = R$, $M^{\otimes 1} = M$ and

$$M^{\otimes n} = \underbrace{M \otimes_R M \otimes_R \cdots \otimes_R M}_{n}$$

for all $n \geq 2$. The tensor algebra has the following universal property proven in Exercise 1.5.1

Proposition 1.5.5. *If A is an algebra, $\lambda_R : R \to A$ is an algebra homomorphism and $\lambda_M : M \to A$ is a R-bimodule homomorphism, then there exists a unique $\mathbb{C}$-algebra homomorphism $\lambda : T_R(M) \to A$ such that the restrictions of λ to R and M are λ_R and λ_M respectively.*

Suppose that $R = \mathbb{C}^n$. Let $Q_0 = \{1, 2, \ldots, n\}$ and define $e_i = (0, \ldots, 0, 1, 0, \ldots, 0)$ as the i-th basis vector. Then $e_1, e_2, \ldots, e_n$ are **orthogonal idempotents**, i.e., $e_i^2 = e_i$ and $e_i e_j = e_j e_i = 0$ for $i \neq j$. Suppose that M is a finite dimensional R-bimodule. We have a direct sum decomposition

$$M = \bigoplus_{i,j \in Q_0} e_j M e_i.$$

An element $a \in M$ is called homogeneous if $a \in e_j M e_i$ for some i and j. In this case, the head of a is $ha = j$ and the tail of a is $ta = i$. Choose a basis Q_1 of M consisting of homogeneous elements. Then $Q = (Q_0, Q_1, h, t)$ is a quiver.

A basis of $M^{\otimes n}$ is given by all paths

$$a_n \otimes a_{n-1} \otimes \cdots \otimes a_1$$

for which $a_n a_{n-1} \cdots a_1$ is a path in Q. It is now clear that $T_R(M)$ is isomorphic to the path algebra $\mathbb{C}Q$.

Exercises.

Exercise 1.5.1. *Prove Proposition* 1.5.5

Exercise 1.5.2. *Suppose that Q is an acyclic quiver. Show that one can label the vertices with $1, 2, \ldots, n$ such that $h(a) < t(a)$ for all arrows $a \in Q_1$. Use this to prove that $\mathbb{C}Q$ is finite dimensional.*

Exercise 1.5.3. *Show, using Exercise* 1.5.2, *that* S_x, $x \in Q_0$ *are (up to isomorphism) the only simple representations if* Q *is acyclic.*

Exercise 1.5.4. *Suppose that* Q *and* Q' *are acyclic quivers. Show that* $\mathbb{C}Q$ *and* $\mathbb{C}Q'$ *are isomorphic* $\mathbb{C}$*-algebras if and only if* Q *is isomorphic to* Q'.

Exercise 1.5.5. *Suppose that* Q *is a quiver and let* $\mathbb{C}$–mod *be the category of finite dimensional* $\mathbb{C}$*-vector spaces. Construct a category* $\mathcal{P}_Q$ *where the objects are the vertices in* Q_0, *and* $\mathrm{Hom}_{\mathcal{P}_Q}(y,x)$ *is the set of all paths in* Q *from* x *to* y *for* $x, y \in \mathbb{Q}_0$. *Show that the category* $\mathrm{Rep}(Q)$ *of quiver representations is equivalent to the category* $(\mathbb{C}\text{–mod})^{\mathcal{P}_Q}$ *of all functors* $\mathcal{P}_Q \to \mathbb{C}$–mod *(and where morphisms are natural transformations).*

1.6. Duality

Duality is a useful notion in category theory. On the level of algebras, the dual corresponds to the opposite algebra.

Definition 1.6.1. If A is an associative algebra, then its **opposite algebra** A^{op} is defined as follows. As a $\mathbb{C}$-vector space, A^{op} is equal to A, but A^{op} is equipped with a multiplication $\circ$ defined by

$$a \circ b := b \cdot a,$$

(here $\cdot$ is the multiplication for A).

Show that if M is a finite dimensional A-module, then the dual space $M^\star$ is an A^{op}-module as follows: For $u \in A^{\mathrm{op}}$ and $f \in M^\star$ we define $u \cdot f$ by $(u \cdot f)(v) = f(uv)$ for all $v \in M$. If $\phi : M \to N$ is a homomorphism of finite dimensional A-modules, then $\phi^\star : N^\star \to M^\star$ is a homomorphism of A^{op}-modules. This gives a canonical isomorphism of $\mathbb{C}$-vector spaces

$$\mathrm{Hom}_A(M,N) \cong \mathrm{Hom}_{A^{\mathrm{op}}}(N^\star, M^\star).$$

We have defined a contravariant functor $\mathcal{F} : A\text{–mod} \to A^{\mathrm{op}}\text{–mod}$. Similarly, we can define a contravariant functor $\mathcal{F}^{\mathrm{op}} : A^{\mathrm{op}}\text{–mod} \to A\text{–mod}$ and $\mathcal{F}^{\mathrm{op}} \circ \mathcal{F}$ and $\mathcal{F} \circ \mathcal{F}^{\mathrm{op}}$ are naturally equivalent to the indentity functors. This shows that the categories A–mod and A^{op}–mod are dual to each other.

If $Q = (Q_0, Q_1, h, t)$ is a quiver, then we define the **opposite quiver** by $Q^{\mathrm{op}} = (Q_0, Q_1, t, h)$, i.e., the heads and tails of the arrows are interchanged.

Exercises.

Exercise 1.6.1. *Show that there is a canonical isomorphism between* $(\mathbb{C}Q)^{\mathrm{op}}$ *and* $\mathbb{C}(Q^{\mathrm{op}})$.

Exercise 1.6.2. *Suppose that* U *is the algebra of upper triangular* $n \times n$ *matrices. Show that* U *and* U^{op} *are isomorphic.*

Exercise 1.6.3. *Let Q be the quiver*

$$\circ \longrightarrow \circ \longleftarrow \circ$$

and let $A = \mathbb{C}Q$. Show that A is not isomorphic to A^{op}.

1.7. The Krull-Remak-Schmidt Theorem

In this section we will prove the Krull-Remak-Schmidt theorem. This result says that every representation of a quiver can be written as a direct sum of indecomposable representations essentially in a unique way. The proof uses the structure of the endomorphism ring $\mathrm{End}_Q(V) = \mathrm{Hom}_Q(V, V)$ of an indecomposable representation V. We will show that such a ring is always local. To see this, we start by studying the generalized eigenspaces of an endomorphism.

Suppose that W is a finite dimensional $\mathbb{C}$-vector space, and $A : W \to W$ is an endomorphism. For $\lambda \in \mathbb{C}$, the **generalized eigenspace** for λ is

$$W_\lambda := \{w \in W \mid \text{for some } N > 0,\ (A - \lambda 1_V)^N(w) = 0\}.$$

It is well-known that W is the direct sum of its generalized eigenspaces:

$$W = \bigoplus_{\lambda \in \mathbb{C}} W_\lambda.$$

A similar decomposition can be done for quiver representations.

Suppose V is a representation of Q, and $\phi : V \to V$ is a morphism of quiver representations. For $\lambda \in \mathbb{C}$ and $x \in Q_0$, let $V(x)_\lambda$ be the generalized eigenspace of $\phi(x) : V(x) \to V(x)$ for eigenvalue λ. Since ϕ is a morphism,

$$\phi(ha)V(a) = V(a)\phi(ta),$$

If $w \in V(ta)_\lambda$, then $(\phi(ta) - \lambda 1)^N w = 0$ for some $N > 0$. From

$$0 = V(a)(\phi(ta) - \lambda 1)^N w = (\phi(ha) - \lambda 1)^N (V(a)w)$$

it follows that $V(a)w \in V(ha)_\lambda$. For every $\lambda \in \mathbb{C}$ we have $V(a)(V(ta)_\lambda) \subseteq V(ha)_\lambda$. So we can define a subrepresentation V_λ of V by $V_\lambda(x) = V(x)_\lambda$ for all $x \in Q_0$ and $\lambda \in \mathbb{C}$. We have an isomorphism

$$V \cong \bigoplus_{\lambda \in \mathbb{C}} V_\lambda. \tag{1.5}$$

Corollary 1.7.1. *If V is an indecomposable representation, and $\phi : V \to V$ is a morphism, then there exists a complex number $\lambda \in \mathbb{C}$ such that $\phi - \lambda 1$ is nilpotent.*

Proof. If V is indecomposable, then $V = V_\lambda$ for some $\lambda \in \mathbb{C}$ by (1.5). □

For a representation V of a quiver Q, the space $\mathrm{Hom}_Q(V, V)$ is a finite dimensional associative $\mathbb{C}$-algebra with 1 (where 1 is the identity).

Corollary 1.7.2. *Suppose that V is an indecomposable representation. For every $\phi \in \operatorname{Hom}_Q(V,V)$, either ϕ is invertible or ϕ is nilpotent. The set of $\mathfrak{m}$ of all noninvertible elements of $\operatorname{Hom}_Q(V,V)$ is the unique maximal (two-sided) ideal of $\operatorname{Hom}_Q(V,V)$ and $\operatorname{Hom}_Q(V,V)/\mathfrak{m} \cong \mathbb{C}$. In particular, $\operatorname{Hom}_Q(V,V)$ is a local ring.*

Proof. If $\phi \in \operatorname{Hom}_Q(V,V)$, then there exists a $\lambda \in \mathbb{C}$ with $\phi - \lambda 1$ nilpotent. If $\lambda = 0$, then ϕ is nilpotent. If $\lambda \neq 0$, then $\phi = (\phi - \lambda 1) + \lambda 1$ is invertible. If $\phi \in \mathfrak{m}$ and $\psi \in \operatorname{Hom}_Q(V,V)$, then ϕ is not invertible, so $\psi\phi$ and $\phi\psi$ are not invertible, and hence in $\mathfrak{m}$. This shows that $\mathfrak{m}$ is an ideal. It is clearly maximal because every element not in $\mathfrak{m}$ is invertible. The composition $\mathbb{C} \to \operatorname{Hom}_Q(V,V) \to \operatorname{Hom}_Q(V,V)/\mathfrak{m}$ is an isomorphism. □

To prepare for the theorem, we need the following technical lemma.

Lemma 1.7.3. *Suppose that V_1, V_2, W_1, W_2 are representations of Q, and*

$$\phi = \begin{pmatrix} \phi_{1,1} & \phi_{1,2} \\ \phi_{2,1} & \phi_{2,2} \end{pmatrix} : V_1 \oplus V_2 \to W_1 \oplus W_2$$

is an isomorphism, where $\phi_{i,j} : V_j \to W_i$. If $\phi_{1,1}$ is an isomorphism, then V_2 is isomorphic to W_2.

Proof. Let

$$\psi = \begin{pmatrix} 1_{W_1} & 0 \\ -\phi_{2,1}\phi_{1,1}^{-1} & 1_{W_2} \end{pmatrix}.$$

Then we have

$$\psi\phi = \begin{pmatrix} \phi_{1,1} & \phi_{1,2} \\ 0 & \phi'_{2,2} \end{pmatrix},$$

where $\phi'_{2,2} = \phi_{2,2} - \phi_{2,1}\phi_{1,1}^{-1}\phi_{1,2}$. Since $\psi\phi$ is invertible, we have that $\phi'_{2,2}$ is an isomorphism. □

Theorem 1.7.4 (Krull-Remak-Schmidt). *Every finite dimensional representation of a quiver Q is isomorphic to a direct sum of indecomposable representations. This decomposition is unique up to isomorphism and permutation of factors. More precisely, if*

$$V_1 \oplus \cdots \oplus V_p \cong W_1 \oplus \cdots \oplus W_r,$$

and $V_1, \ldots, V_p, W_1, \ldots, W_r$ are indecomposable, then $p = r$ and there exists a permutation σ of $\{1, 2, \ldots, p\}$ such that V_i is isomorphic to $W_{\sigma(i)}$ for $i = 1, 2, \ldots, p$.

Proof. We prove by induction on $|\alpha|$ that every representation of dimension α is a direct sum of indecomposable representations. If $|\alpha| = 1$, then every α-dimensional representation is indecomposable. Suppose that V is an

α-dimensional representation with $|\alpha| > 1$. If V is indecomposable, then we are done. Otherwise, $V = W \oplus Z$, where W and Z are nontrivial representations. We have $\alpha = \underline{\dim} V = \underline{\dim} W + \underline{\dim} Z$ and $|\alpha| = |\underline{\dim} W| + |\underline{\dim} Z|$. Since $0 < |\underline{\dim} W|, |\underline{\dim} Z| < |\alpha|$, it follows from the induction hypothesis that W and Z are direct sums of indecomposable representations. Hence V is a direct sum of indecomposable representations. This completes the induction proof.

We now have to show the uniqueness. Suppose that

$$\phi = (\phi_{i,j})_{i,j} : V_1 \oplus \cdots V_p \to W_1 \oplus W_2 \oplus \cdots \oplus W_r$$

is an isomorphism, where $V_1, \ldots, V_p, W_1, \ldots, W_r$ are indecomposable representations, and $\phi_{i,j} \in \operatorname{Hom}_Q(V_j, W_i)$ for all i, j with $1 \leq i \leq r$ and $1 \leq j \leq p$. We show the uniqueness by induction on p. The case $p = 0$ is clear. Define $\psi = \phi^{-1}$ as the inverse of ϕ. Then $\psi = (\psi_{j,i})$, where $\psi_{j,i} \in \operatorname{Hom}_Q(W_i, V_j)$ for all i, j. We have $\sum_i \psi_{p,i}\phi_{i,p} = 1_{V_p}$. Let $\mathfrak{m}$ be the unique maximal ideal of the local ring $\operatorname{Hom}_Q(V_p, V_p)$. Since $1_{V_p} \notin \mathfrak{m}$, we have $\psi_{p,i}\phi_{i,p} \notin \mathfrak{m}$ for some i. After rearranging $W_1, \ldots, W_r$ we may assume that $\psi_{p,r}\phi_{r,p} \notin \mathfrak{m}$. We have that $\psi_{p,r}\phi_{r,p}$ is invertible. It follows that $\phi_{r,p}\psi_{p,r}$ cannot be nilpotent, so it has to be invertible as well. So $\phi_{r,p}$ is an isomorphism. By Lemma 1.7.3, $V_1 \oplus \cdots \oplus V_{p-1} \cong W_1 \oplus \cdots \oplus W_{r-1}$. By the induction hypothesis, we have $p - 1 = r - 1$, so $p = r$. After rearranging $W_1, \ldots, W_{p-1}$ we get that V_i is isomorphic to W_i for $i = 1, 2, \ldots, p$. □

Exercises.

Exercise 1.7.1. *Show that the isomorphism classes of representations of the quiver*

$$1 \xrightarrow{a} 2 \xrightarrow{b} 3$$

are uniquely determined by $\dim V(1)$, $\dim V(2)$, $\dim V(3)$, $\operatorname{rank} V(a)$, $\operatorname{rank} V(b)$ *and* $\operatorname{rank} V(b)V(a)$.

1.8. Bibliographical Remarks

The material of this chapter is introductory. We are dealing only with finite dimensional algebras over $\mathbb{C}$. Most of the results are true in greater generality, either over an algebraically closed field of arbitrary characteristic, for finite dimensional algebras over arbitrary field or even for Artin algebras (which can be infinite dimensional). We refer the reader to several books that have been written on the subject. There is a book [**3**] by Auslander, Reiten and Smalø which contains an excellent exposition of the Auslander-Reiten theory. It is based on the use of stable categories and categories of functors. The book [**43**] by Gabriel and Roiter stresses the language of linear algebra and generalizations to the K-linear categories.

Finally, several years ago the three volumes of the book [**38**] by Assem, Simson and Skowroński started appearing. We recommend especially the first volume of this work which contains all the basics in the most general framework.

Chapter 2

Homological Algebra of Quiver Representations

An algebra is semi-simple if all its modules are direct sums of irreducible modules. By Wedderburn's theorem, finite dimensional semi-simple algebras over $\mathbb{C}$ are products of matrix algebras (see Corollary 3.4.13). For example, group algebras of finite groups are semi-simple (over $\mathbb{C}$) which makes the representation theory of finite groups relatively easy. An algebra A is semi-simple precisely when every A-module is projective, i.e., when global dimension of A is zero.

Hereditary algebras are algebras of global dimension at most one. It is natural to study these as the next step. A path of algebras $\mathbb{C}Q$ for quivers Q without oriented cycles is hereditary. To understand the representation theory of hereditary algebras, it suffices to understand the representation theory of path algebras. In this chapter we develop the homological algebra over path algebras, construct explicitly the projective and injective resolutions of all modules. We also give the definition and all the properties of the Ext vector spaces constructing them as classes of extensions and as derived functors. This material can be found in any homological algebra book.

2.1. Projective and Injective Modules

Let A be a $\mathbb{C}$-algebra. We will work in the category of finite dimensional A-modules and all modules are assumed to be finite dimensional. In this section we give the definitions of projective and injective modules, and discuss some basic properties.

Definition 2.1.1. Let A be a $\mathbb{C}$-algebra.

(1) An A-module P is **projective** if the following equivalent conditions are satisfied:
   (a) the functor $\operatorname{Hom}_A(P, -)$ is exact;
   (b) for every epimorphism $f : N \twoheadrightarrow N'$ and every homomorphism $g : P \to N'$ there exist a homomorphism $h : P \to N$ such that $fh = g$.

$$\begin{array}{ccc} & & P \\ & \overset{h}{\swarrow} & \big\downarrow g \\ N & \underset{f}{\twoheadrightarrow} & N' \end{array}$$

(2) An A-module I is **injective** if the following equivalent conditions are satisfied:
   (a) the functor $\operatorname{Hom}_A(-, I)$ is exact;
   (b) for every monomorphism $f : M' \hookrightarrow M$ and every homomorphism $g : M' \to I$ there exist a homomorphism $h : M \to I$ such that $hf = g$.

$$\begin{array}{ccc} M' & \overset{f}{\hookrightarrow} & M \\ g\big\downarrow & \underset{h}{\swarrow} & \\ I & & \end{array}$$

The following lemma is proven in Exercise 2.1.1.

Lemma 2.1.2.

(1) *If P is projective, then every exact sequence*

$$0 \longrightarrow N \longrightarrow M \longrightarrow P \longrightarrow 0$$

of A-modules splits. In particular, M is isomorphic to $N \oplus P$.

(2) *If I is injective, then every exact sequence*

$$0 \longrightarrow I \longrightarrow M \longrightarrow N \longrightarrow 0$$

splits. In particular, M is isomorphic to $I \oplus N$.

The following lemma is proven in Exercise 2.1.2.

Lemma 2.1.3. *An A-module M is projective if and only if $M^\star$ is an injective A^{op}-module.*

For the remainder of this section, assume that A is finite dimensional. For any A-module M we have $\operatorname{Hom}_A(A, M) = M$. So $\operatorname{Hom}_A(A, -)$ is the identity functor, and it is therefore exact. This shows that A is projective. Also, $A = A^{\mathrm{op}}$ is a projective A^{op}-module, so $A^\star$ is an injective A-module.

If P is a projective module and $P = P' \oplus P''$, then P' and P'' are also projective, and a similar statement is true for injective modules (see Exercise 2.1.3).

Corollary 2.1.4. *Every direct summand of A is projective. Every direct summand of $A^\star$ is injective.*

Also the sum of two projective modules is again projective and the sum of two injective modules is again injective (see Exercise 2.1.3). In particular, a free module A^r is projective, and the module $(A^\star)^r$ is injective. Every direct summand of A^r is projective and every direct summand of $(A^\star)^r$ is injective. The converse is also true.

Lemma 2.1.5. *If P is a projective A module, then P is a direct summand of A^r for some positive integer r. If I is an injective A-module, then I is a direct summand of $(A^\star)^r$ for some positive integer r.*

Proof. Suppose that P is generated by $u_1, \dots, u_r$. Then we define a surjective module homomorphism $\varphi : A^r \twoheadrightarrow P$ defined by $\varphi(a_1, \dots, a_r) = a_1u_1 + \cdots + a_ru_r$. Let K be the kernel of φ. We have an exact sequence

$$0 \longrightarrow K \longrightarrow A^r \longrightarrow P \longrightarrow 0 \,,$$

and $A^r \cong K \oplus P$ by Lemma 2.1.2.

If I is injective, then $I^\star$ is a projective A^{op}-module. So $I^\star$ is isomorphic to a direct summand of A^r for some positive integer r (as A^{op}-modules) and I is a direct summand of $(A^\star)^r$. □

Exercises.

Exercise 2.1.1. *Prove Lemma* 2.1.2.

Exercise 2.1.2. *Prove Lemma* 2.1.3.

Exercise 2.1.3. *Suppose that P, P', P'' are A-modules such that $P = P' \oplus P''$. Show that P is projective if and only if both P' and P'' are projective.*

Exercise 2.1.4.

(1) *Suppose that P is a module such that every exact sequence of A-modules*

$$0 \longrightarrow N \longrightarrow M \longrightarrow P \longrightarrow 0$$

splits. Show that P is projective.

(2) *Suppose that I is a module such that every exact sequence of A-modules*

$$0 \longrightarrow I \longrightarrow M \longrightarrow N \longrightarrow 0$$

splits. Show that I is injective.

2.2. Projective and Injective Quiver Representations

In this section we restrict ourselves to acyclic quivers and their path algebras. There is a 1–1 correspondence between indecomposable projective (respectively injective) representations and vertices of the quiver. The projective and injective representation can be explicitly described in terms of paths.

Let $Q = (Q_0, Q_1, h, t)$ be an acyclic quiver. We investigate the structure of projective and injective resolutions in $\operatorname{Rep}(Q)$. For $x \in Q_0$, define $\mathbb{C}Q$-modules by $P_x := \mathbb{C}Qe_x$ and $I_x = (e_x\mathbb{C}Q)^\star$.

Lemma 2.2.1. *We have a direct sum decompositions of $\mathbb{C}Q$-modules.*

$$\mathbb{C}Q = \bigoplus_{x \in Q_0} P_x \text{ and } (\mathbb{C}Q)^\star = \bigoplus_{x \in Q_0} I_x.$$

Proof. If $u \in \mathbb{C}Q$, then we have

$$u = u \cdot 1 = u \cdot \sum_{x \in Q_0} e_x = \sum_{x \in Q_0} ue_x \in \sum_{x \in Q_0} P_x.$$

Suppose that $\sum_{x \in Q_0} a_x = 0$ with $a_x \in P_x$ for all $x \in Q_0$. For $y \in Q_0$ we have $a_x e_y = a_x e_x e_y = 0$ if $x \neq y$, so

$$0 = \Big(\sum_{x \in Q_0} a_x\Big) e_y = a_y e_y = a_y$$

for all $y \in Q_0$. This shows that the sum is direct. The second direct sum decomposition is proven in a similar way. □

Since P_x is a direct summand of $\mathbb{C}Q$, it is projective by Corollary 2.1.4. Note that P_x is a vector space with a basis given by all paths starting at x. We can view P_x now as a representation of the quiver Q. For any vertex $y \in Q_0$ we have that $P_x(y) = e_y\mathbb{C}Qe_x$ is a vector space with a basis given by all paths from x to y. Since I_x is a direct summand of the injective module $(\mathbb{C}Q)^\star$, it is injective as well. The space $I_x^\star = e_x\mathbb{C}Q$ has a basis given by all paths ending at x. The space $I_x(y)^\star = (e_y I_x)^\star = (e_x\mathbb{C}Qe_y)^\star$ has a basis given by all paths from y to x.

Proposition 2.2.2. *We have canonical isomorphisms*

$$\operatorname{Hom}_Q(P_x, V) \cong V(x) \text{ and } \operatorname{Hom}_Q(V, I_x) \cong V(x)^\star.$$

Proof. Define a linear map $\Psi : \mathrm{Hom}_Q(P_x, V) \to V(x)$ by $\Psi(\phi) = \phi(e_x) \in V(x)$. Define a linear map $\Theta : V(x) \to \mathrm{Hom}_Q(P_x, V)$ as follows; for any $v \in V(x)$ there is a unique morphism $\phi_v : P_x \to V$ of quiver representations given by $\phi_v(p) = V(p)v$. Define $\Theta(v) = \phi_v$. It is easy to see that Ψ and Θ are each others inverses.

For the second isomorphism, note that

$$\mathrm{Hom}_Q(V, I_x) \cong \mathrm{Hom}_{Q^{\mathrm{op}}}(I_x^\star, V^\star) \cong V^\star(x).$$

□

Proposition 2.2.3. *The indecomposable projective representations are (up to isomorphism) exactly all P_x, $x \in Q_0$. The indecomposable injective representations are (up to isomorphism) exactly all I_x, $x \in Q_0$.*

Proof. Note that P_x is indecomposable for all $x \in Q_0$, because $\mathrm{Hom}_Q(P_x, P_x) \cong \mathbb{C}e_x$ is 1-dimensional. If P is a projective representation, then P is a direct summand of $(\mathbb{C}Q)^r = \bigoplus_{x \in Q_0} P_x^r$. By the Krull-Remak-Schmidt theorem, P has to be isomorphic to P_x for some $x \in Q_0$. The second statement in the proposition follows from duality (see Section 1.6 and Lemma 2.1.3). □

Example 2.2.4. Consider the quiver

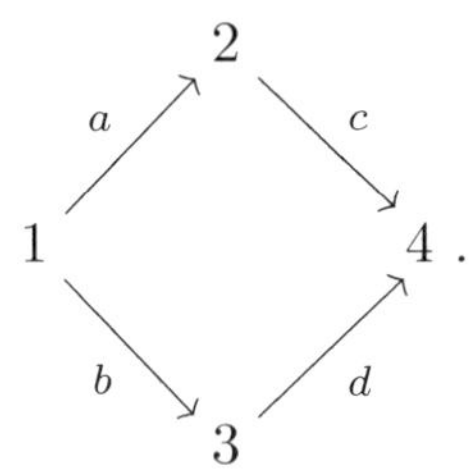

The representation P_1 has dimension vector $(1, 1, 1, 2)$. It has a $\mathbb{C}$-basis $\{e_1, a, b, ca, db\}$. The representation P_2 has dimension vector $(0, 1, 0, 1)$ and $\mathbb{C}$-basis $\{e_2, c\}$. Similarly, the representation P_3 has dimension vector $(0, 0, 1, 1)$ and has $\mathbb{C}$-basis $\{e_3, d\}$. Finally, P_4 has dimension vector $(0, 0, 0, 1)$ and its $\mathbb{C}$-basis is $\{e_4\}$. The simple module S_1 has the following resolution:

$$0 \longrightarrow P_2 \oplus P_3 \xrightarrow{(a,b)} P_1 \longrightarrow S_1 \longrightarrow 0.$$

Exercises.

Exercise 2.2.1. *Show that if Q has an oriented cycle, then there are infinitely many distinct isomorphism classes of simple representations. (See Exercise* 1.4.4.*)*

Exercise 2.2.2. *Let S_x be a simple module corresponding to a vertex $x \in Q_0$. There is a natural map $p_x : P_x \to S_x$ sending the generator e_x to the* *element* $1 \in \mathbb{C} = S_x(x)$. *Prove that the kernel of this map is isomorphic to*

$$\bigoplus_{\substack{a \in Q_1 \\ ta = x}} P_{ha}.$$

The resulting complex

$$0 \longrightarrow \bigoplus_{\substack{a \in Q_1 \\ ta = x}} P_{ha} \longrightarrow P_x \longrightarrow S_x \longrightarrow 0$$

gives a projective resolution of length 1 *of a simple object* S_x.

Exercise 2.2.3. *Let* $e_x^\star \in I_x(x)$ *be dual to* $e_x \in I_x(x)^\star = \mathbb{C}e_x$. *Prove that for a natural embedding* $q_x : S_x \to I_x$ *defined by sending the generator* $1 \in \mathbb{C} = S_x(x)$ *to the basis element* e_x^* *we have*

$$\operatorname{coker}\ q_x = \bigoplus_{\substack{a \in Q_1 \\ ha = x}} I_{ta}.$$

The resulting complex

$$0 \longrightarrow S_x \longrightarrow I_x \longrightarrow \bigoplus_{\substack{a \in Q_1 \\ ha = x}} I_{ta} \longrightarrow 0$$

gives an injective resolution of S_x *of length* 1.

2.3. The Hereditary Property of Path Algebras

Definition 2.3.1. A category of (left) modules over an algebra A is **hereditary** if a submodule of a projective module is projective.

Exercise 2.2.2 suggests that for acyclic quivers Q, the category $\operatorname{Rep}(Q)$ might be hereditary. We will prove this in this section.

Theorem 2.3.2. *Let* Q *be an acyclic quiver. The category* $\operatorname{Rep}(Q)$ *is hereditary.*

Before giving the proof, we need two propositions.

Proposition 2.3.3. *Assume that there are two exact sequences of representations of* Q*:*

$$0 \longrightarrow P_1 \xrightarrow{d} P_0 \longrightarrow V \longrightarrow 0,$$

$$0 \longrightarrow P_1' \xrightarrow{d'} P_0' \longrightarrow V \longrightarrow 0$$

with P_0, P_1, P_0' *projective. Then* $P_1' \oplus P_0$ *is isomorphic to* $P_1 \oplus P_0'$. *In particular* ,P_1' *is projective as well.*

Proof. Since P_0' is projective, the identity map on V can be lifted to the map $f_0 : P_0' \to P_0$ and this map restricts to a map $f_1 : P_1' \to P_1$ making the diagram

$$\begin{array}{ccccccccc} 0 & \longrightarrow & P_1' & \xrightarrow{d'} & P_0' & \xrightarrow{p'} & V & \longrightarrow & 0 \\ & & \downarrow{\scriptstyle f_1} & & \downarrow{\scriptstyle f_0} & & \downarrow{\scriptstyle 1_V} & & \\ 0 & \longrightarrow & P_1 & \xrightarrow{d} & P_0 & \xrightarrow{p} & V & \longrightarrow & 0 \end{array}$$

commutative. Consider the complex

$$\mathcal{C} : 0 \longrightarrow P_1' \xrightarrow{g_2} P_1 \oplus P_0' \xrightarrow{g_1} P_0 \longrightarrow 0,$$

where $g_2 = \begin{pmatrix} -f_1 \\ d' \end{pmatrix}$ and $g_1 = \begin{pmatrix} d & f_0 \end{pmatrix}$. We will show that this complex is exact. Let $H^k(\mathcal{C})$ be the k-th cohomology group of the complex $\mathcal{C}$. The map g_2 is injective because d' is injective, so $H^2(\mathcal{C}) = 0$. Also, g_1 is surjective: if $x_0 \in P_0$, then there exists $x_0' \in P_0'$ with $p'(x_0') = pf_0(x_0') = p(x_0)$. Since $p(x_0 - f_0(x_0')) = 0$, there exists $x_1 \in P_1$ such that $d(x_1) = x_0 - f_0(x_0')$. So $x_0 = f_0(x_0') + d(x_1)$ lies in the image of g_1. Hence $H^0(\mathcal{C}) = 0$. From

$$\begin{aligned} -\dim H^1(\mathcal{C}) &= \dim H^0(\mathcal{C}) - \dim H^1(\mathcal{C}) + \dim H^2(\mathcal{C}) \\ &= \dim P_0 - \dim(P_1 \oplus P_0') + \dim P_1 \\ &= (\dim P_0 - \dim P_1) - (\dim P_0' - \dim P_1') = \dim V - \dim V = 0 \end{aligned}$$

we conclude that $H^1(\mathcal{C}) = 0$. So the complex $\mathcal{C}$ is exact.

Since P_0 is projective our exact sequence splits by Lemma 2.1.2 and we are done. $\square$

For every $\mathbb{C}Q$-module V of Q we define a complex

(2.1)

$$\mathcal{C}(V) : 0 \longrightarrow \bigoplus_{a \in Q_1} V(ta) \otimes P_{ha} \xrightarrow{d^V} \bigoplus_{x \in Q_0} V(x) \otimes P_x \xrightarrow{f^V} V \longrightarrow 0,$$

where

$$f^V(v \otimes p) = p \cdot v$$

and

$$d^V(v \otimes p) = (a \cdot v) \otimes p - v \otimes pa.$$

It is clear that $\mathcal{C}(V)$ is a complex because

$$f^V(d^V(v \otimes p)) = f^V((a \cdot v) \otimes p - v \otimes pa) = p \cdot (a \cdot v) - (pa) \cdot v = 0.$$

Proposition 2.3.4. *The complex $\mathcal{C}(V)$ (2.1) is exact.*

Proof. Clearly, f^V is onto, because $f^V(v \otimes e_x) = e_x \cdot v = v$ for all $v \in V(x) = e_x V$. We claim that d^V is injective. Suppose that

$$d^V\Big(\sum_{a \in Q_1} v_a \otimes p_a\Big) = 0 \text{ and } \sum_{a \in Q_1} v_a \otimes p_a \neq 0,$$

where $v_a \in V(ta)$ and $p_a \in P_{ha}$ for all $a \in Q_1$. We may assume that $Q_0 = \{1, 2, \ldots, n\}$ and that $h(a) < t(a)$ for all $a \in Q_1$ (see Exercise 1.5.2). Choose $y \in Q_0$ maximal such that there exists an arrow a with $v_a \otimes p_a \neq 0$ and $ta = y$. For any arrow a with $ha = y$ we have $ta > ha = y$ and $v_a \otimes p_a = 0$. The component of $d^V(\sum_{a \in Q_1} v_a \otimes p_a)$ in $V(y) \otimes P_y$ is

$$\sum_{i=1}^{r} -v_{a_i} \otimes p_{a_i} a_i = 0,$$

where $a_1, a_2, \ldots, a_r$ are all arrows with tail y. This is a contradiction because $p_{a_1} a_1, \ldots, p_{a_r} a_r$ are independent, and $v_{a_i} \neq 0$ for all i. From Exercise 2.2.2 it follows that

$$\begin{aligned} \dim\Big(\bigoplus_{x \in Q_0} V(x) \otimes P_x\Big) &- \dim\Big(\bigoplus_{a \in Q_1} V(ta) \otimes P_x\Big) \\ &= \sum_{x \in Q_0} \dim V(x)\Big(\dim P_x - \sum_{\substack{a \in Q_1 \\ ta = x}} \dim P_{ha}\Big) = \sum_{x \in Q_0} \dim V(x) = \dim V. \end{aligned}$$

From this follows that the complex $\mathcal{C}(V)$ is exact in the middle as well. □

The complex $\mathcal{C}(V)$ is called the **canonical projective resolution** of V.

Proof of Theorem 2.3.2. Suppose that P_1' is a subrepresentation of a projective representation P_0', and let $V = P_0'/P_1'$. We have a short exact sequence

$$0 \longrightarrow P_1' \longrightarrow P_0' \longrightarrow V \longrightarrow 0.$$

By Proposition 2.3.4 there exists an exact sequence

$$0 \longrightarrow P_1 \longrightarrow P_0 \longrightarrow V \longrightarrow 0,$$

where P_0 and P_1 are projective. So P_1' is projective by Proposition 2.3.3. □

Exercises.

Exercise 2.3.1. *Show for any representation W we have a canonical injective resolution*

$$0 \longrightarrow W \xrightarrow{f_W} \bigoplus_{x \in Q_0} I_x \otimes W(x) \xrightarrow{d_W} \bigoplus_{a \in Q_1} I_{ta} \otimes W(ha) \longrightarrow 0, \tag{2.2}$$

where

$$f_W^\star : \bigoplus_{x \in Q_0} W(x)^\star \otimes I_x^\star \longrightarrow W^\star$$

is defined by

$$f_W^\star(p \otimes w) = p \cdot w$$

and

$$d_W^\star : \bigoplus_{a \in Q_1} W(ha)^\star \otimes I_{ta}^\star \longrightarrow \bigoplus_{x \in Q_0} W(x)^\star \otimes I_x^\star$$

is defined by

$$d_W^\star(w \otimes p) = w \otimes ap - (a \cdot w) \otimes p.$$

2.4. The Extensions Group

Let V and W be two representations of a quiver Q. We introduce the extension group $\mathrm{Ext}_Q(V, W)$. First, we will describe this abelian group as Yoneda extensions. We also describe this group as the derived functor $\mathrm{Ext}_Q^1(V, -)$ of $\mathrm{Hom}_Q(V, -)$ or as the derived functor $\mathrm{Ext}^1(-, W)$ of $\mathrm{Hom}_Q(-, W)$.

An **extension** ξ of V by W is an exact sequence

$$\xi : 0 \longrightarrow W \xrightarrow{i} E \xrightarrow{p} V \longrightarrow 0 .$$

The two extensions

$$\xi : 0 \longrightarrow W \xrightarrow{i} E \xrightarrow{p} V \longrightarrow 0$$

$$\xi' : 0 \longrightarrow W \xrightarrow{i'} E' \xrightarrow{p'} V \longrightarrow 0 .$$

are **equivalent** if there exists an isomorphism $f : E \to E'$ such that the diagram

$$\begin{array}{ccccccccc} 0 & \longrightarrow & W & \xrightarrow{i} & E & \xrightarrow{p} & V & \longrightarrow & 0 . \\ & & \downarrow{\scriptstyle 1_W} & & \downarrow{\scriptstyle f} & & \downarrow{\scriptstyle 1_V} & & \\ 0 & \longrightarrow & W & \xrightarrow{i'} & E' & \xrightarrow{p'} & V & \longrightarrow & 0 \end{array}$$

commutes.

Definition 2.4.1. We define $\mathrm{Ext}_Q(V, W)$ as the class of equivalence classes of extensions of V by W. This is the **Yoneda extension** group.

Consider the following map:

$$\bigoplus_{x \in Q_0} \mathrm{Hom}(V(x), W(x)) \xrightarrow{d_W^V} \bigoplus_{a \in Q_1} \mathrm{Hom}(V(ta), W(ha)), \tag{2.3}$$

where

$$d_W^V(\phi(x), x \in Q_0) = (\phi(ha)V(a) - W(a)\phi(ta), a \in Q_1).$$

The following proposition gives us an easy way to understand $\mathrm{Ext}_Q(V, W)$ and to compute it.

Proposition 2.4.2. *The kernel of d_W^V is $\mathrm{Hom}_Q(V, W)$. The cokernel of d_W^V can be identified with $\mathrm{Ext}_Q(V, W)$. In particular, $\mathrm{Ext}_Q(V, W)$ is a finite dimensional $\mathbb{C}$-vector space.*

Proof. It is clear from the definition of $\mathrm{Hom}_Q(V, W)$ that it is the kernel of d_W^V. We define a map

$$\Theta : \bigoplus_{a \in Q_1} \mathrm{Hom}(V(ta), W(ha)) \to \mathrm{Ext}_Q(V, W).$$

Suppose that

$$\psi = (\psi(a), a \in Q_1) \in \bigoplus_{a \in Q_1} \mathrm{Hom}(V(ta), W(ha)).$$

We define $E(x) := V(x) \oplus W(x)$ and $E(a)$ is given in block form by the matrix

$$\begin{pmatrix} V(a) & 0 \\ \psi(a) & W(a) \end{pmatrix}. \tag{2.4}$$

It is clear that the subspaces $\{W(x)\}_{x \in Q_0}$ form a subrepresentation of E and that the factor is isomorphic to V. Thus we have an extension

$$\xi : 0 \longrightarrow W \longrightarrow E \longrightarrow V \longrightarrow 0.$$

We define $\Theta(\psi)$ as the equivalence class of ξ.

Suppose that

$$\psi' = (\psi'(a), a \in Q_1) \in \bigoplus_{a \in Q_1} \mathrm{Hom}(V(ta), W(ha))$$

and

$$\xi' : 0 \longrightarrow W \longrightarrow E' \longrightarrow V \longrightarrow 0$$

is the extension obtained from ψ'. If ξ and ξ' are equivalent, then there exists an isomorphism $f : E \to E'$ such that the diagram

$$\begin{array}{ccccccccc} 0 & \longrightarrow & W & \xrightarrow{i} & E & \xrightarrow{p} & V & \longrightarrow & 0 \\ & & \downarrow{\scriptstyle 1_W} & & \downarrow{\scriptstyle f} & & \downarrow{\scriptstyle 1_V} & & \\ 0 & \longrightarrow & W & \xrightarrow{i'} & E' & \xrightarrow{p'} & V & \longrightarrow & 0 \end{array} \tag{2.5}$$

commutes. The restriction of $f(x) : V(x) \oplus W(x) \to V(x) \oplus W(x)$ must have the block form

$$f(x) = \begin{pmatrix} 1_V & 0 \\ -\phi(x) & 1_W \end{pmatrix} \tag{2.6}$$

for some $\phi(x) \in \operatorname{Hom}(V(x), W(x))$. Since f is a morphism of quiver representations, we have

$$\begin{pmatrix} 1_{V(ha)} & 0 \\ -\phi(ha) & 1_{W(ha)} \end{pmatrix} \begin{pmatrix} V(a) & 0 \\ \psi(a) & W(a) \end{pmatrix} = \phi(ha)E(a)$$
$$= E'(a)\phi(ta) = \begin{pmatrix} V(a) & 0 \\ \psi'(a) & W(a) \end{pmatrix} \begin{pmatrix} 1_{V(ta)} & 0 \\ -\phi(ta) & 1_{W(ta)} \end{pmatrix}.$$

Looking at the block at position $(2, 1)$ gives us the equation

$$-\phi(ha)V(a) + \psi(a) = \psi'(a) - W(a)\phi(ta)$$

so

$$\psi - \psi' = (\psi(a) - \psi'(a), a \in Q_1) = (\phi(ha)V(a) - W(a)\phi(ta), a \in Q_1) = d_W^V(\phi).$$

Conversely, if $\psi - \psi' = d_W^V(\phi)$ for some $\phi \in \bigoplus_{x \in Q_0} \operatorname{Hom}(V(x), W(x))$, then the diagram (2.5) commutes, where $f : E \to E'$ is the isomorphism defined by (2.6).

Any extension ξ of V by W is equivalent to an extension

$$0 \longrightarrow W \longrightarrow E \longrightarrow V \longrightarrow 0,$$

where $E(x) = V(x) \oplus W(x)$ for all $x \in Q_0$. Then $E(a)$ must be of the form (2.4) for some $\psi = (\psi(a), a \in Q_1)$ and $\Theta(\psi)$ is equivalent to ξ. This shows that Θ is surjective. □

Now, assume that Q is acyclic. Let

$$\mathcal{P}_V : 0 \longrightarrow P_1 \longrightarrow P_0 \longrightarrow V \longrightarrow 0 \tag{2.7}$$

be a projective resolution of V of length 1, i.e., (2.7) is an exact sequence where P_0 and P_1 are projective. Also, let

$$\mathcal{I}^W : 0 \longrightarrow W \longrightarrow I^0 \longrightarrow I^1 \longrightarrow 0 \tag{2.8}$$

be an injective resolution of W of length 1.

We can form the double complex $\operatorname{Hom}_Q(\mathcal{P}_V, \mathcal{I}^W)$. Using the exactness of the functors $\operatorname{Hom}_Q(P_0, -), \operatorname{Hom}_Q(P_1, -), \operatorname{Hom}_Q(-, I^0), \operatorname{Hom}_Q(-, I^1)$, the left-exactness of the functor $\operatorname{Hom}_Q(V, -)$ and the right-exactness of the functor $\operatorname{Hom}_Q(-, W)$, we get the following diagram with exact rows and columns:
(2.9)

$$\begin{array}{ccccccccc}
 & & 0 & & 0 & & 0 & & \\
 & & \downarrow & & \downarrow & & \downarrow & & \\
0 & \longrightarrow & \operatorname{Hom}_Q(V, W) & \longrightarrow & \operatorname{Hom}_Q(V, I^0) & \longrightarrow & \operatorname{Hom}_Q(V, I^1) & & \\
 & & \downarrow & & \downarrow & & \downarrow & & \\
0 & \longrightarrow & \operatorname{Hom}_Q(P_0, W) & \longrightarrow & \operatorname{Hom}_Q(P_0, I^0) & \longrightarrow & \operatorname{Hom}_Q(P_0, I^1) & \longrightarrow & 0 \\
 & & \downarrow & & \downarrow & & \downarrow & & \\
0 & \longrightarrow & \operatorname{Hom}_Q(P_1, W) & \longrightarrow & \operatorname{Hom}_Q(P_1, I^0) & \longrightarrow & \operatorname{Hom}_Q(P_1, I^1) & \longrightarrow & 0 \\
 & & & & \downarrow & & \downarrow & & \\
 & & & & 0 & & 0 & &
\end{array}$$

Lemma 2.4.3. *The cokernel of* $\operatorname{Hom}_Q(V, I^0) \to \operatorname{Hom}_Q(V, I^1)$ *and the kernel of* $\operatorname{Hom}_Q(P_0, W) \to \operatorname{Hom}_Q(P_1, W)$ *are both isomorphic to* $\operatorname{Ext}_Q(V, W)$.

Proof. From the snake lemma applied to the diagram (2.9) it follows that the cokernel of $\operatorname{Hom}_Q(P_0, W) \to \operatorname{Hom}_Q(P_1, W)$ and the cokernel of $\operatorname{Hom}_Q(V, I^0) \to \operatorname{Hom}_Q(V, I^1)$ are isomorphic. If we choose $P_0 = \bigoplus_{x \in Q_0} V(x) \otimes P_x$ and $P_1 = \bigoplus_{a \in Q_1} V(ta) \otimes P_{ha}$ as in (2.1), then the cokernel of $\operatorname{Hom}_Q(P_0, W) \to \operatorname{Hom}_Q(P_1, W)$ is $\operatorname{Ext}_Q(V, W)$ by Exercise 2.4.1. Therefore, the cokernel of $\operatorname{Hom}(V, I^0) \to \operatorname{Hom}(V, I^1)$ is isomorphic to $\operatorname{Ext}_Q(V, W)$. $\square$

In the category of quiver representations, one can define groups $\operatorname{Ext}^i_Q(V, W)$ for all $i \geq 0$ using derived functors of $\operatorname{Hom}_Q(V, -)$ or $\operatorname{Hom}_Q(-, W)$. We have $\operatorname{Ext}^0_Q(V, W) = \operatorname{Hom}_Q(V, W)$. Because of the previous lemma, we have $\operatorname{Ext}^1_Q(V, W) = \operatorname{Ext}_Q(V, W)$. Also, $\operatorname{Ext}^i_Q(V, W) = 0$ for all $i > 1$, because every representation has a projective resolution of length 1.

Exercises.

Exercise 2.4.1. *Show that if we apply the functor* $\operatorname{Hom}_Q(-, W)$ *to* (2.1) *or the functor* $\operatorname{Hom}_Q(V, -)$ *to* (2.2) *then we obtain the map*

$$d^V_W = \operatorname{Hom}_Q(d^V, W) = \operatorname{Hom}_Q(V, d_W)$$

in (2.3).

Exercise 2.4.2. *We have identified* $\mathrm{Ext}_Q(V,W)$ *with the cokernel of* d_W^V *and this gives* $\mathrm{Ext}_Q(V,W)$ *the structure of a finite dimensional* $\mathbb{C}$*-vector space. But the vector space structure of* $\mathrm{Ext}_Q(V,W)$ *can also be directly defined in terms of extensions.*

(1) *Show that the split extensions of* V *by* W *form exactly the equivalence class which corresponds to the* 0 *element in* $\mathrm{Ext}_Q(V,W)$*.*

(2) *Suppose that*

$$\xi : 0 \longrightarrow W \xrightarrow{\ i\ } E \xrightarrow{\ p\ } V \longrightarrow 0$$

is an extension of V *by* W *and denote its equivalence class by* $[\xi]$*. For* $\lambda \in \mathbb{C}$*, define* $E_\lambda := \{(u,v) \in E \oplus V \mid p(u) = \lambda v\}$*. Define* $p_\lambda : E_\lambda \to V$ *as the projection onto* V *and define* $i_\lambda : W \to E_\lambda$ *by* $i_\lambda(w) = (i(w), 0)$*. Show that*

$$\xi_\lambda : 0 \longrightarrow W \xrightarrow{\ i_\lambda\ } E_\lambda \xrightarrow{\ p_\lambda\ } V \longrightarrow 0$$

is exact and that $\lambda \cdot [\xi] = [\xi_\lambda]$*.*

(3) *Suppose that*

$$\xi' : 0 \longrightarrow W \xrightarrow{\ i'\ } E' \xrightarrow{\ p'\ } V \longrightarrow 0$$

is another extension of V *by* W*. Let*

$$F = \{(u,u') \in E \oplus E' \mid p(u) = p'(u')\} \subset E \oplus E'$$

be the pull-back in the diagram

$$\begin{array}{ccc} F & \dashrightarrow & E' \\ \downarrow & & \downarrow \\ E & \longrightarrow & V \end{array}$$

and let

$$G = \{(i(w), -i'(w)) \mid w \in W\} \subseteq F.$$

Define $E'' = F/G$*. The map* $F \to V$ *contains* G *in the kernel, so it induces a morphism* $p'' : E'' \to W$*. Define* $i'' : W \to E''$ *by* $i''(w) = (i(w), 0) + G \in F/G = E''$*. Show that the sequence*

$$\xi'' : 0 \longrightarrow W \xrightarrow{\ i''\ } E'' \xrightarrow{\ p''\ } V \longrightarrow 0$$

is exact and that $[\xi] + [\xi'] = [\xi'']$*.*

Exercise 2.4.3. *Consider the subsquare*

$$\begin{array}{ccc} \mathrm{Hom}_Q(P_0, I^0) & \xrightarrow{d_0'} & \mathrm{Hom}_Q(P_0, I^1) \\ {\scriptstyle d_0}\downarrow & & \downarrow{\scriptstyle d_1} \\ \mathrm{Hom}_Q(P_1, I^0) & \xrightarrow[d_1']{} & \mathrm{hom}_Q(P_1, I^1) \end{array}$$

of (2.9). *We can form the total complex*

$$\mathcal{C}^{\mathrm{tot}}\colon \mathrm{Hom}_Q(P_0, I^0) \longrightarrow \mathrm{Hom}_Q(P_1, I^0) \oplus \mathrm{Hom}_Q(P_0, I^1) \longrightarrow \mathrm{Hom}(P_1, I^1),$$

where the two maps are given by

$$\begin{pmatrix} d_0 \\ -d_0' \end{pmatrix} \text{ and } \begin{pmatrix} d_1' & d_1 \end{pmatrix}.$$

Show that the homology groups are as follows:

$$\begin{aligned} H_0(\mathcal{C}^{\mathrm{tot}}) &= \mathrm{Hom}_Q(V, W), \\ H_1(\mathcal{C}^{\mathrm{tot}}) &= \mathrm{Ext}_Q(V, W), \\ H_2(\mathcal{C}^{\mathrm{tot}}) &= 0. \end{aligned}$$

2.5. The Euler Form

Definition 2.5.1. The Euler characteristic of two representations V, W of Q is

$$\chi(V, W) = \dim \mathrm{Hom}_Q(V, W) - \dim \mathrm{Ext}_Q(V, W).$$

We define the **Euler form** or **Ringel form** on $\mathbb{R}^{Q_0}$ by

$$\langle \alpha, \beta \rangle = \sum_{x \in Q_0} \alpha(x)\beta(x) - \sum_{a \in Q_1} \alpha(ta)\beta(ha).$$

Note that this bilinear form may not be symmetric (unless Q has no arrows) or skew-symmetric.

Proposition 2.5.2. *If* $\alpha = \underline{\dim}\, V$ *and* $\beta = \underline{\dim}\, W$, *then*

$$\chi(V, W) = \langle \alpha, \beta \rangle.$$

In particular, $\chi(V, W)$ *only depends on the dimension vectors of* V *and* W.

Proof. We have an exact sequence

$$0 \to \mathrm{Hom}_Q(V, W) \to \bigoplus_{x \in Q_0} \mathrm{Hom}(V(x), W(x))$$
$$\to \bigoplus_{x \in Q_1} \mathrm{Hom}(V(ta), W(ha)) \to \mathrm{Ext}_Q(V, W) \to 0,$$

so

$$\begin{aligned}
&\dim \operatorname{Hom}_Q(V,W) - \dim \operatorname{Ext}_Q(V,W) \\
&\quad = \dim \bigoplus_{x\in Q_0} \operatorname{Hom}(V(x),W(x)) - \dim \bigoplus_{a\in Q_1} \operatorname{Hom}(V(ta),W(ha)) \\
&\qquad\qquad = \sum_{x\in Q_0} \alpha(x)\beta(x) - \sum_{a\in Q_1} \alpha(ta)\beta(ha) = \langle \alpha,\beta\rangle.
\end{aligned}$$

□

We can think about the Euler form in terms of matrices.

Definition 2.5.3. Let Q be a quiver with $Q_0 = \{1,2,\dots,n\}$. The **Euler matrix** E is an $n\times n$ matrix with the entries given by the formulas

$$E_{i,j} = \delta_{i,j} - |\{a\in Q_1 \mid ta = i, ha = j\}|,$$

where $\delta_{i,j}$ is the **Kronecker symbol** defined by

$$\delta_{i,j} = \begin{cases} 1 & \text{if } i = j, \\ 0 & \text{if } i \neq j. \end{cases}$$

Identifying the dimension vectors with column vectors, we have

$$\langle \alpha,\beta\rangle = \alpha^t E\beta.$$

If Q is an acyclic quiver, then we can relabel the vertices of Q by $1,2,\dots,n$ such that $ha < ta$ for all $a\in Q_1$. In this case, the Euler matrix is lower triangular with 1's on the diagonal.

Example 2.5.4. Consider the quiver

$$1 \longleftarrow 2 \longleftarrow 3\ .$$

The Euler matrix is

$$E = \begin{pmatrix} 1 & 0 & 0 \\ -1 & 1 & 0 \\ 0 & -1 & 1 \end{pmatrix}.$$

Exercises.

Exercise 2.5.1. *Suppose that Q is an acyclic quiver with $Q_0 = \{1,2,\dots,n\}$. Show that $\chi(P_i,S_j) = \delta_{i,j} = \chi(S_i,I_j)$. Show that the columns of $(E^{-1})^t$ are the dimension vectors of the indecomposable projective representations, and the columns of E^{-1} are the dimension vectors of the indecomposable injective representations. Also show that $(E^{-1})_{i,j}$ is exactly the number of paths from i to j.*

Exercise 2.5.2. *Suppose that Q and Q' are acyclic quivers. Use Exercise* 1.5.3 *to show that the categories* $\mathrm{Rep}(Q)$ *and* $\mathrm{Rep}(Q')$ *are equivalent if and only if Q and Q' are isomorphic quivers (i.e., one is obtained from the other by relabeling of the vertices/arrows.)*

2.6. Bibliographical Remarks

The material of this chapter is standard. We choose to develop it from scratch so the book would be accessible to readers not familiar with homological algebra. We only deal with the hereditary case needed to deal with path algebras $\mathbb{C}Q$ for acyclic quivers.

The path algebras $\mathbb{C}Q$ (and even KQ where K is an arbitrary field) are hereditary even when Q is a finite quiver with oriented cycles. In fact, for arbitrary Q and an ideal I in KQ, contained in the square of the ideal generated by the arrows, the category of KQ/I-modules is hereditary if and only if $I = 0$. This is proved in [**43**, Section 8.2]. The general theory of derived functors $\mathrm{Ext}^n(V, W)$ of functors $\mathrm{Hom}(V, W)$ for categories of modules is described for example in a book [**70**] of Rotman. A more sophisticated homological algebra approach, related to derived categories, which will be needed when dealing with cluster categories, can be found in the book [**44**] of Gelfand and Manin.

Chapter 3

Finite Dimensional Algebras

In this section we will investigate how quivers can be used to understand arbitrary finite dimensional algebras. If a finite dimensional algebra A is of the form $\mathbb{C}Q/I$ where I is a two-sided ideal, then we can view A-modules as quiver representations satisfying the relations that come from the ideal I. We will consider finite dimensional algebras $A = \mathbb{C}Q/I$ where I is a so-called admissible ideal. In that case, the quiver Q encodes important properties of the algebra A. As we will see, not every finite dimensional algebra is of this form. However, for every finite dimensional algebra A there exists another finite dimensional algebra B such that A and B are Morita equivalent. This means that the category of finite dimensional left A-modules and the category of finite dimensional left B modules are equivalent. Along the way, we will also develop some important notions of algebras such as the Jacobson radical.

3.1. Quivers with Relations

In this section we give some basic results on homological algebra for quivers with relations. Suppose that Q is a quiver and denote the two-sided ideal of $\mathbb{C}Q$ generated by all arrows by $\mathfrak{m}$.

Definition 3.1.1. An element $r \in \mathbb{C}Q$ is homogeneous with head $hr = y$ and tail $tr = x$ if r is a linear combination of paths from x to y.

If $r \in \mathbb{C}Q$ is homogeneous with $hr = y$ and $tr = x$, then we define $V(r) : V(tr) \to V(hr)$ as the restriction of left multiplication $r\cdot : V \to V$. If $V(r) = 0$, then we say that V satisfies the relation r.

Suppose that J is an ideal of $\mathbb{C}Q$. If $r \in J$, then $e_y r e_x \in J$ is homogeneous. Since

$$r = 1 \cdot r \cdot 1 = \Big(\sum_{y \in Q_0} e_y\Big) r \Big(\sum_{x \in Q_0} e_x\Big) = \sum_{y \in Q_0} \sum_{x \in Q_0} e_y r e_x,$$

we see that every ideal is generated by homogeneous elements.

Definition 3.1.2. An ideal $J \subseteq \mathbb{C}Q$ is **admissible** if

$$\mathfrak{m}^k \subseteq J \subseteq \mathfrak{m}^2$$

for some integer $k \geq 2$.

Suppose that J is an admissible ideal. Then we have $\mathfrak{m}^k \subseteq J$ for some $k \geq 2$. The space $\mathbb{C}Q/\mathfrak{m}^k$ is spanned by all paths of length $< k$, so $\mathbb{C}Q/\mathfrak{m}^k$ and $A := \mathbb{C}Q/J$ are finite dimensional. Let $\operatorname{Rep}(Q, J)$ be the full subcategory of $\operatorname{Rep}(Q)$ whose objects are all representations V of Q which satisfy all relations $r \in J$. The categories A–mod and $\operatorname{Rep}(Q, J)$ are equivalent. We will now see that the projectives and injectives of A are again parametrized by the vertices of Q.

For simplicity, we will just write p instead of $p + J \in A$ for any path p in Q if there is no confusion. Note that $e_x, x \in Q_0$ are orthogonal idempotents in A. We have

$$A = \bigoplus_{x \in Q_0} P_x$$

and

$$A^\star = \bigoplus_{x \in Q_0} I_x,$$

where $P_x := Ae_x$ and $I_x = (e_x A)^\star$. By Corollary 2.1.4, P_x is projective and I_x is injective for all $x \in Q_0$.

For $x, y \in Q_0$, $I_y(x)^\star = P_x(y) = e_y A e_x = e_y \mathbb{C}Q e_x / e_y I e_x$ is the space of paths from x to y modulo the space of relations from x to y.

Proposition 3.1.3. *For any A-module V we have* $\operatorname{Hom}_A(P_x, V) = V(x)$, *and* $\operatorname{Hom}_A(V, I_x) = V(x)^\star$.

Proof. The proof goes exactly like the proof of 2.2.2. $\square$

Let $\mathfrak{m}_A = \mathfrak{m} + I \subseteq A$. Since $\mathfrak{m}^k \subseteq I$, we have $\mathfrak{m}_A^k = 0$ in A.

Lemma 3.1.4 (Nakayama lemma, special case)**.** *If V is an A-module with $\mathfrak{m}_A V = V$, then $V = 0$.*

Proof. If $\mathfrak{m}_A V = V$, then we have by induction that $0 = \mathfrak{m}_A^k V = V$. $\square$

Proposition 3.1.5. *The indecomposable projective representations up to isomorphism are exactly $P_x, x \in Q_0$ and the injective indecomposable representations up to isomorphism are $I_x, x \in Q_0$.*

Proof. It is clear that $P_x/\mathfrak{m}_A P_x$ is spanned by e_x. So $P_x/\mathfrak{m}_A P_x \cong S_x$ is 1-dimensional and P_x must be indecomposable. Moreover, since S_x and S_y are nonisomorphic for $x \neq y$ we have that P_x, $x \in Q_0$ are distinct. Using duality we also see that I_x, $x \in Q_0$ are indecomposable and pairwise nonisomorphic. Now the proposition follows from Proposition 2.1.5. □

Lemma 3.1.6. *The simple A-modules are S_x, $x \in Q_0$.*

Proof. If V is simple, then $\mathfrak{m}_A V$ is a proper submodule of V by Lemma 3.1.4, so $\mathfrak{m}_A V = 0$. It follows that every arrow acts trivially and V must be isomorphic to S_x for some $x \in Q_0$. □

Proposition 3.1.7. *Assume that Q is acyclic and that it has n vertices. Then every object in* $\operatorname{Rep}(Q, J)$ *has a projective resolution of length $\leq n-1$.*

Proof. We may relabel the vertices of Q such that $ha < ta$ for all $a \in Q_1$. Define representations $V_0, V_1, \dots, V_n$ as follows. We define $V_0 = V$. For $i > 0$ we have a canonical surjection

$$f_i : P_{(i)} \longrightarrow V_i,$$

where $P_{(i)} = \bigoplus_{j=1}^n V_i(j) \otimes P_i$ and f_i is defined by

$$f_i(v \otimes p) = pv.$$

Define V_{i+1} as the kernel of f_i. By induction on i we see that $V_i(j) = 0$ if $i + j > n$ and

$$f_i(j) : P_i(j) \to V_i(j)$$

is an isomorphism if $i + j = n$. In particular, $V_n(j) = 0$ for $j = 1, 2, \dots, n$. So we have a resolution

$$0 \longrightarrow P_{(n-1)} \longrightarrow P_{(n-2)} \longrightarrow \cdots \longrightarrow P_{(0)} \longrightarrow V \longrightarrow 0.$$

□

Example 3.1.8. Consider the quiver

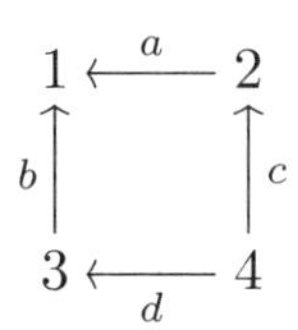

with the relation $ac - bd$. Let $J = (ac - bd)$ and $A = \mathbb{C}Q/J$. The projective P_4 is spanned by $e_4, c, d, ac = bd$ and has dimension vector $(1,1,1,1)^t$. A resolution of S_4 is given by

$$0 \longrightarrow P_1 \longrightarrow P_2 \oplus P_3 \longrightarrow P_4 \longrightarrow S_4 \longrightarrow 0.$$

Exercises.

Exercise 3.1.1. *Consider the quiver*

$$1 \xleftarrow{a_1} 2 \xleftarrow{a_2} 3 \qquad \cdots \xleftarrow{a_{n-1}} n,$$

where J is the ideal generated by $a_1a_2, a_2a_3, \ldots, a_{n-2}a_{n-1}$. Describe the projective modules for $A = \mathbb{C}Q/J$ and give a projective resolution of S_n.

Exercise 3.1.2. *Consider the quiver*

$$1 \circlearrowright a$$

with $J = (a^n)$ and $A = \mathbb{C}Q/J = \mathbb{C}[a]/(a^n)$. What are the projective A-modules? Give an infinite projective resolution of S_1.

3.2. The Jacobson Radical

Suppose that A is a $\mathbb{C}$-algebra, and $J \subseteq A$ is an ideal.

Proposition 3.2.1. *The following statements are equivalent:*

(1) *J is the intersection of all maximal right ideals;*
(2) *$b \in J$ if and only if for every $a \in A$, $1 - ba$ has a right inverse;*
(3) *$b \in J$ if and only if for every $a \in A$, $1 - ba$ is a unit;*
(4) *J is the intersection of all maximal left ideals;*
(5) *$b \in J$ if and only if for every $a \in A$, $1 - ab$ has a left inverse;*
(6) *$b \in J$ if and only if for every $a \in A$ $1 - ab$ is a unit.*

Proof. For a maximal right ideal $\mathfrak{m}$ and $b \in A$ we have

$$\begin{aligned} & b \notin \mathfrak{m} \\ \Leftrightarrow\ & bA + \mathfrak{m} = A \\ \Leftrightarrow\ & 1 \in bA + \mathfrak{m} \\ \Leftrightarrow\ & \text{there exists an } a \in A \text{ such that } 1 - ba \in \mathfrak{m}. \end{aligned}$$

If J is the intersection of all maximal right ideals, then

$$\begin{aligned} & b \notin J \\ \Leftrightarrow\ & b \notin \mathfrak{m} \text{ for some right maximal ideal } \mathfrak{m} \\ \Leftrightarrow\ & \text{there exists } a \in A \text{ such that } 1 - ba \subseteq \mathfrak{m} \\ & \text{for some right maximal ideal } \mathfrak{m} \\ \Leftrightarrow\ & 1 - ba \text{ has no right inverse for some } a \in A. \end{aligned}$$

This shows (1)$\Leftrightarrow$(2).

Suppose that (2) is true. Then $1 - ba$ has a right inverse, say c. So we have $(1-ba)c = 1$ and $c = 1 - b(-ac)$. So c has a right inverse, say d. Then we have $d = (1 - ba)cd = 1 - ba$. So $1 - ba$ is a unit with inverse c. This shows that (2)$\Rightarrow$(3). The implication (3)$\Rightarrow$(2) is obvious, so (1), (2) and (3) are equivalent. Similarly (4), (5) and (6) are equivalent.

Suppose that $1 - ab$ is a unit. If $c = (1 - ab)^{-1}$, then

$$(1+bca)(1-ba) = 1-ba+bca-bcaba = 1-ba+b(c(1-ab))a = 1-ba+ba = 1.$$

Similarly, we have $(1 - ba)(1 + bca) = 1$ and therefore $1 - ba$ is a unit. By reversing the roles of a and b we get that $1 - ab$ is a unit if and only if $1 - ba$ is a unit. We conclude that (3)$\Leftrightarrow$(6). □

Definition 3.2.2. The ideal J for which statements (1)–(6) in Proposition 3.2.1 are true is called the **Jacobson radical** and it is denoted by $\operatorname{rad}(A)$.

It is clear from Proposition 3.2.1 that $\operatorname{rad}(A)$ is a two-sided ideal.

Lemma 3.2.3. *An element $b \in A$ lies in* $\operatorname{rad}(A)$ *if and only if $bV = 0$ for every simple module V.*

Proof. Suppose that $b \in \operatorname{rad}(A)$ and V is a simple module. If $v \in V$ is nonzero and $\mathfrak{m} = \{a \in A \mid av = 0\}$, then $\mathfrak{m}$ is a maximal left ideal. So $b \in \operatorname{rad}(A) \subseteq \mathfrak{m}$ and $bv = 0$.

Conversely, if $bV = 0$ for every simple module V, then $b(A/\mathfrak{m})$ for every maximal left ideal $\mathfrak{m}$, so b lies in the intersection of all maximal left ideals, which is $\operatorname{rad}(A)$. □

Lemma 3.2.4 (Nakayama's lemma). *If M is an A-module with* $\operatorname{rad}(A)M = M$, *then we have $M = 0$.*

Proof. If $M \neq 0$, then we can choose a maximal submodule N of M such that $V = M/N$ is simple. From $\operatorname{rad}(A)V = 0$ follows that $\operatorname{rad}(A)M \subseteq N$; contradiction. □

An easy and useful corollary is the following (see Exercise 3.2.6).

Corollary 3.2.5. *Suppose that $f : M \to N$ is a module homomorphism and f induces a map $M/\operatorname{rad}(A)M \to N/\operatorname{rad}(A)N$ that is surjective. Then f is surjective.*

Some results for quiver representations also apply to finite dimensional modules of a finitely generated $\mathbb{C}$-algebra.

Remark 3.2.6. Suppose A is a $\mathbb{C}$-algebra generated by finitely many elements $a_1, \ldots, a_r$. Define Q as the quiver with r loops, say $b_1, b_2, \ldots, b_r$. Then $\mathbb{C}Q$ is the free algebra $\mathbb{C}\langle b_1, \ldots, b_r\rangle$ and we can define a surjective algebra homomorphism $\psi : \mathbb{C}Q \to A$ defined by $\psi(b_i) = a_i$. Every A-module is a $\mathbb{C}Q$-module, and for A-modules M and N we have $\mathrm{Hom}_Q(M, N) = \mathrm{Hom}_A(M, N)$. Also a module M is indecomposable as an A-module if and only if it is indecomposable as a representation of Q. Then Corollary 1.7.2 and the Krull-Remak-Schmidt theorem (Theorem 1.7.4) also apply to A-modules, because these results immediately are reduced to the same statement for the quiver Q.

Lemma 3.2.7. *Suppose that $A = \mathbb{C}Q/J$, where Q is a quiver and J is an admissible ideal. Then* $\mathrm{rad}(A)$ *is equal to the ideal $\mathfrak{m}_A$ generated by all arrows.*

Proof. Since J is admissible, every element of $\mathfrak{m}_A$ is nilpotent. So $1 - au$ is a unit for every arrow a and every $u \in A$. This shows that $\mathfrak{m}_A \subseteq \mathrm{rad}(A)$. Since $V = A/\mathfrak{m}_A = \bigoplus_{x \in Q_0} S_x$ is a sum of simple modules, we have $\mathrm{rad}(A)V = 0$. This proves that $\mathrm{rad}(A) \subseteq \mathfrak{m}_A$. □

Lemma 3.2.8. *If A is finite dimensional, then* $\mathrm{rad}(A)$ *is nilpotent.*

Proof. Since

$$A \supseteq \mathrm{rad}(A) \supseteq \mathrm{rad}(A)^2 \supseteq \cdots$$

we must have $\mathrm{rad}(A)^n = \mathrm{rad}(A)^{n+1}$ for some n. From $\mathrm{rad}(A)\,\mathrm{rad}(A)^n = \mathrm{rad}(A)^n$ it follows that $\mathrm{rad}(A)^n = 0$. □

Definition 3.2.9. An A-module is called **semi-simple** if it is a direct sum of irreducible modules. An finite dimensional algebra A is called semi-simple if A is semi-simple as a left A module.

Lemma 3.2.10. *A finite dimensional algebra is semi-simple if and only if* $\mathrm{rad}(A) = 0$.

Proof. This follows from Exercises 3.2.4 and 3.2.5 below. □

Wedderburn's theorem states that every finite dimensional semi-simple $\mathbb{C}$-algebra is a product of matrix algebras. We will prove this in Corollary 3.4.13.

Exercises.

Exercise 3.2.1. *Let $A = \mathrm{Mat}_{n,n}(\mathbb{C})$ be the algebra of $n \times n$ matrices. Prove that* $\mathrm{rad}(A) = 0$.

Exercise 3.2.2. *Prove that* $\mathrm{rad}(A \times B) = \mathrm{rad}(A) \times \mathrm{rad}(B)$.

Exercise 3.2.3. *Prove that the following statements are equivalent for a finite dimensional algebra A:*

(1) *A is semi-simple.*

(2) *Every finite dimensional A-module is semi-simple.*

(3) *Every finite dimensional indecomposable A-module is irreducible.*

Exercise 3.2.4. *Suppose that A is a finite dimensional semi-simple algebra. Then there exists a module M such that $A = \mathrm{rad}(A) \oplus M$ as a left A-module.*

(1) *Show that* $\mathrm{rad}(A)M = 0$.

(2) *Show that* $\mathrm{rad}(A)^2 = \mathrm{rad}(A)$.

(3) *Prove that* $\mathrm{rad}(A) = 0$.

Exercise 3.2.5. *Suppose that A is a finite dimensional A-algebra with* $\mathrm{rad}(A) = 0$. *Suppose that M is an indecomposable summand of A and let $m \in M$. Right multiplication with m defines a left-module homomorphism $f : M \to M$.*

(1) *Show that f is an isomorphism or nilpotent.*

(2) *Show that $M = Am$ or m is nilpotent.*

(3) *Show that M is irreducible.*

Exercise 3.2.6. *Use Nakayama's lemma to prove Corollary* 3.2.5.

3.3. Basic Algebras

Definition 3.3.1. A finite dimensional algebra A is called **basic** if the algebra $A/\mathrm{rad}(A)$ is isomorphic to $\mathbb{C}^n$ for some n.

For example, from Lemma 3.2.7 it follows that an algebra of the form $\mathbb{C}Q/J$, where Q is a quiver and J is an admissible ideal, is basic. The following theorem shows that these are all basic algebras.

Theorem 3.3.2. *Every basic algebra A is isomorphic to $\mathbb{C}Q/J$ for some quiver Q and an admissible ideal. Moreover, Q is uniquely determined.*

The remainder of this section is dedicated to the proof of this theorem. We want to construct a surjective algebra homomorphism $\lambda : \mathbb{C}Q \to A$ where the kernel J is admissible. First we need to determine the images of $\varphi(e_x)$, $x \in Q_0$. These will be orthogonal idempotents in A, and must be lifts of the orthogonal idempotents of $A/\mathrm{rad}(A) \cong \mathbb{C}^n$. In Proposition 3.3.3 we will show that these idempotents in $A/\mathrm{rad}(A)$ can indeed be lifted. After this we have to determine the images of $\lambda(a)$, $a \in Q_1$. This is a problem of lifting the elements of $\mathrm{rad}(A)/\mathrm{rad}(A)^2$ to arrows in Q_1 (Lemma 3.3.5). Finally, $\mathbb{C}Q$ can be viewed as a tensor algebra as in Section 1.5, and using the universal property we can construct λ.

Proposition 3.3.3 (lifting of the idempotents). *Suppose that A is an algebra, and $J \subseteq A$ is an ideal consisting of nilpotent elements. If $e_1, \ldots, e_n \in A/J$ are orthogonal idempotents, then there exist orthogonal idempotents $\overline{e}_1, \ldots, \overline{e}_n \in A$ such that $\overline{e}_i + J = e_i$ for all i.*

Proof. Choose $f_n \in A$ such that $f_n + J = e_n$. Then $f_n^2 - f_n \in J$, and $(f_n^2 - f_n)^k = 0$ for some $k > 0$. If $p(x) = (1-(1-x)^k)^k$, then $p(x)$ is divisible by x^k and $1 - p(x)$ is divisible by $(1-x)^k$, so $p(x) - p(x)^2$ is divisible by $(x - x^2)^k$. For $x = f_n$ we have $p(f_n) - p(f_n)^2 = 0$. So $\overline{e}_n := p(f_n)$ is idempotent. Also, $p(x) - x^k$ is divisible by $x(1-x)$, so $\overline{e}_n - f_n^k \in J$ and $\overline{e}_n + J = e_n^k = e_n$. Let $A' = (1-\overline{e}_n)A(1-\overline{e}_n)$ and $J' = (1-\overline{e}_n)J(1-\overline{e}_n)$. Note that $e_1, \ldots, e_{n-1}$ lie in $A'/J' = (1-e_n)(A/J)(1-e_n)$ because $e_i = (1-e_n)e_i(1-e_n)$ for all i. By induction, we can lift $e_1, \ldots, e_{n-1}$ to orthogonal idempotents $\overline{e}_1, \ldots, \overline{e}_{n-1} \in A'$. We also have $\overline{e}_i\overline{e}_n = \overline{e}_n\overline{e}_i = 0$ for all $i < n$ because $\overline{e}_i \in A' = (1-\overline{e}_n)A(1-\overline{e}_n)$. □

Suppose that A is a basic algebra. Let $\operatorname{rad}(A)$ be the Jacobson radical. Then $R = A/\operatorname{rad}(A)$ is isomorphic to $\mathbb{C}^n$ for some integer n. For some integer k we have $\operatorname{rad}(A)^k = 0$ by Lemma 3.2.8.

Lemma 3.3.4. *Let $\pi_R : A \to R$ be the quotient homomorphism. Then there exists a ring homomorphism $\lambda_R : R \to A$ such that $\pi_R\lambda_R = 1_R$.*

Proof. Let $e_1, \ldots, e_n$ be the basis vectors of $R = \mathbb{C}^n = A/\operatorname{rad}(A)$. Then $e_1, \ldots, e_n$ are orthogonal idempotents. We can lift $e_1, \ldots, e_n$ to orthogonal idempotents $\overline{e}_1, \ldots, \overline{e}_n \in A$ by Proposition 3.3.3. We now define $\lambda_R : R \to A$ by $\lambda_R(e_i) = \overline{e}_i$. □

Let $M = \operatorname{rad}(A)/\operatorname{rad}(A)^2$. Then M is an R-bimodule. The following lemma is obvious.

Lemma 3.3.5 (lifting of the arrows). *Let $\pi_M : \operatorname{rad}(A) \to M$ be the quotient homomorphism. There exists a bimodule homomorphism $\lambda_M : M \to \operatorname{rad}(A)$ such that $\pi_M\lambda_M = 1_M$.*

We may view M as a sub-bimodule of A. There now is a unique algebra homomorphism

$$\lambda : T_R(M) \to A$$

such that the restrictions of λ to R and M are equal to λ_R and λ_M, respectively, by Proposition 1.5.5.

Lemma 3.3.6. *The map λ is surjective.*

Proof. We have

$$A \subseteq \operatorname{im} \lambda + \operatorname{rad}(A)$$

and

$$\operatorname{rad}(A) \subseteq \operatorname{im}\lambda + \operatorname{rad}(A)^2$$

by construction. Taking the i-th power gives

$$\operatorname{rad}(A)^i \subseteq \operatorname{im}\lambda + \operatorname{rad}(A)^{i+1}.$$

So we have

$$A \subseteq \operatorname{im}\lambda + \operatorname{rad}(A) \subseteq \operatorname{im}\lambda + \operatorname{rad}(A)^2 \subseteq \cdots \subseteq \operatorname{im}\lambda + \operatorname{rad}(A)^k = \operatorname{im}\lambda.$$

□

We identify $T_R(M)$ with a path algebra $\mathbb{C}Q$. So $A \cong \mathbb{C}Q/J$, where J is the kernel of λ. Since λ induces an isomorphism $T_R(M)/\mathfrak{m} \cong A/\operatorname{rad}(A)$, we have $J \subseteq \mathfrak{m}$. Since λ also induces an isomorphism $M \cong \mathfrak{m}/\mathfrak{m}^2 \to \operatorname{rad}(A)/\operatorname{rad}(A)^2$ we see that $J \subseteq \mathfrak{m}^2$. We also have $\mathfrak{m}^k \subseteq J$ because $\lambda(\mathfrak{m}^k) \subseteq \operatorname{rad}(A)^k = 0$. So J is an admissible ideal.

The number of arrows from i to j is $\dim e_j M e_i$. So the quiver Q is uniquely determined by A.

Exercises.

Exercise 3.3.1. *Show that a basic algebra is hereditary if and only if it is a path algebra.*

Exercise 3.3.2. *Suppose that A is a finite dimensional algebra and $M_1, \ldots, M_n$ are pairwise nonisomorphic indecomposable finite dimensional A-modules. Define $M = M_1 \oplus \cdots \oplus M_n$. Show that endomorphism ring $\operatorname{End}_A(M) = \operatorname{Hom}_A(M, M)$ is a basic algebra.*

3.4. Morita Equivalence

Definition 3.4.1. We say that two finite dimensional algebras A and B are **Morita equivalent** if the categories A–mod and B–mod are equivalent as $\mathbb{C}$-categories.

The main result of this section is Theorem 3.4.11 which gives us a characterization of Morita equivalence on the level of algebras. As a consequence, we will see that every finite dimensional algebra is Morita equivalent to a basic algebra (Corollary 3.4.13). Also, we use the main result to give a proof of Wedderburn's theorem for finite dimensional $\mathbb{C}$-algebras (Corollary 3.4.12).

The following easy, but useful lemma, we will use repeatedly.

Lemma 3.4.2. *Suppose that that $\mathcal{F}_1, \mathcal{F}_2 : \mathcal{C}_1 \to \mathcal{C}_2$ are two additive functors between two abelian categories, and $\gamma : \mathcal{F}_1 \to \mathcal{F}_2$ is a natural transformation. If A and C are objects such that A is a direct summand of C^l for some $l > 1$ and $\gamma(C) : \mathcal{F}_1(C) \to \mathcal{F}_2(C)$ is an isomorphism, then $\gamma(A)$ is an isomorphism.*

Proof. Let $A \oplus B = C^l$. We have

$$\gamma(A \oplus B) = \begin{pmatrix} \gamma(A) & 0 \\ 0 & \phi(B) \end{pmatrix} = \gamma(C^l) = \begin{pmatrix} \gamma(C) & & \\ & \ddots & \\ & & \gamma(C) \end{pmatrix}$$

is an isomorphism because $\gamma(C)$ is an isomorphism. So $\gamma(A)$ and $\gamma(B)$ are isomorpisms. □

Let P be an A-module. The space of endomorphisms $\mathrm{End}_A(P) := \mathrm{Hom}_A(P,P)$ is an associative ring. For an A-module M, $\mathrm{Hom}_A(P,M)$ is a right $\mathrm{End}_A(P)$-module. If we set $B = \mathrm{End}_A(P)^{\mathrm{op}}$, then $\mathrm{Hom}_A(P,-)$ is a functor from A–mod to B–mod. Under certain assumptions on P, this functor will be an equivalence of categories, and A and B will be Morita equivalent algebras.

In what follows, A and A^{op} naturally appear as A-modules. So we make the following cautionary remark.

Remark 3.4.3. Note that as A–A-bimodules, A and A^{op} are isomorphic. However, A and A^{op} may not be isomorphic as algebras. In particular, $\mathrm{End}_A(A) \cong A^{\mathrm{op}}$ is isomorphic to A as an A–A-bimodule, but they may not be isomorphic as algebras.

Let us assume that P is a projective A-module. So P is a direct summand of A^l for some $l > 0$. We define $\widetilde{P} := \mathrm{Hom}_A(P,A)$ and $B = \mathrm{End}_A(P)^{\mathrm{op}}$. Now P is a (A,B)-bimodule, and $\widetilde{P}$ is a (B,A)-bimodule.

Lemma 3.4.4. *For every A-module M we have an isomorphism*

$$\widetilde{P} \otimes_A M \cong \mathrm{Hom}_A(P,M)$$

of left B-modules.

Proof. There is a natural transformation

$$\gamma : \mathrm{Hom}_A(-,A) \otimes_A M \to \mathrm{Hom}_A(-,M),$$

where

$$\gamma(N) : \mathrm{Hom}_A(N,A) \otimes_A M \to \mathrm{Hom}_A(N,M)$$

maps $f \otimes m$ to the function $n \mapsto f(n)m$. Since $\gamma(A)$ is an isomorphism, and P is a direct summand of A^l, it follows from Lemma 3.4.2 that

$$\gamma(P) : \widetilde{P} \otimes_A M \to \mathrm{Hom}_A(P,M)$$

is an isomorphism. Note that $\gamma(P)$ is a B-module homomorphism. □

The reader may verify that γ is also natural in M. So the functors $\widetilde{P} \otimes_A -$ and $\mathrm{Hom}_A(P,-)$ are naturally isomorphic.

Lemma 3.4.5. *If M is a left B-module, N is an (A, B)-bimodule and U is a left A-module, then we have an isomorphism*

$$\mathrm{Hom}_B(M, \mathrm{Hom}_A(N, U)) \cong \mathrm{Hom}_A(N \otimes_B M, U).$$

Proof. See Exercise 3.4.1. □

We call P a **progenerator** if every indecomposable projective module appears in the decomposition of P. This means that A is a direct summand of P^k for some positive integer k. From now on, we assume that P is a progenerator. In Theorem 3.4.11 we will see that the functors $\mathcal{F} = \mathrm{Hom}_A(P, -)$ and $\mathcal{G} = \mathrm{Hom}_B(\widetilde{P}, -)$ give equivalences of categories: $\mathcal{G} \circ \mathcal{F}$ and $\mathcal{F} \circ \mathcal{G}$ are naturally equivalent to the identity of the categories A–mod and B–mod, respectively.

The following lemma shows that $\widetilde{P}$ inherits the progenerator property from P.

Lemma 3.4.6. *The B-module $\widetilde{P}$ is projective, and B is a direct summand of $\widetilde{P}^l$. So $\widetilde{P}$ is a progenerator in the category B–mod.*

Proof. Since A is a direct summand of P^k, $\widetilde{P} = \mathrm{Hom}_A(P, A)$ is a direct summand of the free module $\mathrm{Hom}_A(P, P^k) \cong B^k$. Hence $\widetilde{P}$ is projective. Since P is a direct summand of A^l, $B = \mathrm{Hom}_A(P, P)$ is a direct summand of $\mathrm{Hom}_A(P, A^l) \cong \widetilde{P}^l$. □

The following two lemmas show that the functor $\mathcal{F}$ and $\mathcal{G}$ give equivalences.

Lemma 3.4.7. *For every projective A-module P', we have an isomorphism*

$$P \otimes_B \mathrm{Hom}_A(P, P') \cong P'$$

of left A-modules. In particular,

$$P \otimes_B \widetilde{P} = P \otimes_B \mathrm{Hom}_A(P, A) \cong A.$$

Proof. We define a natural transformation

$$\gamma : P \otimes_B \mathrm{Hom}_A(P, -) \to \mathrm{id}$$

as follows. For an A-module M,

$$\gamma(M) : P \otimes_B \mathrm{Hom}_A(P, M) \to M$$

maps $v \otimes f$ to $f(v)$.

Since P' is a direct summand of P^s for some s, it suffices to show that $\gamma(P)$ is an isomorphism because of Lemma 3.4.2. We have

$$P \otimes_B \mathrm{Hom}_A(P, P) \cong P \otimes_B B \cong P,$$

so $\gamma(P)$ is indeed an isomorphism. □

Lemma 3.4.8. *For every A-module M we have an isomorphism of A-modules*

$$\operatorname{Hom}_B(\widetilde{P}, \operatorname{Hom}_A(P, M)) \cong M.$$

Proof. By Lemma 3.4.5 we have

$$\operatorname{Hom}_B(\widetilde{P}, \operatorname{Hom}_A(P, M)) = \operatorname{Hom}_A(P \otimes_B \widetilde{P}, M) = \operatorname{Hom}_A(A, M) \cong M.$$

□

Corollary 3.4.9. *We have an isomorphism of K-algebras*

$$\operatorname{Hom}_B(\widetilde{P}, \widetilde{P}) \cong A^{\mathrm{op}}.$$

Proof. By Lemma 3.4.8, we have

$$\operatorname{Hom}_B(\widetilde{P}, \widetilde{P}) \cong \operatorname{Hom}_B(\widetilde{P}, \operatorname{Hom}_A(P, A)) \cong A$$

as left A-modules. Since $\operatorname{Hom}_B(\widetilde{P}, \widetilde{P})$ contains A^{op} as a subalgebra, we have $\operatorname{Hom}_B(\widetilde{P}, \widetilde{P}) \cong A^{\mathrm{op}}$. □

Corollary 3.4.10. *We have an isomorphism of A-modules*

$$\widetilde{\widetilde{P}} := \operatorname{Hom}_B(\widetilde{P}, B) \cong P.$$

Proof. From Lemma 3.4.8 it follows that

$$\operatorname{Hom}_B(\widetilde{P}, B) \cong \operatorname{Hom}_B(\widetilde{P}, \operatorname{Hom}_A(P, P)) \cong P.$$

□

Theorem 3.4.11. *Two finite dimensional algebras A and B are Morita equivalent if and only if $B = \operatorname{End}_A(P)^{\mathrm{op}}$ for some progenerator P in A–mod.*

Proof. Suppose that the A-module P is a progenerator. Define

$$\mathcal{F} = \operatorname{Hom}_A(P, -) : A\text{–mod} \to B\text{–mod}$$

and

$$\mathcal{G} = \operatorname{Hom}_B(\widetilde{P}, -) : B\text{–mod} \to A\text{–mod}.$$

For every A-module M we have

$$\mathcal{G}(\mathcal{F}(M)) = \operatorname{Hom}_B(\widetilde{P}, \operatorname{Hom}_A(P, M)) \cong M$$

by Lemma 3.4.8. This gives a natural isomorphism between the functors $\mathcal{G} \circ \mathcal{F}$ and the identity functor. Since the B-module $\widetilde{P}$ is a progenerator, $A = \operatorname{Hom}_B(\widetilde{P}, \widetilde{P})$ and $P = \operatorname{Hom}_B(\widetilde{P}, A)$, we can reverse the roles of A and B, and of $\mathcal{F}$ and $\mathcal{G}$. We see that there exists a natural isomorphism between $\mathcal{F} \circ \mathcal{G}$ and the identity functor. So the categories A–mod and B–mod are equivalent, and A and B are Morita equivalent.

Conversely, suppose that A and B are Morita equivalent. Let $\mathcal{F}$: A–mod $\to$ B–mod and $\mathcal{G}$: B–mod $\to$ A–mod be such that $\mathcal{G} \circ \mathcal{F}$ and

$\mathcal{F} \circ \mathcal{G}$ are naturally equivalent to the identity. Since $\operatorname{End}_B(B) \cong B^{\mathrm{op}}$, it follows that $\operatorname{End}_A(\mathcal{G}(B)) \cong B^{\mathrm{op}}$. Since B is a progenerator, $\mathcal{G}(B)$ is a progenerator. □

Corollary 3.4.12. *Every finite dimensional algebra is Morita equivalent to a basic algebra.*

Proof. Suppose A is a finite dimensional algebra. Decompose A into projectives: $A = P_1^{m_1} \oplus \cdots \oplus P_n^{m_n}$ with $P_1, \ldots, P_n$ pairwise nonisomorphic. and define $P = P_1 \oplus P_2 \oplus \cdots \oplus P_n$. Then $B = \operatorname{Hom}_A(P, P)^{\mathrm{op}}$ is a basic algebra (see Exercise 3.3.2) and A and B are Morita equivalent. □

Recall that an A-module is called semi-simple if it is the direct sum of irreducible modules. A finite dimensional $\mathbb{C}$-algebra A is called semi-simple if it is semi-simple as a left A-module.

Corollary 3.4.13 (Wedderburn's theorem for $\mathbb{C}$)**.** *If A is a finite dimensional semi-simple $\mathbb{C}$ algebra, then A is a finite product of matrix algebras.*

Proof. By Exercise 3.2.3, the algebra A is semi-simple if and only if every object in the category A–mod is semi-simple. The algebra A is Morita equivalent to $B = \mathbb{C}Q/J$ where J is an admissible ideal. So B is semi-simple as well. It follows that $Q_1 = \emptyset$ and $J = 0$. Now $A = \operatorname{Hom}_B(P, P)^{\mathrm{op}}$ for some B-module P. Because P is semi-simple, A is a product of matrix algebras. □

Exercises.

Exercise 3.4.1. *Proof Lemma* 3.4.5

Exercise 3.4.2. *Show that a finite dimensional algebra is hereditary if and only if it is Morita equivalent to a path algebra. (See Exercise* 3.3.1.*)*

3.5. Bibliographical Remarks

The quivers with relations are convenient when dealing with module categories defined in terms of linear algebra. As proved in this chapter, the module category of a finite dimensional K-algebra is equivalent to the category of modules over some quiver with relations. As before, we restrict ourselves to path algebras of acyclic quivers over the complex numbers. More general versions of the results of this chapter can be found in the books [**3**], [**43**] and [**38**].

The quivers with relations of the form $\mathbb{C}Q/I$ where Q is acyclic are special, as they have finite global dimension.

One of the important unsolved problems about quivers with relations (or more generally Artin algebras) is the Auslander finitistic global dimension conjecture. It says that if $\mathbb{C}Q/I$ is a quiver with relations (Q is an arbitrary finite quiver, I is an admissible ideal), then the maximum of projective dimensions of modules of finite projective dimensions is finite.

Chapter 4

Gabriel's Theorem

This chapter is devoted to Gabriel's theorem which was the turning point in the development of quiver representations. The classification of hereditary algebras of finite representation type corresponds to the Cartan classification of root systems with roots of one length, and the dimension vectors of indecomposables correspond to positive roots. These results pointed to deep connections to Lie theory. It is interesting to observe that for algebras of finite representation type the dimension vectors of indecomposables do not depend on the orientation of arrows in a quiver Q. This led to many developments. The connection of Euler form with the scalar product on the root space coming from Killing form of a the algebra led to a quick proof of half of the theorem. The idea of reflection functors and Coxeter functors provided connection to a Weyl group of a given root system. Still, reflections functors are possible only at the sink and source vertices.

Gabriel's theorem was quickly generalized to the extended Dynkin quivers where the explicit classification of indecomposables was still possible. we will discuss it in Chapter 7. Later Victor Kac generalized these results to arbitrary quivers providing a deep connection to Kac-Moody Lie algebras and corresponding infinite root systems. This will be discussed in Chapter 8.

At the same time Gabriel's theorem inspired researchers to look for notions related to quiver representations that would not depend on the orientation of arrows of a quiver as such notions would be more likely to be connected to root systems. This led to the notion of preprojective algebras (see Section 6.7). This aspect was also reflected in the structure of derived categories and cluster categories (see Chapter 12).

4.1. Quivers of Finite Representation Type

Definition 4.1.1. A quiver Q has **finite representation type** if the category $\operatorname{Rep}(Q)$ contains finitely many isomorphism classes of indecomposable objects.

The goal of this chapter is to classify the quivers of finite representation type. An important tool is the quadratic form called the Tits form.

Definition 4.1.2. The **Tits form** for Q is defined by

$$B_Q(\alpha) = \langle \alpha, \alpha \rangle = \sum_{x \in Q_0} \alpha(x)\alpha(x) - \sum_{a \in Q_1} \alpha(ta)\alpha(ha)$$
$$= \dim \mathrm{GL}_\alpha - \dim \operatorname{Rep}_\alpha(Q).$$

The **Cartan form** is defined by

$$(\alpha, \beta) = \langle \alpha, \beta \rangle + \langle \beta, \alpha \rangle.$$

We have that $(\alpha, \alpha) = 2B_Q(\alpha)$.

For a quiver, let $\Gamma(Q)$ be the (nonoriented) graph obtained from Q by forgetting about the orientations of the arrows. Note that B_Q and $(\cdot, \cdot)$ only depend on the graph $\Gamma(Q)$, and not on the orientation of Q.

Lemma 4.1.3. *If Q has finite representation type, then $B_Q(\alpha) \geq 1$ for every nonzero dimension vector α.*

Proof. Suppose that Q has finite representation type and that $\alpha \in \mathbb{N}^{Q_0}$ is nonzero. Then there are only finitely many isomorphism classes of representations of dimension α. So there are only finitely many GL_α-orbits in $\operatorname{Rep}_\alpha(Q)$. In particular, one of these orbits, say $\mathrm{GL}_\alpha \cdot V$ is Zariski open and has dimension $\dim \operatorname{Rep}_\alpha(Q)$. The stabilizer $(\mathrm{GL}_\alpha)_V$ of V is at least 1 dimensional by Remark 1.3.4. We conclude that

$$B_Q(\alpha) = \dim \mathrm{GL}_\alpha - \dim \operatorname{Rep}_\alpha(Q)$$
$$= \dim \mathrm{GL}_\alpha - \dim \mathrm{GL}_\alpha \cdot V = \dim(\mathrm{GL}_\alpha)_V \geq 1.$$

□

Exercises.

Exercise 4.1.1. *Suppose that Q has finite representation type. Show that B_Q is positive definite using the following steps.*

(1) *Show that $B_Q(\alpha) \geq 1$ for every nonzero $\alpha \in \mathbb{Z}^{Q_0}$.*

(2) *Show that $B_Q(\alpha) > 0$ for every $\alpha \in \mathbb{Q}^{Q_0}$.*

(3) *Show that* $B_Q(\alpha) \geq 0$ *for every* $\alpha \in \mathbb{R}^{Q_0}$, *i.e.,* B_Q *is positive semi-definite.*

(4) *Finally, prove that* B_Q *is positive definite.*

4.2. Dynkin Graphs

The simple Lie algebras correspond to irreducible root systems and Dynkin diagrams. The Dynkin diagrams are $\mathbf{A}_n$ $(n \geq 1)$, $\mathbf{B}_n$ $(n \geq 2)$, $\mathbf{C}_n$ $(n \geq 3)$, $\mathbf{D}_n$ $(n \geq 4)$, $\mathbf{E}_6$, $\mathbf{E}_7$, $\mathbf{E}_8$, $\mathbf{F}_4$ and $\mathbf{G}_2$. The diagrams of type **B**, **C**, **F** and **G** correspond to root systems with roots of different lengths. These diagrams will not appear in the classification of quivers of finite representation type.

In this section we will prove half of Gabriel's theorem. Namely, if a quiver Q has finite representation type, then the underlying graph $\Gamma(Q)$ must be a union of Dynkin graphs of type ADE (Theorem 4.2.4). The **Dynkin graphs** of type ADE are:

$\mathbf{A}_n(n \geq 1)$

$\mathbf{D}_n(n \geq 4)$

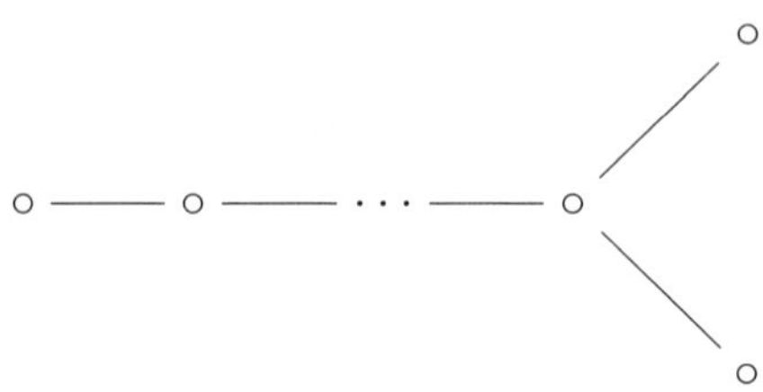

$\mathbf{E}_6$

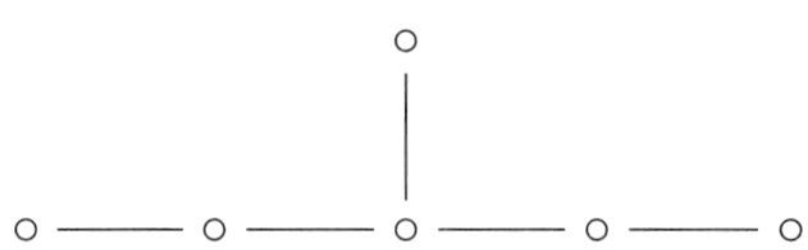

$\mathbf{E}_7$

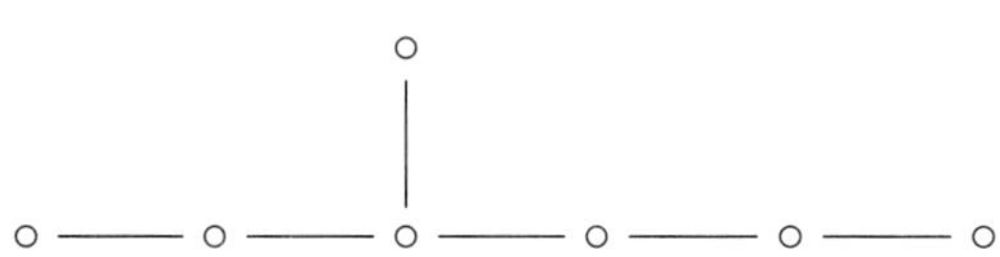

$\mathbf{E}_8$

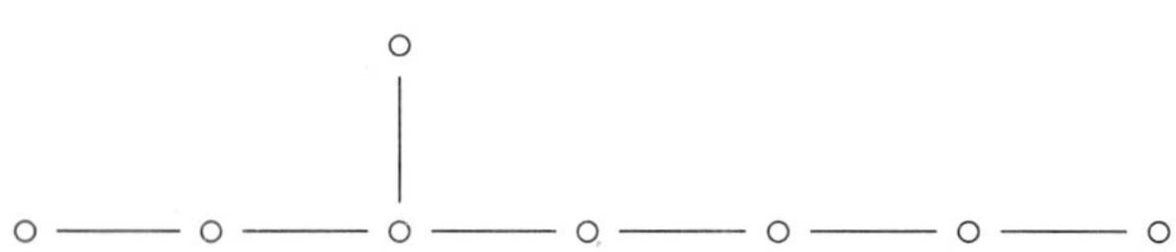

Below we introduce the extended Dynkin graphs. As we will see later these diagrams are critical in the following sense: A quiver has finite representation type if and only if the underlying graph contains an extended Dynkin diagram of type ADE as a subgraph. The **extended Dynkin graphs** of type ADE are:

$\widehat{\mathbf{A}}_n$

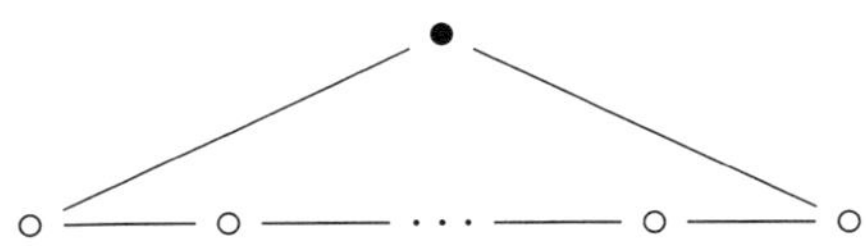

$\widehat{\mathbf{D}}_n$

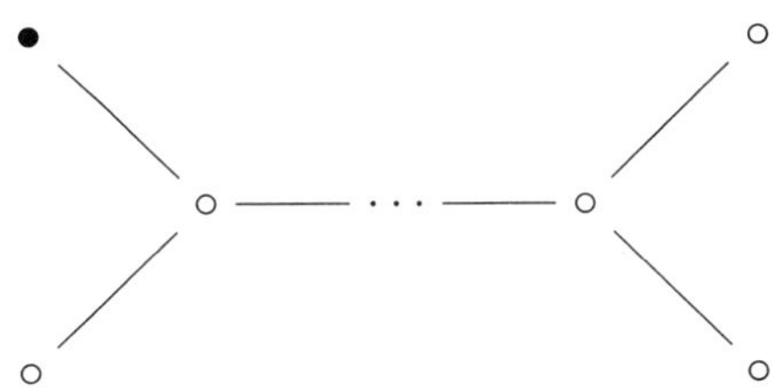

$\widehat{\mathbf{E}}_6$

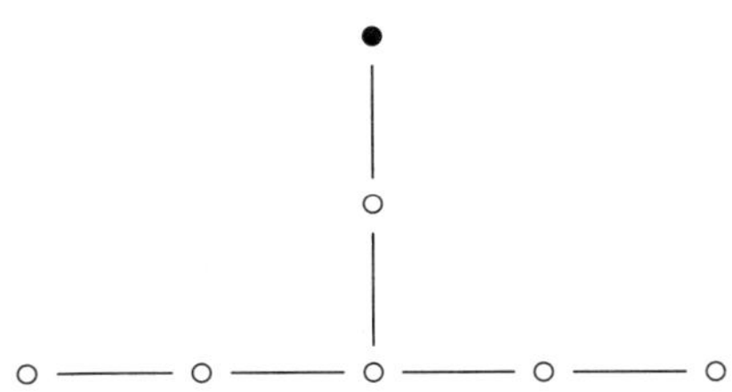

$\widehat{\mathbf{E}}_7$

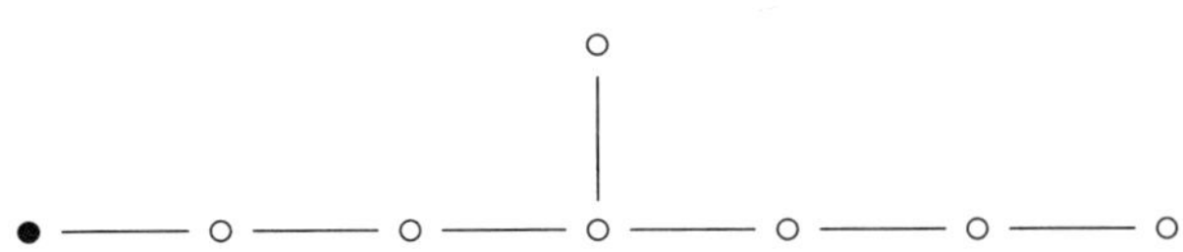

$\widehat{\mathbf{E}}_8$

We use the convention that $\widehat{A}_0$ and $\widehat{A}_1$ are the graphs

respectively.

Lemma 4.2.1. *If $\Gamma(Q)$ is an extended Dynkin graph of type ADE, then $B_Q(\alpha) = 0$ for some nonzero vector $\alpha \in \mathbb{N}^n$.*

Proof. For each of the graphs we can take choose the following vector α such that $B_\Gamma(\alpha) = 0$:

(4.1)

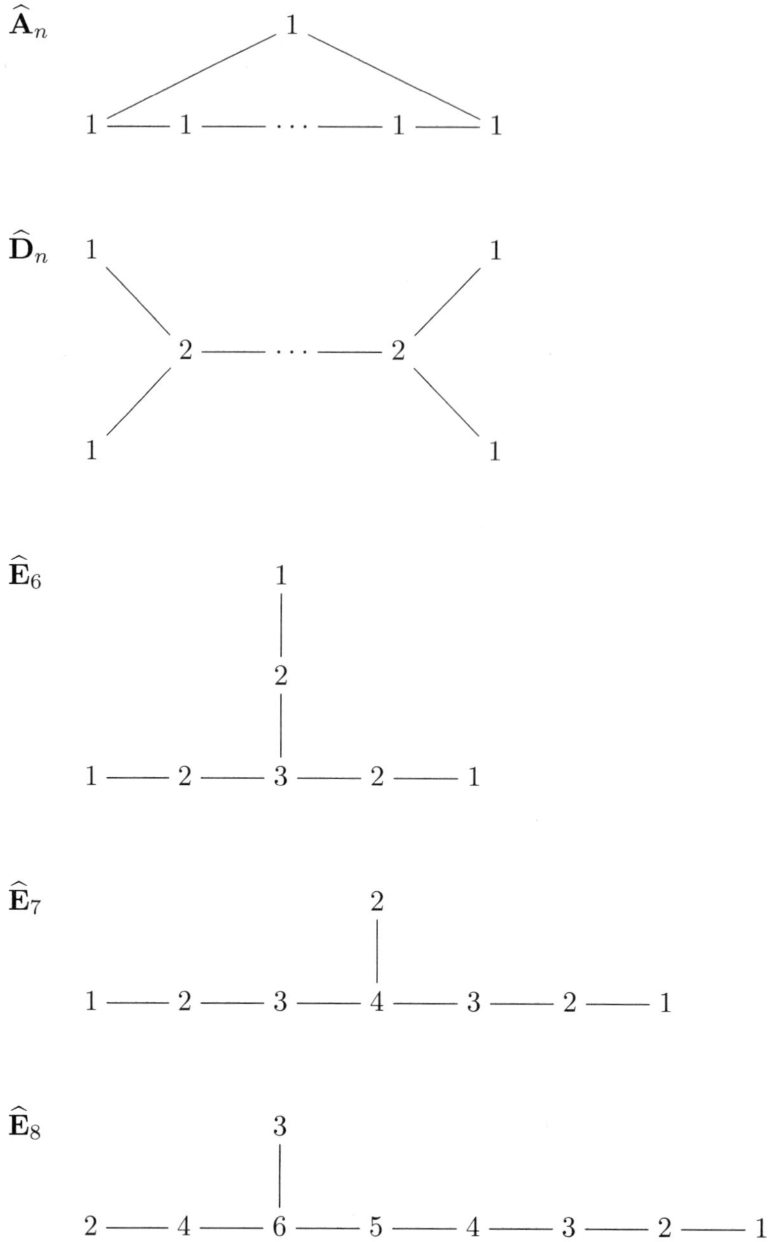

□

Lemma 4.2.2. *If $\Gamma(Q)$ is an extended Dynkin diagram, then B_Q is nonnegative definite, and $B_Q(\beta) = 0$ if and only if β is a multiple of the vector given in* (4.1).

Proof. The proof is Exercise 4.2.3 □

Lemma 4.2.3. *If Q' is a subquiver of Q and $B_Q(\alpha) \geq 1$ for every nonzero dimension vector $\alpha \in \mathbb{N}^{Q_0}$, then $B_{Q'}(\alpha) \geq 1$ for every nonzero dimension vector $\alpha \in \mathbb{N}^{Q'_0}$.*

Proof. If $\alpha \in \mathbb{N}^{Q'_0}$ is nonzero, then we define $\overline{\alpha} \in \mathbb{N}^{Q_0}$ by $\overline{\alpha}(x) = \alpha(x)$ if $x \in Q'_0$ and $\overline{\alpha}(x) = 0$ if $x \in Q_0 \setminus Q'_0$. Then it follows that $B_{Q'}(\alpha) \geq B_Q(\overline{\alpha}) \geq 1$. □

Theorem 4.2.4. *If Q has finite representation type, then $\Gamma(Q)$ is a union of Dynkin graphs of type ADE.*

Proof. Suppose that Q has finite representation type. For every subquiver Q' of Q and every nonzero dimension vector α we have $B_{Q'}(\alpha) \geq 1$. So $\Gamma(Q')$ cannot be an extended Dynkin graph of type ADE. So no subgraph of $\Gamma(Q)$ is isomorphic to an extended Dynkin graph of type ADE. Suppose that $\Gamma \subseteq \Gamma(Q)$ is a connected component of $\Gamma(Q)$. The graph $\widehat{\mathbf{A}}_0$ is a loop, so Γ does not contain loops. Since Γ cannot contain the graph $\widehat{\mathbf{A}}_1$, it does not have multiple edges. Since Γ cannot contain the graphs $\widehat{\mathbf{A}}_n$ for $n \geq 2$, it has no cycles. Hence Γ is a tree. Since Γ does not contain $\widehat{\mathbf{D}}_3$, every vertex has degree ≤ 3. If every vertex of Γ has degree ≤ 2, then Γ is isomorphic to $\widehat{\mathbf{A}}_n$ for some n. Otherwise, Γ has at least 1 vertex with degree 3. It cannot have 2 vertices with degree 3 because then it would contain $\widehat{\mathbf{D}}_n$ for some $n \geq 4$. So Γ is of the following form:

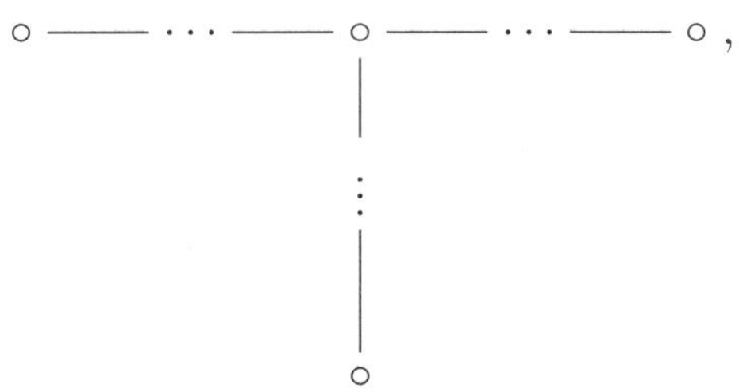

where the arms (including the central vertex) have lengths p, q, r, respectively, with $2 \leq p \leq q \leq r$. The total number of vertices is $n = p+q+r-2$. We have $p = 2$ because if $p \geq 3$, then Γ contains $\widehat{\mathbf{E}}_6$. We also have $q \leq 3$ because if $q \geq 4$, then Γ contains $\widehat{\mathbf{E}}_7$. If $q = 2$, then Γ us isomorphic to $\mathbf{D}_n$. If $q = 3$, then $r \leq 5$ because otherwise Γ contains $\widehat{\mathbf{E}}_8$. If $r = 3, 4, 5$, then Γ is isomorphic to $\mathbf{E}_6, \mathbf{E}_7, \mathbf{E}_8$, respectively. □

Exercises.

Exercise 4.2.1. *Show directly that if Q' is a subquiver of Q, and Q has finite representation type, then Q' has finite representation type.*

Exercise 4.2.2. *Show that if Γ is Dynkin graph of type ADE, then B_Γ is positive definite.*

Exercise 4.2.3. *Show that if Γ is an extended Dynkin diagram, then B_Q is nonnegative definite. Show that $B_\Gamma(\beta) = 0$ if and only if β is a multiple of the vector in* (4.1).

Exercise 4.2.4. *Show that if Γ is one of the graphs below, then B_Γ is* not *nonnegative definite.*

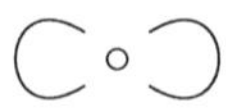

$\widehat{\widehat{\mathbf{A}}}_n (n \geq 0)$

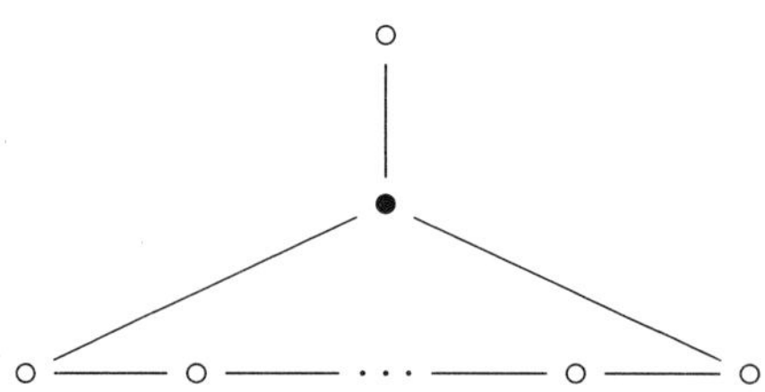

$\widehat{\widehat{\mathbf{D}}}_n (n \geq 4)$

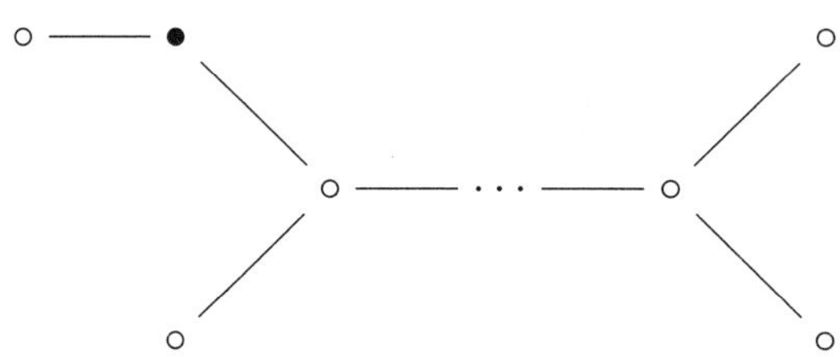

$\widehat{\widehat{\mathbf{E}}}_6$

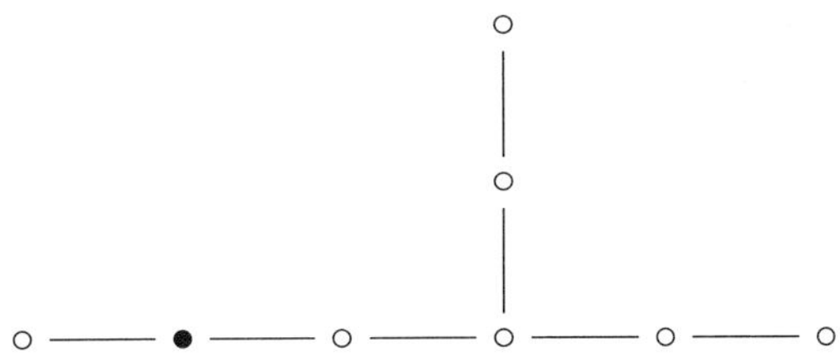

$\widehat{\widehat{\mathbf{E}}}_7$

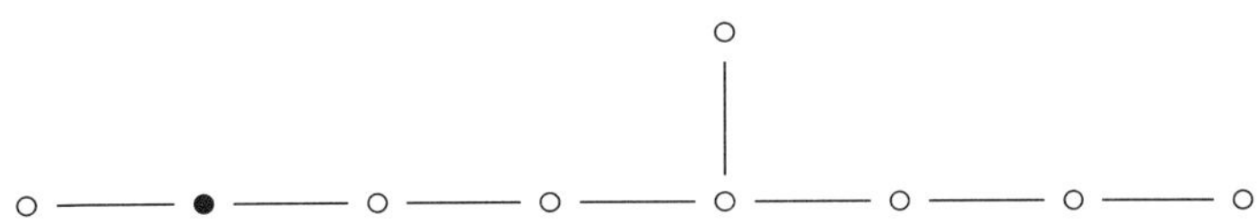

$\widehat{\widehat{\mathbf{E}}}_8$

Our convention is that $\widehat{\widehat{\mathrm{A}}}_0$ *and* $\widehat{\widehat{\mathrm{A}}}_1$ *are the graphs*

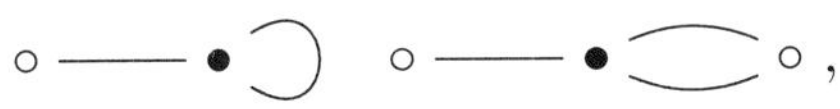

respectively.

Exercise 4.2.5. *Show that if* Γ *is a graph for which* B_Γ *is nonnegative definite, then* Γ *is a union of Dynkin graphs and extended Dynkin graphs of type ADE.*

4.3. The Reflection Functors

In a root system, every root can be obtained from a simple root by a sequence of simple reflections. We will use this idea to construct indecomposable representations from simple representations by using a sequence of reflection. We have to develop a notion of a reflection of a quiver representation. First, we define the action of reflections on the quiver Q itself.

Let $Q = (Q_0, Q_1, h, t)$ be a quiver.

Definition 4.3.1. For $x \in Q_0$, define $\sigma_x(Q)$ as the quiver in which the direction of the arrows connecting to x is reversed. More precisely, define

$\sigma_x(Q) = (Q_0, Q_1', h, t)$, where

$$Q_1' = \{a \in Q_1 \mid ha \neq x \text{ and } ta \neq x\} \cup \{a^\star \mid a \in Q_1,\ ha = x \text{ or } ta = x\},$$

and $ha^\star = ta$, $ta^\star = ha$.

Note that $\sigma_x^2(Q)$ is isomorphic to Q.

Definition 4.3.2. If $\alpha \in \mathbb{R}^{Q_0}$, then we define $\sigma_x(\alpha) \in \mathbb{R}^{Q_0} = \mathbb{R}^{\sigma_x(Q)_0}$ by

$$\sigma_x(\alpha)(y) = \begin{cases} \alpha(y) & \text{if } y \neq x, \\ \displaystyle\sum_{\substack{a \in Q_1 \\ ha = x}} \alpha(ta) + \sum_{\substack{a \in Q_1 \\ ta = x}} \alpha(ha) - \alpha(x) & \text{if } y = x. \end{cases}$$

We have $\sigma_x^2(\alpha) = \alpha$.

Example 4.3.3. If Q is the quiver

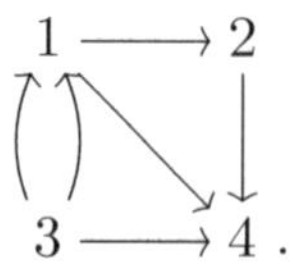

Then $\sigma_1(Q)$ is

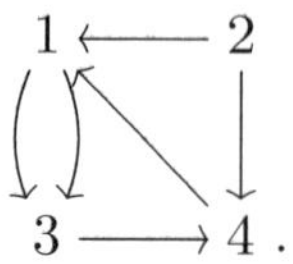

Next we will define reflections of a quiver representation. First we introduce some notation. If $x_1, x_2, \ldots, x_k \in Q_0$ are vertices and V is a representation of Q, and

$$x = \begin{pmatrix} x_1 \\ x_2 \\ \vdots \\ x_k \end{pmatrix},$$

then we use the convention $V(x) = \bigoplus_{i=1}^k V(x_i)$. If $\phi : V \to V'$ is a morphism of representations, then we use the convention

$$\phi(x) = \begin{pmatrix} \phi(x_1) & & & \\ & \phi(x_2) & & \\ & & \ddots & \\ & & & \phi(x_k) \end{pmatrix} : V(x) \to V'(x).$$

If

$$y = \begin{pmatrix} y_1 \\ y_2 \\ \vdots \\ y_l \end{pmatrix} \in Q_0^l$$

and

$$p = \begin{pmatrix} p_{1,1} & p_{1,2} & \cdots & p_{1,k} \\ p_{2,1} & p_{2,2} & \cdots & p_{2,k} \\ \vdots & \vdots & \ddots & \vdots \\ p_{l,1} & p_{l,2} & \cdots & p_{l,k} \end{pmatrix},$$

where $p_{i,j}$ is a path with $tp_{i,j} = x_j$ and $hp_{i,j} = y_i$, then we use the convention

$$V(p) = \begin{pmatrix} V(p_{1,1}) & V(p_{1,2}) & \cdots & V(p_{1,k}) \\ V(p_{2,1}) & V(p_{2,2}) & \cdots & V(p_{2,k}) \\ \vdots & \vdots & \ddots & \vdots \\ V(p_{l,1}) & V(p_{l,2}) & \cdots & V(p_{l,k}) \end{pmatrix}.$$

Definition 4.3.4. A vertex $x \in Q_0$ is a **sink** if there are no arrows $a \in Q_0$ such that $ta = x$. Similarly, $x \in Q_0$ is a **source** if there are no arrows $a \in Q_1$ with $ha = x$.

sink:

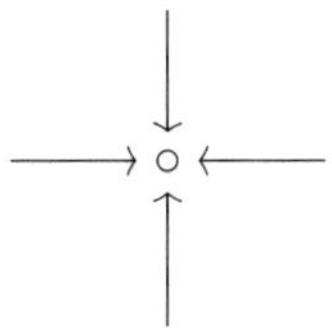

source:

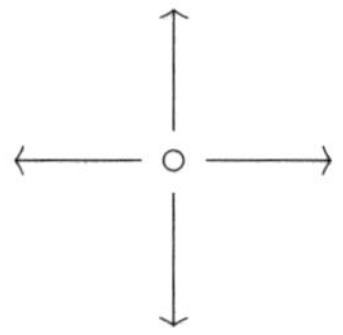

We will only define here reflections of quiver representations at vertices that are a sink or a source. Within a more general context of cluster categories and quivers with potentials it is possible to define reflections at arbitrary vertices (see Chapter 12).

Definition 4.3.5. Suppose that $x \in Q_0$ is a sink. Let $\{a_1, \dots, a_k\}$ be the set of arrows in Q whose head equals x. Let $V \in \operatorname{Rep}_\alpha(Q)$. We define a representation $C_x^+(V) = W$ of $\sigma_x(Q)$ as follows. For $y \neq x$ we define $W(y) = V(y)$, and for $b \in Q_1 \setminus \{a_1, \dots, a_k\}$ we define $W(b) = V(b)$. Let

$$\operatorname{in}_x = \begin{pmatrix} ta_1 \\ ta_2 \\ \vdots \\ ta_k \end{pmatrix},$$

$$a = \begin{pmatrix} a_1 & a_2 & \cdots & a_k \end{pmatrix} : \operatorname{in}_x \to x,$$

and

$$a^\star = \begin{pmatrix} a_1^\star \\ a_2^\star \\ \vdots \\ a_k^\star \end{pmatrix} : x \to \mathrm{in}_x.$$

We define $W(x)$ and

$$W(a_i^\star) : W(x) \to W(ha_i^\star) = W(ta_i) = V(ta_i)$$

such that

$$W(a^\star) = \begin{pmatrix} W(a_1^\star) \\ \vdots \\ W(a_k^\star) \end{pmatrix} : W(x) \to V(\mathrm{in}_x)$$

is the kernel of

$$V(a) = \begin{pmatrix} V(a_1) & \cdots & V(a_k) \end{pmatrix} : V(\mathrm{in}_x) \to V(x).$$

If $\phi : V \to V'$ is a morphism of quiver representations and $W' = C_x^+(V')$, then we define $\psi = C_x^+(\phi) : W \to W'$ as follows. For $y \neq x$ we define $\psi(y) = \phi(y) : W(y) = V(y) \to W'(y) = V'(y)$. The diagram

$$\begin{array}{ccc} V(\mathrm{in}_x) & \xrightarrow{V(a)} & V(x) \\ {\scriptstyle \phi(\mathrm{in}_x)}\downarrow & & \downarrow{\scriptstyle \phi(x)} \\ V'(\mathrm{in}_x) & \xrightarrow[V'(a)]{} & V'(x) \end{array}$$

commutes.

The map $\psi(x)$ is the unique linear map $W(x) \to W'(x)$ such that

$$\begin{array}{ccc} W(x) & \xrightarrow{W(a^\star)} & V(\mathrm{in}_x) \\ {\scriptstyle \psi(x)}\downarrow & & \downarrow{\scriptstyle \phi(\mathrm{in}_x)} \\ W'(x) & \xrightarrow[W'(a^\star)]{} & V'(\mathrm{in}_x) \end{array}$$

commutes. The reader now can verify that C_x^+ is a functor (see Exercise 4.3.2).

Lemma 4.3.6 (see Exercise 4.3.3). *Use the notation of Definition 4.3.5. Then V has a direct summand isomorphic to S_x if and only if $V(a)$ is not onto.*

Definition 4.3.7. Suppose that $x \in Q_0$ is a source. Let $\{b_1, \ldots, b_l\}$ be the set of arrows in Q whose tail equals x. Let $V \in \mathrm{Rep}_\alpha(Q)$. We define a representation $C_x^-(V) = W$ of $\sigma_x(Q)$ as follows. For $y \neq x$ we define $W(y) = V(y)$, and for $a \in Q_1 \setminus \{b_1, \dots, b_l\}$ we define $W(a) = V(a)$. Let

$$\mathrm{out}_x = \begin{pmatrix} hb_1 \\ hb_2 \\ \vdots \\ hb_l \end{pmatrix},$$

$$b = \begin{pmatrix} b_1 \\ b_2 \\ \vdots \\ b_l \end{pmatrix} : x \to \mathrm{out}_x,$$

and

$$b^\star = \begin{pmatrix} b_1^\star & b_2^\star & \cdots & b_l^\star \end{pmatrix} : \mathrm{out}_x \to x.$$

We define $W(x)$ and

$$W(b_j^\star) : W(tb_i^\star) = W(hb_j) = V(hb_j) \to W(x)$$

such that

$$W(b^\star) = \begin{pmatrix} W(b_1^\star) & \cdots & W(b_l^\star) \end{pmatrix} : V(\mathrm{out}_x) \to W(x)$$

is the cokernel of

$$V(b) = \begin{pmatrix} V(b_1) \\ \vdots \\ V(b_l) \end{pmatrix} : V(x) \to V(\mathrm{out}_x).$$

If $\phi : V \to V'$ is a morphism of quiver representations and $W' = C_x^-(V')$, then we define $\psi = C_x^-(\phi) : W \to W'$ as follows. For $y \neq x$ we define $\psi(y) = \phi(y) : W(y) = V(y) \to W'(y) = V'(y)$. The diagram

$$\begin{array}{ccc} V(x) & \xrightarrow{V(b)} & V(\mathrm{out}_x) \\ {\scriptstyle \phi(x)}\downarrow & & \downarrow{\scriptstyle \phi(\mathrm{out}_x)} \\ V'(x) & \xrightarrow[V'(b)]{} & V'(\mathrm{out}_x) \end{array}$$

commutes. The map $\psi(x) : W(x) \to W'(x)$ is the unique linear map such that the diagram

$$\begin{array}{ccc} V(\mathrm{out}_x) & \xrightarrow{W(b^\star)} & W(x) \\ {\scriptstyle \phi(\mathrm{out}_x)}\downarrow & & \downarrow{\scriptstyle \psi(x)} \\ V(\mathrm{out}_x) & \xrightarrow[W'(b^\star)]{} & W'(x) \end{array}$$

commutes. The reader now can verify that C_x^- is a functor (see Exercise 4.3.2).

Lemma 4.3.8 (see Exercise 4.3.3). *Use the notation of Definition 4.3.7. Then V has a direct summand isomorphic to S_x if and only if $V(b)$ is not injective.*

If we want to use the functors C_x^+ and C_x^- to construct indecomposable representations from simple representations, then we would like them to send indecomposable representations to indecomposable representations. This is almost true. The theorem below gives a precise statement.

Theorem 4.3.9 (Bernstein-Gelfand-Ponomarev). *Let us preserve the notation of Definitions 4.3.5 and 4.3.7.*

(1) *Let $x \in Q_0$ be a sink. Let $V \in \operatorname{Rep}(Q)$ be an indecomposable object of dimension α.*
   (a) *We have $V \cong S_x$ if and only if $C_x^+(V) = 0$.*
   (b) *If $V \ncong S_x$, then $C_x^+(V)$ is indecomposable of dimension $\sigma_x(\alpha)$ and $C_x^- C_x^+(V) \cong V$.*

(2) *Let $x \in Q_0$ be a source. Let $V \in \operatorname{Rep}(Q)$ be an indecomposable object of dimension α.*
   (a) *We have $V \cong S_x$ if and only if $C_x^-(V) = 0$.*
   (b) *If $V \ncong S_x$, then $C_x^-(V)$ is indecomposable of dimension $\sigma_x(\alpha)$ and $C_x^+ C_x^-(V) \cong V$.*

Proof. Parts 1(a) and 2(a) are obvious.

Suppose that x is a sink and V is a representation of Q. Let $W = C_x^+(V)$. Then $Z := C_x^-(W)$ is a representation of Q again.

Note that $W(\text{out}_x) = V(\text{in}_x)$ because in the quiver $\sigma_x(Q)$, the outgoing arrows at x are $a_1^\star, \ldots, a_k^\star$ and $ha_i^\star = ta_i$ for all i. We have an exact sequence

$$0 \longrightarrow W(x) \xrightarrow{W(a^\star)} V(\text{in}_x) \xrightarrow{V(a)} V(x) \,. \tag{4.2}$$

Since

$$Z(a) : V(\text{in}_x) = W(\text{out}_x) \to Z(x)$$

is the cokernel of $W(a^\star)$, there exists a unique linear map $\pi(x) : V(x) \to Z(x)$ such that the diagram

$$\begin{array}{ccc} V(\text{in}_x) & \xrightarrow{V(a)} & V(x) \\ {\scriptstyle Z(a)}\big\downarrow & \swarrow{\scriptstyle \pi(x)} & \\ Z(x) & & \end{array}$$

commutes. For $y \neq x$ we have $Z(y) = W(y) = V(y)$ and we define $\pi(y) = 1_{V(y)}$. It is easy to verify that $\pi_V = \pi : V \to Z = C_x^- C_x^+(V)$ is a morphism of quiver representations.

Suppose that V is indecomposable and $V \ncong S_x$. Then $V(a)$ is onto and $\pi(x)$ is an isomorphism. So $\pi : V \to Z = C_x^- C_x^+(V)$ is an isomorphism. Since (4.2) is also right exact, we have

$$\dim W(x) = \dim V(\mathrm{in}_x) - \dim V(x) = \sum_{i=1}^{k} \alpha(ta_i) - \alpha(x) = \sigma_x(\alpha)(x),$$

so $\underline{\dim} W = \sigma_x(\alpha)$.

Since $W(a^\star)$ is injective, W does not have any summands isomorphic to S_x by Lemma 4.3.8. If $W = W_1 \oplus \cdots \oplus W_r$ with W_i indecomposable, then

$$V \cong C_x^-(W) \cong C_x^-(W_1) \oplus \cdots \oplus C_x^-(W_r).$$

Since W_i is not isomorphic to S_x, we have $C_x^-(W_i) \neq 0$ by 2(a). We conclude that $r = 1$ and W is indecomposable. This proves 1(b). The proof of 2(b) goes the same. $\square$

Remark 4.3.10. Note that $\pi : V \mapsto C_x^- C_x^+(V)$ gives a natural transformation $\pi : \mathrm{id} \to C_x^- C_x^+$ between two functors. Dually, there is a natural transformation $\iota : C_x^+ C_x^- \to \mathrm{id}$. (See Exercise 4.3.4).

Exercises.

Exercise 4.3.1. *Verify that the Cartan form $(-,-)$ is the same for Q and for $\sigma_x(Q)$. Note that $(\sigma_x(\alpha), \sigma_x(\beta)) = (\alpha, \beta)$.*

Exercise 4.3.2. *Prove that C_x^+ and C_x^- are functors.*

Exercise 4.3.3. *Prove Lemmas 4.3.6 and 4.3.8.*

Exercise 4.3.4. *Verify Remark 4.3.10 that there exist natural transformations $\pi : \mathrm{id} \to C_x^- C_x^+$ and $\iota : C_x^+ C_x^- \to \mathrm{id}$.*

Exercise 4.3.5. *Suppose that V is representation of a quiver Q with $V(x) = 0$. If x is a sink, then V is injective if and only if $C_x^+ V$ is injective. If x is a source, then V is projective if and only if $C_x^- V$ is projective.*

Exercise 4.3.6. *Suppose that V is a representation of Q and W is a representation of $\sigma_x(Q)$ where $x \in Q_0$ is a sink. Show that we have isomorphisms $\mathrm{Hom}_Q(W, C_x^+ V) \cong \mathrm{Hom}_{\sigma_x Q}(C_x^- W, V)$ and $\mathrm{Ext}_Q(C_x^+ V, W) \cong \mathrm{Ext}_{\sigma_x Q}(V, C_x^- W)$.*

4.4. The Coxeter Functor

In a Weyl group, a Coxeter element is a composition of all the simple reflections in some order. In a similar way we will construct Coxeter functors C^+ and C^- as compositions of reflection functors. We have to choose the ordering of the reflections such that we are always reflecting at a source or a sink. This leads to the notion of an admissible sequence. We will show that we can construct the indecomposable representations from indecomposable projectives (or injectives) by applying the Coxeter functor C^- (respectively C^+) a number of times.

Definition 4.4.1. A sequence $x_1, \dots, x_m$ of vertices of Q is **an admissible sequence of sinks (resp. sources)** if x_{i+1} is a sink (resp. a source) in $\sigma_{x_i} \cdots \sigma_{x_1}(Q)$ for $i = 0, 1, \dots, m-1$.

Lemma 4.4.2. *Let Q be a quiver and let $x_1, \dots, x_m$ be an admissible sequence of sinks, and define $Q' = \sigma_{x_m} \cdots \sigma_{x_1}(Q)$. Define functors $C^+ = C^+_{x_m} \cdots C^+_{x_1} : \operatorname{Rep}(Q) \to \operatorname{Rep}(Q')$ and*

$$C^- = C^-_{x_1} \cdots C^-_{x_m} : \operatorname{Rep}(Q') \to \operatorname{Rep}(Q).$$

(1) *For every indecomposable representation V of Q, we have $C^+(V) = 0$ or $C^+(V)$ is indecomposable. If $C^+(V) = 0$, then V is isomorphic to $C^-_{x_1} \cdots C^-_{x_{i-1}}(S_{x_i})$ for some i. If $C^+(V)$ is indecomposable, then $C^-(C^+(V))$ is isomorphic to V.*

(2) *For every indecomposable representation V' of Q', we have $C^-(V') = 0$ or $C^-(V')$ is indecomposable. If $C^-(V') = 0$, then V' is isomorphic to $C^+_{x_m} \cdots C^+_{x_{i+1}}(S_{x_i})$ for some i. If $C^-(V')$ is indecomposable, then $C^+(C^-(V'))$ is isomorphic to V'.*

(3) *Let $\operatorname{Ind}(Q)$ (resp. $\operatorname{Ind}(Q')$) denote the sets of isomorphism classes of indecomposable representations of Q (resp. Q'). Let $\mathcal{M} \subseteq \operatorname{Ind}(Q)$ be the set of isomorphism classes of the representations $C^-_{x_1} \dots C^-_{x_{i-1}}(S_{x_i})$, $1 \leq i \leq m$ and let $\mathcal{M}'$ be the set of isomorphism classes of $C^+_{x_k} \dots C^+_{x_{i+1}}(S_{x_i})$, $1 \leq i \leq m$. Then the functors C^+ and C^- set up a bijection between $\operatorname{Ind}(Q) \backslash \mathcal{M}$ and $\operatorname{Ind}(Q') \backslash \mathcal{M}'$.*

Proof. Parts (1) and (2) follow by induction from Theorem 4.3.9. Part (3) follows from (1) and (2). □

Lemma 4.4.3. *Suppose that Q is a quiver with n vertices and without cycles (i.e., $\Gamma(Q)$ is a tree). There exists an admissible sequence of sinks $x_1, \dots, x_n$ with $\{x_1, x_2, \dots, x_n\} = \{1, 2, \dots, n\}$.*

Proof. Without loss of generality we may assume that $Q_0 = \{1, 2, \dots, n\}$ and $ha < ta$ for all $a \in Q_1$ by Exercise 1.5.2. Then $1, 2, \dots, n$ is an admissible sequence of sinks. □

In Lemma 4.4.3 we have $\sigma_{x_n} \cdots \sigma_{x_1}(Q) = Q$, because every arrow is reversed exactly twice.

The reflection functors allow us to compare representation theory of quivers with different orientations of arrows.

Lemma 4.4.4. *Let Q be a quiver without cycles. Let Q' be a quiver that differs from Q only by changing the orientation of some arrows. Then there exists an admissible sequence $x_1, \dots, x_k \in Q_0$ of sinks such that $Q' = \sigma_{x_k} \cdots \sigma_{x_1}(Q)$.*

Proof. We may assume that $Q = (Q_0, Q_1, h, t)$, $Q' = (Q_0, Q_1, h', t')$ where $\{ha, ta\} = \{h'a, t'a\}$ for all $a \in Q_1$. We prove the lemma by induction on the number of vertices of Q. if $|Q_0| = 1$ then the lemma is obvious. Suppose that $n = |Q_0| > 1$. Choose a leaf $z \in Q_0$ of $\Gamma(Q)$, i.e., z is only connected to one other vertex w in Q, via an arrow a, i.e., we have $\{z, w\} = \{ha, ta\} = \{h'a, t'a\}$. Let $\overline{Q}$ (respectively $\overline{Q}'$) be the quiver Q (respectively Q') with z and a omitted. By induction, there exists an admissible sequence of sinks $x_1, \dots, x_k$ for $\overline{Q}$ such that $\overline{Q}' = \sigma_{x_k}\sigma_{x_{k-1}} \cdots \sigma_{x_1}(\overline{Q})$. There exists an admissible sequence of sinks $v_1, \dots, v_{n-1}$ for Q' such that $\{v_1, \dots, v_{n-1}\} = \overline{Q}'_0 = Q_0 \setminus \{z\}$, and $\sigma_{v_{n-1}} \cdots \sigma_{v_1}\sigma_{x_k} \cdots \sigma_{x_1}(\overline{Q}) = \overline{Q}'$ by Lemma 4.4.3. So without loss of generality, we may assume that $w \in \{x_1, \dots, x_k\}$. By replacing "$w$" with "$z, w$" in the sequence $x_1, \dots, x_k$ at various places, we obtain an admissible sequence $y_1, \dots, y_l$. Note that z is a sink in $\sigma_{y_l} \cdots \sigma_{y_1}(Q) = Q'$ because there is no z after the last w in the sequence $y_1, \dots, y_l$. If z is a sink in Q', then we have $\sigma_{y_l} \cdots \sigma_{y_1}(Q) = Q'$. If z is a source in Q', then $y_1, \dots, y_l, z$ is an admissible sequence of sinks, and $\sigma_z\sigma_{y_l} \cdots \sigma_{y_1}(Q) = Q'$. □

Corollary 4.4.5. *If $\Gamma(Q)$ is a tree, then whether Q is of finite representation type only depends on $\Gamma(Q)$ and not on the orientation of Q.*

Proof. This follows from Lemmas 4.4.4 and 4.4.2. □

Lemma 4.4.6. *Suppose that $x_1, \dots, x_k$ and $y_1, \dots, y_k$ are admissible sequences of sinks, such that $y_1, \dots, y_k$ is a rearrangement of $x_1, \dots, x_k$. Then the functors $C^+_{x_k} \cdots C^+_{x_1}$ and $C^+_{y_k} \cdots C^+_{y_1}$ are naturally equivalent.*

Proof. If $v, w \in Q_0$ are two sinks not joined by an arrow, then $C^+_v C^+_w$ and $C^+_w C^+_v$ are naturally equivalent. We prove the lemma by induction on k. The case $k = 1$ is trivial. Suppose that $x_1 = y_m$ for some m with $1 \leq m \leq k$. Choose m minimal. Suppose that there is an arrow a between x_1 and y_i for some i with $1 \leq i \leq m - 1$. Since x_1 is a sink, the arrow a goes from y_i to x_1. Because $y_1, \dots, y_{m-1}$ is an admissible sequence, and $x_1 \notin \{y_1, \dots, y_{m-1}\}$ by the minimality of m, applying $\sigma_{y_1}, \dots, \sigma_{y_{m-1}}$

to Q in that order will not reverse the arrow a. In particular, y_i is not a sink in the quiver $\sigma_{y_{i-1}} \cdots \sigma_{y_1}(Q)$, and $y_1, \dots, y_i$ is not admissible; contradiction. So $C^+_{x_1}$ commutes with $C^+_{y_1}, \dots, C^+_{y_m}$. It follows that $C^+_{y_k} \cdots C^+_{y_1}$ and $C^+_{y_k} \cdots C^+_{y_{m+1}} C^+_{y_{m-1}} \cdots C^+_{y_1} C^+_{x_1}$ are naturally equivalent. By induction we have that $C^+_{y_k} \cdots C^+_{y_{m+1}} C^+_{y_{m-1}} \cdots C^+_{y_1}$ and $C^+_{x_k} \cdots C^+_{x_2}$ are naturally equivalent. Therefore, the functors

$$C^+_{x_k} \cdots C^+_{x_1}, C^+_{y_k} \cdots C^+_{y_{m+1}} C^+_{y_{m-1}} \cdots C^+_{y_1} C^+_{x_1}, C^+_{y_k} \cdots C^+_{y_1}$$

are all naturally equivalent to each other. □

Definition 4.4.7. Let Q be a quiver with n vertices and without oriented cycles. Suppose that $x_1, \dots, x_n$ is an admissible sequence of sinks such that $Q_0 = \{x_1, \dots, x_n\}$. Define

$$C^+ := C^+_{x_n} \dots C^+_{x_1}, C^- := C^-_{x_1} \dots C^-_{x_n}.$$

The functors $C^+, C^- : \mathrm{Rep}(Q) \to \mathrm{Rep}(Q)$ are called **Coxeter functors**. By Lemma 4.4.6, the definitions of C^+, C^- do not depend on the choice of $x_1, \dots, x_n$.

Corollary 4.4.8.

(1) *A representation V is projective if and only if $C^+(V) = 0$.*

(2) *A representation V is injective if and only if $C^-(V) = 0$.*

(3) *If V does not have any projective summands, then $C^-C^+(V)$ is isomorphic V.*

(4) *If V does not have any injective summands, then $C^+C^-(V)$ is isomorphic to V.*

Proof. We apply Lemma 4.4.2 where $m = n$ is the number of vertices and $x_1, x_2, \dots, x_m$ are the vertices of Q, arranged so that they form an admissible sequence of sinks. By Exercise 4.4.2, $\mathcal{M}$ consists of the projective indecomposable representations, and $\mathcal{M}'$ consists of the injective indecomposable representations. For an indecomposable representation V we have $C^+(V) = 0$ if and only if V is projective, and $C^-(V)$ if and only if V is injective. Parts (1) and (2) follow. If V is indecomposable, then (3) and (4) follow from Lemma 4.4.2. The case where V is decomposable easily follows. □

Suppose that $\Gamma(Q)$ is a union of Dynkin graphs of type ADE. Denote dimension vector of the simple representation S_i by ϵ_i. The Weyl group $\mathcal{W}$ is the group generated by $\sigma_x, x \in Q_0$. The group $\mathcal{W}$ acts on $\mathbb{Z}^{Q_0}$ and $\mathbb{R}^{Q_0}$. We say that $\alpha \in \mathbb{Z}^{Q_0}$ is a root if it lies in the $\mathcal{W}$-orbit of ϵ_i for some i. We say that an element $\alpha \in \mathbb{R}^{Q_0}$ is nonnegative (denoted $\alpha \geq 0$) if $\alpha(x) \geq 0$ for all $x \in Q_0$.

Proposition 4.4.9.

(1) *The Tits form B_Q is positive definite.*
(2) *If α is a root, then $B_Q(\alpha) = 1$.*
(3) *There only finitely many roots.*
(4) *$\mathcal{W}$ is finite.*
(5) *For every root α, we have $\alpha \geq 0$ or $\alpha \leq 0$.*

Proof. (1) This follows from Exercise 4.2.2. (2) If $w \in \mathcal{W}$ and $\alpha \in \mathbb{R}^{Q_0}$, then $B_Q(w(\alpha)) = B_Q(\alpha)$. In particular, for every $\alpha = w(\epsilon_i) \in \mathbb{R}^{Q_0}$ we have $B_Q(\alpha) = B_Q(\epsilon_i) = 1$. (3) The set of roots is contained in the intersection of $\mathbb{Z}^{Q_0}$ and the compact set $\{\alpha \in \mathbb{R}^{Q_0} \mid B_Q(\alpha) \leq 1\}$. This set is discrete and compact, and hence finite. (4) An element $w \in \mathcal{W}$ is completely determined by the action on the roots, since $\epsilon_1, \dots, \epsilon_n$ is an $\mathbb{R}$-basis. So $\mathcal{W}$ is contained in the permutation group on the set of roots, and hence it is finite. (5) If α is a root, then we can write $\alpha = \alpha^+ - \alpha^-$, where $\alpha^+, \alpha^- \in \mathbb{N}^{Q_0}$ are dimension vectors and $\alpha^+(x)\alpha^-(x) = 0$ for all $x \in Q_0$. We have

$$1 = B_Q(\alpha) = B_Q(\alpha^+) + B_Q(\alpha^-) - (\alpha^+, \alpha^-) \geq B_Q(\alpha^+) + B_Q(\alpha^-).$$

Since B_Q is positive definite, we must have either $\alpha^+ = 0$ or $\alpha^- = 0$. □

Definition 4.4.10. Let Q be a quiver without loops. Let $Q_0 = \{x_1, \dots, x_n\}$. An element $c := \sigma_{x_n} \dots \sigma_{x_1}$ is called a **Coxeter transformation**. The Coxeter transformation depends on the ordering of vertices.

Corollary 4.4.11. *Suppose that $x_1, x_2, \dots, x_n$ is an admissible sequence of sinks, and let $c = \sigma_{x_n}\sigma_{x_{n-1}} \cdots \sigma_{x_1}$ be the corresponding Coxeter transformation. If V is a representation without projective summands with dimension vector α, then $C^+(V)$ is a representation with dimension vector $c(\alpha)$.*

Proof. This follows from Corollary 4.4.8 and Theorem 4.3.9. □

Lemma 4.4.12.

(1) *There are no nonzero vectors in $\mathbb{R}^{Q_0}$ fixed by c,*
(2) *For every nonzero $\alpha \in \mathbb{R}^{Q_0}$ there exists $k \in \mathbb{N}$ such that $c^k(\alpha)$ has a negative coordinate.*

Proof. Let $0 \neq \alpha \in \mathbb{R}^{Q_0}$ be a vector such that $\alpha = c(\alpha) = \sigma_{x_1} \dots \sigma_{x_n}(\alpha)$. The transformation $\sigma_{x_2} \dots \sigma_{x_n}$ does not change the coordinate $\alpha(x_1)$, so $\sigma_{x_1}(\alpha)(x_1) = c(\alpha)(x_1) = \alpha(x_1)$. This means that $\sigma_{x_1}(\alpha) = \alpha$. Similarly $\sigma_{x_2}(\alpha) = \alpha, \dots, \sigma_{x_n}(\alpha) = \alpha$. Since $(\epsilon_i, \alpha) = \alpha(x_i) - \sigma_{x_i}(\alpha)(x_i) = 0$ for all i, and $(-,-)$ is nondegenerate, it follows that $\alpha = 0$ and (1) is proved.

The element c has a finite order because $\mathcal{W}$ is finite, say h (the so-called **Coxeter number**). The vector $\alpha + c(\alpha), \dots + c^{h-1}(\alpha)$ is fixed by c, so

$\alpha + c(\alpha) + \cdots + c^{h-1}(\alpha) = 0$. If α is nonzero, then one of the vectors $\alpha, c(\alpha), \dots, c^{h-1}(\alpha)$ must have a negative coordinate. This proves (2). □

Now we can prove Gabriel's theorem.

Theorem 4.4.13 (Gabriel). *A quiver Q is of finite representation type if and only if $\Gamma(Q)$ is a union of Dynkin graphs of type ADE. If Q is of finite representation type, then the map $V \mapsto \underline{\dim}\, V$ gives a bijection between* $\operatorname{Ind}(Q)$ *and the set of positive roots.*

Proof. If Q has finite representation type, then $\Gamma(Q)$ is a union of Dynkin graphs of type ADE by Theorem 4.2.4. Without loss of generality, we may assume that $Q_0 = \{1, 2, \dots, n\}$ and that $ha < ta$ for all $a \in Q_1$.

Suppose that $\Gamma(Q)$ is a union of Dynkin graphs of type ADE. Let $c = \sigma_n \sigma_{n-1} \cdots \sigma_1$ be the Coxeter transformation. Suppose that V is an indecomposable representation of Q, and $\alpha = \underline{\dim}\, V$. By Lemma 4.4.12, $c^i(\alpha)$ has a negative coordinate for some i. Choose l minimal such that $c^l(\alpha)$ has a negative coordinate. Choose j minimal such that $\sigma_j \cdots \sigma_1 c^{l-1}(\alpha)$ has a negative coordinate. It follows that $C_j^+ \cdots C_1^+ (C^+)^{l-1}(V) = 0$, and $C_{j-1}^+ \cdots C_1^+ (C^+)^{l-1}(V) \neq 0$. Then V is isomorphic to $(C^-)^{l-1} C_1^- \cdots C_{j-1}^+(S_j)$, and $\alpha = c^{l-1} \sigma_1 \cdots \sigma_{j-1}(\epsilon_j)$ is a root.

Suppose that α is a root. Again, choose l minimal such that $c^l(\alpha)$ has a negative coordinate. Choose j minimal such that $\sigma_j \cdots \sigma_1 c^{l-1}(\alpha)$ has a negative coordinate. Then $\beta = \sigma_{j-1} \cdots \sigma_1 c^{l-1}(\alpha)$ is a positive root and $\sigma_j(\beta)$ is a negative root. It follows that $\beta = \epsilon_j$. It follows that $V = (C^-)^{l-1} C_1^- \cdots C_{j-1}^-(S_j)$ is an indecomposable representation of dimension α. □

Exercises.

Exercise 4.4.1. *Suppose that $x_1, x_2, \dots, x_m \in Q_0$ are distinct vertices. Prove that $x_1, x_2, \dots, x_m$ is an admissible sequence of sinks if and only if there is no arrow from x_i to x_j with $j > i$.*

Exercise 4.4.2. *Suppose that $x_1, \dots, x_m$ is an admissible sequence of sinks for the quiver Q, and $x_1, \dots, x_m$ are distinct. Show that $C_{x_1}^- \cdots C_{x_{m-1}}^-(S_{x_m})$ is a projective representation (where S_{x_m} is the simple representation of the quiver $\sigma_{x_{m-1}} \cdots \sigma_{x_1} Q$ corresponding to the vertex x_m). Similarly, if $x_1, \dots, x_m$ is an admissible sequence of sources for the quiver Q, then show that $C_{x_1}^+ \cdots C_{x_{m-1}}^+(S_{x_m})$ is an injective representation. (Use Exercise 4.3.5.)*

Exercise 4.4.3. *Suppose that Q is a quiver of finite representation type. For every ordering $x_1, \dots, x_n$ of the vertices, we get a Coxeter transformation $c = \sigma_{x_n} \cdots \sigma_{x_1}$. Show that the conjugacy class of c is independent of the* *choice of the ordering of the vertices. Is this still true if Q is a quiver that is not of finite representation type?*

Exercise 4.4.4. *Consider the following quivers:*

$$Q: \begin{array}{ccc} \circ & \longrightarrow & \circ \\ \uparrow & & \downarrow \\ \circ & \longrightarrow & \circ \end{array} \qquad Q': \begin{array}{ccc} \circ & \longrightarrow & \circ \\ \downarrow & & \downarrow \\ \circ & \longrightarrow & \circ \end{array}.$$

Show that there is no sequence of reflections that turns Q into a quiver isomorphic to Q'.

Exercise 4.4.5. *Show that the Coxeter transformation c has finite order if and only if Q is of finite representation type.*

Exercise 4.4.6. *Compute the characteristic polynomial $p(x)$ of the Coxeter element c for the type $\widehat{\widehat{\mathbf{E}}}_8$ (see Exercise 4.2.4. What are the absolute values of the eigenvalues? (The polynomial $p(x)$ is the polynomial with the smallest known Mahler measure > 1.)*

Exercise 4.4.7. *Suppose that Q is a quiver with $Q_0 = \{1, 2, \dots, n\}$ such that $ha < ta$ for every arrow $a \in Q_1$. Let $c = \sigma_n \sigma_{n-1} \cdots \sigma_1$ be a Coxeter transformation.*

1. *Show that $c(\underline{\dim}(P_k)) = -\underline{\dim}(I_k)$ for all k.*
2. *Show that the matrix of c is given by $-E^{-1}E^t$, where E is the Euler matrix as defined in Definition 2.5.3.*

Exercise 4.4.8. *Suppose that V and W are representations without projective summands. Show that C^+ gives an isomorphism between $\operatorname{Hom}_Q(V, W)$ and $\operatorname{Hom}_Q(C^+(V), C^+(W))$.*

4.5. Examples

Example 4.5.1. Consider the quiver

$$1 \xrightarrow{a} 2 .$$

There are three indecomposable objects $K \to 0$, $K \to K$, $0 \to K$. We have

$$\begin{aligned} C^+(K \to 0) &= C_1^+ C_2^+(K \to 0) = C_1^+(K \leftarrow K) = (0 \to K), \\ C^+(K \to K) &= C_1^+ C_2^+(K \to K) = C_1^+(K \leftarrow 0) = 0, \\ C^+(0 \to K) &= C_1^+ C_2^+(0 \to K) = 0. \end{aligned}$$

We also have

$$\begin{aligned} C^-(K \to 0) &= C_2^- C_1^-(K \to 0) = 0, \\ C^-(K \to K) &= C_2^- C_1^-(K \to K) = C_2^-(0 \leftarrow K) = 0, \\ C^-(0 \to K) &= C_2^- C_1^-(0 \to K) = C_2^-(K \leftarrow K) = (K \to 0). \end{aligned}$$

Example 4.5.2. Consider the equioriented $\mathbf{A}_n$ quiver

$$1 \xrightarrow{a_1} 2 \xrightarrow{a_2} \cdots \xrightarrow{a_{n-1}} n.$$

The indecomposable representations are $E_{i,j}$ with dimension vector $\epsilon_i + \ldots + \epsilon_j$ with $1 \leq i \leq j \leq n$. We have

$$C^+(E_{i,j}) = E_{i+1,j+1} \text{ if } j < n, \ C^+(E_{i,n}) = 0,$$
$$C^-(E_{i,j}) = E_{i-1,j-1} \text{ if } i > 0, \ C^-(E_{0,j}) = 0.$$

We see that the functors C^+ (resp. C^-) act as left (resp. right) translation.

Example 4.5.3. Consider the quiver

$$1 \xrightarrow{a} 2 \xleftarrow{b} 3 .$$

An admissible sequence of sinks is $2, 3, 1$. There are six indecomposable objects $E_{1,1}, E_{1,2}, E_{1,3}, E_{2,2}, E_{2,3}, E_{3,3}$. We have (here we do not indicate the orientations of intermediate quivers)

$$C^+E_{1,1} = C_1^+C_3^+C_2^+E_{1,1} = C_1^+C_3^+E_{1,2} = C_1^+E_{1,3} = E_{2,3},$$
$$C^+E_{1,2} = C_1^+C_3^+C_2^+E_{1,2} = C_1^+C_3^+E_{1,1}C_1^+E_{1,1} = 0,$$
$$C^+E_{1,3} = C_1^+C_3^+C_2^+E_{1,3} = C_1^+C_3^+E_{1,3} = C_1^+E_{1,2} = E_{2,2},$$
$$C^+E_{2,2} = C_1^+C_3^+C_2^+E_{2,2} = 0,$$
$$C^+E_{2,3} = C_1^+C_3^+C_2^+E_{2,3} = C_1^+C_3^+E_{3,3} = 0,$$
$$C^+E_{3,3} = C_1^+C_3^+C_2^+E_{3,3} = C_1^+C_3^+E_{2,3} = C_1^+E_{2,2} = E_{1,2}.$$

Similarly we have $C^-E_{1,1} = 0$, $C^-E_{1,2} = E_{3,3}$, $C^-E_{1,3} = 0$, $C^-E_{2,2} = E_{1,3}$, $C^-E_{2,3} = E_{1,1}$ and $C^-E_{3,3} = 0$.

Example 4.5.4. Consider the quiver

$$\begin{array}{ccc} 1 \xrightarrow{a} & 4 & \xleftarrow{b} 2 \\ & \uparrow{\scriptstyle c} & \\ & 3. & \end{array}$$

This is the quiver of type $\mathbf{D}_4$ so it has 12 indecomposable objects. The orbits under C^+ are

$$\begin{aligned} &(1000) \to (0111) \to (1001) \\ &(0100) \to (1011) \to (0101) \\ &(1111) \to (1112) \to (0001) \\ &(0010) \to (1101) \to (0011). \end{aligned}$$

Exercises.

Exercise 4.5.1. *Consider the quiver*

$$Q:\ 1 \xrightarrow{a} 2 \xleftarrow{b} 3 \xrightarrow{c} 4\,.$$

Determine how C^+ and C^- act on the set of indecomposable representations.

Exercise 4.5.2. *Consider the quiver*

$$\begin{array}{ccccc} Q:\ 1 & \xrightarrow{a} & 4 & \xleftarrow{b} & 2 \\ & & {\scriptstyle c}\downarrow & & \\ & & 3\,. & & \end{array}$$

Determine how C^+ and C^- act on the set of indecomposable representations.

4.6. Bibliographical Remarks

Gabriel's theorem was the starting point of a long period of interest in quiver representations. The connection with root systems attracted many mathematicians to the subject. The proof that Q is Dynkin if and only if B_Q is positive definite can be found in the book of Humphreys [**50**], Chapter III. The dimension argument of Tits based on this fact proves half of the result: if the connected quiver is not Dynkin, then it cannot have a finite representation type. The other part is more involved. There are two approaches: the one using the reflection functors and the approach of Auslander and Reiten that defines Coxeter functor in homological terms. In this chapter we follow the proof based on [**8**] where the reflection functors were introduced. The Auslander-Reiten theory will be discussed in the next two chapters.

If Q is a Dynkin quiver, then for every dimension vector α the representation space $\operatorname{Rep}_\alpha(Q)$ has finitely many $\operatorname{GL}_\alpha(Q)$-orbits. There are natural questions about the geometry and singularities of the orbit closures for these actions. Typical interesting questions are about normality, the Cohen-Macaulay property, and rational singularities for the orbit closures.

The first result in this direction was the paper of Abeasis, Del Fra and Kraft [**1**] where they proved rational singularities for the orbit closures for the equioriented $\mathbf{A}_n$ quiver. Then Lakshmibai and Magyar [**61**] proved that result over a field of arbitrary characteristic, using normality and rational singularities of Schubert varieties.

In his groundbreaking paper [**78**] Zwara proved a general result allowing to compare orbit closures for two algebras A and B for which there existed a certain type of functor (Hom-controlled functor) from mod-A to mod-B.

Using this result Bobiński and Zwara ([**9**], [**10**]) proved rational singularities of orbit closures for quivers of type $\mathbf{A}_n$, $\mathbf{D}_n$. The method of Hom-controlled functors allows us to reduce the problem of singularities of orbit closures for a given quiver Q of type $\mathbf{A}_n$ or $\mathbf{D}_n$ to the singularities of $\mathbf{A}_n$ or $\mathbf{D}_n$ with particular orientation, and then use the paper [**61**] (for type $\mathbf{A}_n$) and some results of Brion ([**11**]) for type $\mathbf{D}_n$.

For the quivers of type $\mathbf{E}_6$ $\mathbf{E}_7$, $\mathbf{E}_8$ the normality and rational singularities of orbit closures are not known.

Chapter 5

Almost Split Sequences

This and the next chapter are devoted to Auslander-Reiten theory. In this chapter we study irreducible morphisms between modules of a finite dimensional algebra A. Irreducible morphisms are roughly speaking morphisms that cannot be factored in a nontrivial way. Understanding the structure of the irreducible morphisms will give a lot of information about the module category. The irreducible morphisms are intimately connected with the so-called almost split sequences, and the Auslander-Reiten transform that will be introduced in the next chapter.

The irreducible morphisms, almost split sequences and Auslander-Reiten transform can neatly be summarized in the Auslander Reiten quiver of the algebra A. This is a quiver whose vertices are indecomposable modules and for which the arrows are irreducible morphisms.

In this chapter A is a finite dimensional algebra over $\mathbb{C}$ and $\mathfrak{m} = \mathrm{rad}(A)$ denotes the Jacobson radical.

5.1. Ideals of Morphisms in the Module Categories

Irreducible morphisms are similar to the notion of minimal generators of an ideal. For example, a subset $\{f_1, \dots, f_r\} \subseteq \mathfrak{m}$ of the Jacobson radical is a minimal set of generators of $\mathfrak{m}$ if and only if $f_1 + \mathfrak{m}, \dots, f_r + \mathfrak{m}$ form a basis of the space $\mathfrak{m}/\mathfrak{m}^2$. Instead of working with ideals in the algebra A itself, we will work with ideals in the category of A-modules. We will first introduce the notion of ideals of morphisms in a category, and then define a particular ideal, the radical rad_A (which is similar to the Jacobson radical of an algebra). Irreducible morphisms are morphisms that lie in rad_A

but not in rad_A^2. The arrows in the Auslander-Reiten quiver that we will construct later will correspond to irreducible morphisms that form a basis of $\operatorname{rad}_A / \operatorname{rad}_A^2$.

Definition 5.1.1. A collection $\mathcal{I}$ of morphisms in A–mod is **a two-sided ideal of morphisms** if the following conditions are satisfied.

(1) For every representation V the zero map $0_V : V \to V$ is in $\mathcal{I}(V, V)$,
(2) If $f, g : V \to W$ are in $\mathcal{I}(V, W)$, then $f + g \in \mathcal{I}(V, W)$,
(3) If $f \in \mathcal{I}$, and g is a homomorphism for which $g \circ f$ is defined, then $g \circ f \in \mathcal{I}$,
(4) If $f \in \mathcal{I}$, and h is a homomorphism for which $f \circ h$ is defined, then $f \circ h \in \mathcal{I}$.

If $\mathcal{I}$ is a two-sided ideal of morphisms, $f \in \mathcal{I}(V, W)$ and $\lambda \in \mathbb{C}$, then $\lambda f = (\lambda 1) \circ f \in \mathcal{I}(V, W)$. This shows that $\mathcal{I}(V, W)$ is a linear subspace.

Remark 5.1.2. The collection $\mathcal{I}$ is **a two-sided ideal** in A–mod if and only if the correspondance

$$(V, W) \mapsto \mathcal{I}(V, W)$$

defines a subfunctor of two variables

$$\mathcal{I}(-, -) : (A\text{–mod}) \times (A\text{–mod}) \to \mathbb{C}\text{–mod}$$

of the functor $\operatorname{Hom}_A(-, -)$.

Lemma 5.1.3. *The following classes of morphisms are two-sided ideals of morphisms:*

(1) *The ideal $\mathcal{P}$ of morphisms $f : V \to W$ admitting a factorization $V \to P \to W$ where P is a projective module.*
(2) *The ideal $\mathcal{Q}$ of morphisms $f : V \to W$ admitting a factorization $V \to Q \to W$ where Q is a injective module.*

Proof. It is enough to prove part (1) because part (2) is dual to it. The points (1), (3), (4) of Definition 5.1.1 are obvious. To prove (2) notice that if we have factorizations $f : V \to P \to W$, $g : V \to P' \to W$, then $f + g$ factors through $P \oplus P'$. □

Definition 5.1.4. Let $\mathcal{I}$ be a two-sided ideal in A–mod. We define **the quotient category** $(A\text{–mod})/\mathcal{I}$ as follows. The objects of $(A\text{–mod})/\mathcal{I}$ are the finite dimensional A-modules, and for every two finite dimensional A-modules V, W, we define

$$\operatorname{Hom}_{(A-\mathrm{Mod})/\mathcal{I}}(V, W) := \operatorname{Hom}_A(V, W)/\mathcal{I}(V, W).$$

We have the obvious functor

$$\pi : (A\text{–mod}) \to (A\text{–mod})/\mathcal{I}.$$

The category $(A\text{–mod})/\mathcal{I}$ is an additive $\mathbb{C}$-category (i.e., morphisms from $\mathbb{C}$-vector spaces, with compositions being bilinear).

Definition 5.1.5. If $\mathcal{I}$ is a two-sided ideal of morphisms, then we define $\mathcal{I}^m$ such that $\mathcal{I}^m(V, W)$ is the linear span of all morphisms $f : V \to W$ that are a composition of m morphisms in $\mathcal{I}$.

By Exercise 5.1.1, $\mathcal{I}^m$ is again a two-sided ideal of morphisms. The following lemma is proven in Exercise 5.1.2.

Lemma 5.1.6. *If $f \in \operatorname{Hom}_A(V, W)$ and $g \in \operatorname{Hom}_A(W, V)$, then $1_W - fg$ is invertible if and only if $1_V - gf$ is invertible.*

Definition 5.1.7. The radical rad_A of the category A–mod is the class of morphisms defined as follows:

$$\begin{aligned}
&\operatorname{rad}_A(V, W)\\
&\quad = \{f \in \operatorname{Hom}_A(V, W) \mid \forall g \in \operatorname{Hom}_A(W, V)\ 1_V - gf \text{ is invertible}\}\\
&\quad\quad = \{f \in \operatorname{Hom}_A(V, W) \mid \forall g \in \operatorname{Hom}_A(W, V)\ 1_V - fg \text{ is invertible}\}.
\end{aligned}$$

Proposition 5.1.8. *The radical rad_A is a two-sided ideal of morphisms.*

Proof. We verify the 4 properties of a two-sided ideal of morphisms (see Definition 5.1.1):

(1) Clearly $0_V : V \to V$ lies in $\operatorname{rad}_A(V, V)$ for all modules V.

(2) Suppose that $f, f' \in \operatorname{rad}_A(V, W)$. Let $h = (1 - gf)^{-1}$. Then

$$h(1 - g(f + f')) = h((1 - gf) + gf') = 1 + hgf' = 1 - (-hg)f'$$

is invertible, so $1 - g(f + f')$ is invertible. This shows that $f + f' \in \operatorname{rad}_A(V, W)$.

(3) If $f \in \operatorname{rad}_A(V, W)$, $g \in \operatorname{Hom}_A(W, Z)$ and $h \in \operatorname{Hom}_A(Z, V)$, then $1 - (hgf) = 1 - h(gf)$ is invertible. We get $gf \in \operatorname{rad}_A(V, Z)$, because h can be arbitrary. Also, $1 - (gf)h = 1 - g(fh)$ is invertible because of Lemma 5.1.6. Therefore $fh \in \operatorname{rad}_A(Z, W)$ because g can be arbitrary. So this proves (4). □

Proposition 5.1.9. *A morphism*

$$f = \begin{pmatrix} f_{1,1} & f_{1,2} & \cdots & f_{1,m} \\ f_{2,1} & f_{2,2} & \cdots & f_{2,m} \\ \vdots & \vdots & \ddots & \vdots \\ f_{n,1} & f_{n,2} & \cdots & f_{n,m} \end{pmatrix} : V_1 \oplus \ldots \oplus V_m \to W_1 \oplus \ldots \oplus W_n$$

is in rad_A if and only if all components $f_{i,j} \in \operatorname{rad}_A(V_j, W_i)$ for all i, j.

Proof. Let $V = V_1 \oplus \cdots \oplus V_m$ and $W = W_1 \oplus \cdots \oplus W_n$ and let $p_i^V : V \to V_i$, $p_i^W : W \to W_i$ be the projections and $u_i^V : V_i \to V$ and let $u_i^W : W_i \to W$ be the inclusions for all i.

If $f \in \operatorname{rad}_A(V, W)$, then we have $f_{i,j} = p_i^W f u_j^V \in \operatorname{rad}_A(V_j, W_i)$. Conversely, if $f_{i,j} \in \operatorname{rad}_A(V_j, W_i)$ for all i, j, then $f = \sum_{i,j} u_i^W f_{i,j} p_j^V \in \operatorname{rad}_A(V, W)$. □

Recall that a homomorphism of A-modules is a **section** (resp. **retraction**) if it admits left (resp. right) inverse.

Remark 5.1.10.

(1) A homomorphism $f : L \to M$ is a section if and only if it is a monomorphism and the exact sequence

$$0 \longrightarrow L \xrightarrow{f} M \longrightarrow \operatorname{coker}(f) \longrightarrow 0$$

splits.

(2) A homomorphism $g : M \to N$ is a retraction if and only if it is an epimorphism and the exact sequence

$$0 \longrightarrow \ker(g) \longrightarrow M \xrightarrow{g} N \longrightarrow 0$$

splits.

Proposition 5.1.11.

(1) *If V is an indecomposable module, then* $\operatorname{rad}_A(V, W)$ *consists of all nonsections.*

(2) *If W is an indecomposable module, then* $\operatorname{rad}_A(V, W)$ *consists of all nonretractions.*

(3) *Suppose that V and W are both indecomposable. If V and W are nonisomorphic, then* $\operatorname{rad}_A(V, W) = \operatorname{Hom}_A(V, W)$ *and* $\operatorname{rad}_A(V, V)$ *is the set of all nonisomorphisms, the Jacobson radical of* $\operatorname{End}_A(V)$.

Proof. (1) If $f : V \to W$ is a section, then $gf = 1_V$ for some $g : W \to V$ and $0 = 1 - gf$ is not invertible. Therefore, $f \notin \operatorname{rad}_A(V, W)$. Conversely, suppose that $f : V \to W$ does not lie in the radical. Then there exists a $g \in \operatorname{Hom}_A(W, V)$ such that $1 - gf$ is not invertible. It follows that 1 is an eigenvalue of gf. By Corollary 1.7.2 and Remark 3.2.6, gf can only have 1 eigenvalue. So gf is invertible and f is a section.

(2) Is dual to (1) and (3) is clear because every section or retraction from V to W is an isomorphism if V and W are indecomposable. □

Exercises.

Exercise 5.1.1. *Prove that, if $\mathcal{I}$ is a two-sided ideal of morphisms, then $\mathcal{I}^m$ is a two-sided ideal of morphisms.*

Exercise 5.1.2. *Prove Lemma* 5.1.6. *(Hint: Look at the proof of Proposition* 3.2.1*).*

5.2. Irreducible Morphisms

In this section we will study irreducible morphisms and their properties.

Definition 5.2.1. A homomorphism $f : V \to W$ is **irreducible** if

(1) f is neither a section nor a retraction, and

(2) for any factorization

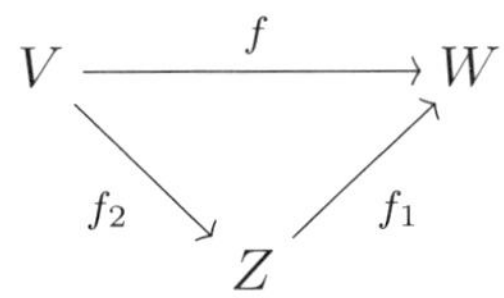

either f_1 is a retraction or f_2 is a section.

Irreducibility is a self-dual notion.

Lemma 5.2.2. *If $f : V \to W$ is an irreducible homomorphism, then f is either a monomorphism or an epimorphism (but not both).*

Proof. Suppose that f is not an epimorphism. Then we can factor f through $\operatorname{im}(f)$:

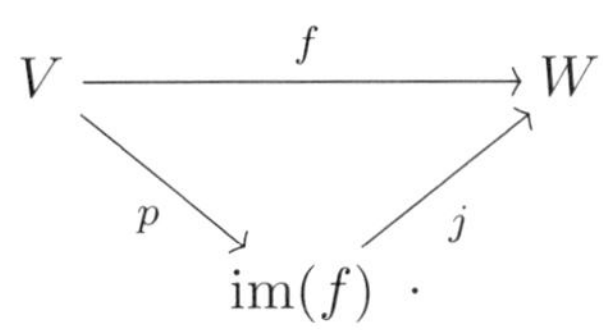

Factoring it through $\operatorname{im}(f)$ we get $f = jp$, where j is not a retraction. This implies that p is a section, so f is a proper monomorphism. □

Now we relate the notion of irreducible maps to the radical of the module category.

Lemma 5.2.3. *Assume that V and W are indecomposable modules. Then $f : V \to W$ is irreducible if and only if $f \in \operatorname{rad}_A(V, W) \setminus \operatorname{rad}^2_A(V, W)$.*

Proof. Assume f is irreducible. Then $f \in \operatorname{rad}_A(V, W)$ as otherwise it would be an isomorphism. If $f \in \operatorname{rad}^2_A(V, W)$, then f could be written as

$f = \sum_{i=1}^t g_i h_i$, where $g_i \in \mathrm{rad}_A(Z_i, W)$, $h_i \in \mathrm{rad}_A(V, Z_i)$. Without loss of generality we may assume that $Z_1, \dots, Z_t$ are indecomposable. Define

$$h = \begin{pmatrix} h_1 \\ \vdots \\ h_t \end{pmatrix}, \quad g = \begin{pmatrix} g_1 & \cdots & g_t \end{pmatrix}.$$

By Proposition 5.1.9 $h \in \mathrm{rad}_A(V, Z)$ and $g \in \mathrm{rad}_A(Z, W)$. For every $u \in \mathrm{Hom}_A(Z, V)$, $uh \in \mathrm{rad}_A(V, V)$. So $uh \neq 1_V$ and h has no left inverse. So h is not a section, and dually we see that g is not a retraction. This contradicts the fact that f is irreducible.

To prove the reverse implication, assume that

$$f \in \mathrm{rad}_A(V, W) \setminus \mathrm{rad}_A^2(V, W).$$

Since V, W are indecomposable, f is not an isomorphism, so it is neither a section nor a retraction. Assume that $f = gh$, where $h : V \to Z$, $g : Z \to W$. Decompose $Z = \bigoplus_{i=1}^t Z_i$ and write h and g in the matrix form

$$h = \begin{pmatrix} h_1 \\ \vdots \\ h_t \end{pmatrix}, \quad g = \begin{pmatrix} g_1 & \cdots & g_t \end{pmatrix}.$$

We have $f = \sum_{i=1}^t g_i h_i$. Because $f \notin \mathrm{rad}_A^2(V, W)$, we have $g_i h_i \notin \mathrm{rad}_A^2(V, W)$ for some i. It follows that $g_i \notin \mathrm{rad}_A(V, W)$ or $h_i \notin \mathrm{rad}_A(V, W)$. Therefore, g_i or h_i is an isomorphism by Proposition 5.1.11. Thus g is a retraction or h is a section. □

We understand the irreducible maps between indecomposables. The next two lemmas characterize when a morphism $f : V \to W$ is irreducible when either V or W is indecomposable.

Lemma 5.2.4. *Suppose that V and W are indecomposable representations.*

(1) *Suppose that*

$$f = \begin{pmatrix} f_1 \\ f_2 \\ \vdots \\ f_t \end{pmatrix} : V \to W^t$$

is a monomorphism. Then f is irreducible if and only if $f_1, \dots, f_t \in \mathrm{rad}_A(V, W)$ are linearly independent modulo $\mathrm{rad}_A^2(V, W)$.

(2) *Suppose that*

$$f = \begin{pmatrix} f_1 & f_2 & \cdots & f_t \end{pmatrix} : V^t \to W$$

is an epimorphism. Then f is irreducible if and only if $f_1, \dots, f_t \in \mathrm{rad}_A(V, W)$ are linearly independent modulo $\mathrm{rad}_A^2(V, W)$.

Proof. (1) Suppose that f is irreducible. If $f_i \not\in \operatorname{rad}_A(V,W)$, then f_i is an isomorphism and hence f is a section. Contradiction. So $f_1,\dots,f_t \in \operatorname{rad}_A(V,W)$. Suppose that $\sum_{i=1}^t \lambda_i f_i \in \operatorname{rad}_A^2(V,W)$ with $\lambda_i \neq 0$ for some i. Without loss of generality we may assume $\lambda_1 \neq 0$ and if we replace f_1 by $\sum_{i=1}^t \lambda_i f_i$ we may assume that $f_1 \in \operatorname{rad}_A^2(V,W)$. Write $f_1 = g_1 h_1$ where $h_1 \in \operatorname{rad}_A(V,Z)$ and $g_1 \in \operatorname{rad}_A(Z,W)$. We have a decomposition $f = gh$ where

$$g = \begin{pmatrix} g_1 & 0 & \cdots & 0 \\ 0 & 1 & \cdots & 0 \\ \vdots & \vdots & & \vdots \\ 0 & 0 & \cdots & 1 \end{pmatrix} \text{ and } h = \begin{pmatrix} h_1 \\ f_2 \\ \vdots \\ f_t \end{pmatrix}.$$

Now h is not a section because $h_1, f_2, \cdots, f_t \in \operatorname{rad}_A$. The map g is not a retraction because g_1 is not a retraction (because it lies in rad_A). So f is reducible. Contradiction.

Suppose now that $f_1,\dots,f_t \in \operatorname{rad}_A(V,W)$ are linearly independent modulo $\operatorname{rad}_A^2(V,W)$. Suppose that $f = gh$, where $h : V \to Z$ and $g : Z \to W$. Assume that h is not a section. Then $h \in \operatorname{rad}_A(V,Z)$. Let us write $Z = W^s \oplus Y$ such that Y does not have any indecomposable summand isomorphic to W. We can write

$$g = \begin{pmatrix} g_1 & g_2 \end{pmatrix}, \quad h = \begin{pmatrix} h_1 \\ h_2 \end{pmatrix},$$

where $h_1 : V \to W^s$, $h_2 : V \to Y$, $g_1 : W^s \to W^t$, $g_2 : Y \to W^t$. Since no direct summand of Y is isomorphic to W, we have $g_2 \in \operatorname{rad}_A$. Modulo $\operatorname{rad}_A(W,W)$, we may view g_1 as a $t \times s$ matrix $\overline{g}_1$ with complex entries. Suppose that $\overline{g}_1$ has rank $< t$. Then we have

$$\begin{pmatrix} \lambda_1 & \cdots & \lambda_t \end{pmatrix} \overline{g}_1 = 0$$

for some $\lambda_1,\dots,\lambda_t \in \mathbb{C}$, not all 0. It follows that $\lambda g_1 \in \operatorname{rad}_A(W^s,W)$, where

$$\lambda = \begin{pmatrix} \lambda_1 & \cdots & \lambda_t \end{pmatrix} : W \to W^t,$$

and $\lambda g \in \operatorname{rad}_A(Z,W)$. Therefore, we get $\lambda gh = \lambda f \in \operatorname{rad}_A^2(V,W)$. This contradicts the assumption that $f_1,\dots,f_t$ are independent modulo $\operatorname{rad}_A^2(V,W)$. We conclude that $\overline{g}_1$ has rank t. This implies that g is a retraction.

(2) is dual to (1). □

Lemma 5.2.5.

(1) *Suppose that V is indecomposable,*

$$f = \begin{pmatrix} f_1 \\ f_2 \end{pmatrix} : V \to W_1 \oplus W_2$$

is a monomorphism and that W_1 and W_2 do not have any indecomposable summand in common. Then f is irreducible if and only if both f_1 and f_2 are irreducible.

(2) *Suppose that W is indecomposable and*

$$f = \begin{pmatrix} f_1 & f_2 \end{pmatrix} : V_1 \oplus V_2 \to W$$

is an epimorphism and W_1 and W_2 do not have any indecomposable summand in common. Then f is irreducible if and only if both f_1 and f_2 are irreducible.

Proof. (1) Suppose that

$$f = gh,$$

where $g = \begin{pmatrix} g_1 \\ g_2 \end{pmatrix}$, $g_i : Z \to W_i$ and $h : V \to Z$.

Suppose that f_1, f_2 are both irreducible. Suppose that h is not a section. Since $f_i = g_i h$, we have that g_1, g_2 are retractions. So let $u_i : W_i \to Z$ such that $u_i g_i = 1$. So W_1, W_2 are direct summands of Z. Consider

$$u = \begin{pmatrix} u_1 & u_2 \end{pmatrix} : W_1 \oplus W_2 \to Z.$$

Then

$$gu = \begin{pmatrix} 1_{W_1} & v_{1,2} \\ v_{2,1} & 1_{W_2} \end{pmatrix},$$

where $v_{1,2}, v_{2,1} \in \operatorname{rad}_A$ because W_1 and W_2 have no direct summand in common. It follows that gu is an isomorphism and g is a retraction. This shows that f is irreducible.

Conversely, suppose that f is irreducible. If $f_1 = g_1 h_1$, then we have a decomposition

$$f = gh,$$

where

$$g = \begin{pmatrix} g_1 & 0 \\ 0 & 1 \end{pmatrix}, \quad h = \begin{pmatrix} h_1 \\ f_2 \end{pmatrix}.$$

If h_1 is not a section, then $h_1, f_2 \in \operatorname{rad}_A$ and h is not a section. Then g must be a retraction and therefore g_1 is a retraction.

(2) is dual to (1). □

The following two results show that the irreducible monomorphisms (resp. epimorphisms) can be characterized by means of their cokernels (resp. kernels).

Lemma 5.2.6. *Let*

$$0 \longrightarrow L \xrightarrow{f} M \xrightarrow{g} N \longrightarrow 0$$

be a nonsplit exact sequence in A–mod.

(1) *The homomorphism f is irreducible if and only if for every homomorphism $v : V \to N$ there exists $w_1 : V \to M$ such that $v = gw_1$, or there exists $w_2 : M \to V$ such that $g = vw_2$.*

(2) *The homomorphism g is irreducible if and only if for every homomorphism $u : L \to U$ there exists $z_1 : M \to U$ such that $u = z_1 f$, or there exists $z_2 : U \to M$ such that $f = z_2 u$.*

Proof. We prove only part (1). Part (2) follows by duality. Assume f is irreducible and that $v : V \to N$ is arbitrary. Define $E := M \times_N V \subseteq M \oplus V$ by $E = \{(x, y) \mid g(x) = v(y)\}$ and let $u : E \to M$ and $g' : E \to V$ be the restrictions of the projections $M \oplus V \to M$ and $M \oplus V \to V$, respectively. Then we have a commutative diagram

$$\begin{array}{ccccccccc} 0 & \longrightarrow & L & \xrightarrow{f'} & E & \xrightarrow{g'} & V & \longrightarrow & 0 \\ & & \downarrow{\scriptstyle 1_L} & & \downarrow{\scriptstyle u} & & \downarrow{\scriptstyle v} & & \\ 0 & \longrightarrow & L & \xrightarrow[f]{} & M & \xrightarrow[g]{} & N & \longrightarrow & 0 \end{array}$$

where the square on the right is a pull-back diagram.

We get a factorization $f = uf'$. Thus f' is a section or u is a retraction.

If f' is a section, then g' is a retraction, so there exists $w_1' : V \to E$ such that $g'w_1' = 1_V$. We can take $w_1 = uw_1'$ and we get $gw_1 = guw_1' = vg'w_1' = v$.

If u is a retraction, then there exists a homomorphism $w_2' : M \to E$ such that $uw_2' = 1_M$. Let $w_2 = g'w_2'$. Then we have $vw_2 = vg'w_2' = guw_2' = g$.

Conversely, suppose that for every homomorphism $v : V \to N$ there exists $w_1 : V \to M$ such that $v = gw_1$, or there exists $w_2 : M \to V$ such that $g = vw_2$. Since the given sequence is not split, f is neither a section nor a retraction. Assume f factors $f = f_1 f_2$, where $f_1 : U \to M$, $f_2 : L \to U$. Since f is a monomorphism, we have a commutative diagram

$$\begin{array}{ccccccccc} 0 & \longrightarrow & L & \xrightarrow{f_2} & U & \xrightarrow{u} & V & \longrightarrow & 0 \\ & & \downarrow{\scriptstyle 1_L} & & \downarrow{\scriptstyle f_1} & & \downarrow{\scriptstyle v} & & \\ 0 & \longrightarrow & L & \xrightarrow[f]{} & M & \xrightarrow[g]{} & N & \longrightarrow & 0 \end{array}$$

where $V = \operatorname{coker}(f_2)$. We show that the module U has to be isomorphic to the pull-back $V \times_N M$. The map

$$\begin{pmatrix} u \\ f_1 \end{pmatrix} : U \to V \oplus M$$

is injective: If $u(x) = f_1(x) = 0$ for $x \in U$, then $x = f_2(y)$ for some $y \in L$ and $0 = f_1(x) = f_1(f_2(y)) = f(y)$. Hence $y = 0$ because f is injective and $x = f_2(y) = 0$. It follows that $U \subseteq V \oplus M$ and since $\dim U = \dim V + \dim L = \dim V + \dim M - \dim N = \dim M \times_N V$ we conclude that $U = M \times_N V$.

Now, if there exists a map $w_1 : V \to M$ such that $v = gw_1$, from the commutative diagram

$$\begin{array}{ccc} V & \xrightarrow{1_V} & V \\ {\scriptstyle w_1}\downarrow & & \downarrow{\scriptstyle v} \\ M & \xrightarrow[g]{} & N \end{array}$$

and the universal property of the pull-back it follows that there exists a morphism $h : V \to U$ such that $uh = 1_V$. So u is a retraction and f_2 is a section.

Suppose that there exists $w_2 : M \to V$ such that $g = vw_2$. From the commutative diagram

$$\begin{array}{ccc} M & \xrightarrow{w_2} & V \\ {\scriptstyle 1_M}\downarrow & & \downarrow{\scriptstyle v} \\ M & \xrightarrow[g]{} & N \end{array}$$

and the universal property of the pull-back it follows that there exists a morphism $k : M \to U$ such that $f_1 k = 1_M$, so f_1 is a retraction.

We conclude that f is irreducible. □

Corollary 5.2.7.

(1) *If $f : L \to M$ is an irreducible monomorphism, then $N := \operatorname{coker}(f)$ is indecomposable.*

(2) *If $g : M \to N$ is an irreducible epimorphism, then $\ker(g)$ is indecomposable.*

Proof. We prove part (1). Part (2) follows by duality. Assume $N = N_1 \oplus N_2$ with N_1, N_2 nonzero. Let $q_i : N_i \to N$ be inclusions. Since q_i is not surjective, g cannot factor through q_i. Therefore, q_i must factor through g by Lemma 5.2.6. We can write $q_i = gw_i$ for $i = 1, 2$. But this means that

$$g \begin{pmatrix} w_1 & w_2 \end{pmatrix} = \begin{pmatrix} q_1 & q_2 \end{pmatrix} = 1_N,$$

so g splits. Therefore, f is a section which contradicts the irreducibility of f. □

Definition 5.2.8. We call a morphism $f : V \to W$ special if there exist indecomposable modules $M_0 = V, M_1, \ldots, M_l = W$ and a sequence of irreducible morphisms $f_i : M_{i-1} \to M_i$, $i = 1, 1, \ldots, l$ such that $f = f_l f_{l-1} \cdots f_1$. (The identity $1_V : V \to V$ is also considered special.)

The following proposition will be useful for understanding morphisms in the module category in terms of the Auslander-Reiten quiver that is defined in the next section.

Proposition 5.2.9. *Suppose that A has finite representation type. Then there exists an integer k such that for every two indecomposable modules V and W we have $\operatorname{rad}_A^k(V, W) = 0$. Then $\operatorname{Hom}_A(V, W)$ is spanned by special homomorphisms.*

Proof. Let $N_1, \ldots, N_r$ be all indecomposable representations and set $N = N_1 \oplus \cdots \oplus N_r$. We have that $\mathfrak{m} = \operatorname{rad}_A(N, N)$ is the Jacobson radical of the algebra $\operatorname{End}_A(N)$. We have $\mathfrak{m}^k = 0$ for some integer k. Suppose that $M_0 = V, M_1, \ldots, M_l = W$ are indecomposables and $f_i : M_{i-1} \to M_i$ is a nonisomorphism for all i. We can write $f_i = p_i f_i' u_{i-1}$ where $u_i : M_i \to N$ is the inclusion $f_i' : N \to N$ lies in the radical $\mathfrak{m}$ and $p_i : N \to M_i$ is the projection. Then we have

$$f_l f_{l-1} \cdots f_1 = p_l f_l' \cdots f_1' u_0 = 0$$

because $f_l' \cdots f_1' \in \mathfrak{m}^k = 0$. This shows that $\operatorname{rad}_A^k(V, W) = 0$.

By induction on l we show that $\operatorname{Hom}_A(V, W)$ is spanned by all special morphisms and $\operatorname{rad}_A(V, W)^l$. The cases $l = 0$ and $l = 1$ are obvious. Suppose that $\operatorname{Hom}_A(V, W)$ is spanned by $\operatorname{rad}_A(V, W)^l$ and all special morphisms from V to W. Now $\operatorname{rad}_A^l(V, W)$ is spanned by morphisms f that have a decomposition $f = f_l f_{l-1} \cdots f_1$ and each f_i is a nonisomorphism between two indecomposable modules. If each f_i is irreducible, then f is special. If not, then $f_i \in \operatorname{rad}_A^2$ for some i and $f \in \operatorname{rad}_A(V, W)^{l+1}$. This shows that $\operatorname{rad}_A^l(V, W)$ is contained in the span of $\operatorname{rad}_A^{l+1}(V, W)$ and all special morphisms. We conclude that $\operatorname{Hom}_A(V, W)$ is spanned by $\operatorname{rad}_A(V, W)^{l+1}$ and all special morphisms from V to W. This concludes the induction proof.

Taking $l = k$ gives us the desired result. □

5.3. The Auslander-Reiten Quiver

We will now introduce the **Auslander-Reiten Quiver** (or **AR Quiver**) for the algebra A.

Definition 5.3.1. The vertices of the AR Quiver correspond to the indecomposable modules. For two indecomposable modules V and W we draw $\dim \operatorname{rad}_A(V,W)/\operatorname{rad}^2_A(V,W)$ arrows from V to W.

There is a lot more structure to the AR Quiver which we will discover in Section 6.6. In particular, we will see that the Auslander-Reiten quiver is closely related to almost split sequences and the Auslander-Reiten transform. From Proposition 5.2.9 we see that for indecomposable modules V and W of a finite dimensional algebra A of finite representation type, the space $\operatorname{Hom}_A(V,W)$ is spanned by the morphisms corresponding to paths from V to W. There may be linear relations among those morphisms. These relations can be understood in terms of almost split sequences that we will discuss in the next section.

Example 5.3.2. For the quiver

$$\circ \longrightarrow \circ \longrightarrow \circ$$

the AR quiver is

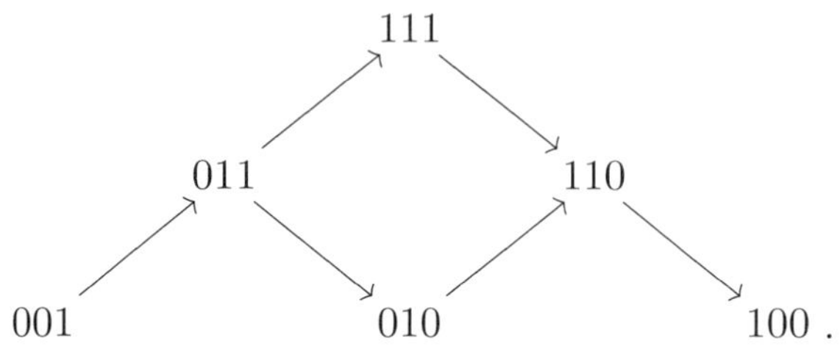

For example, $\operatorname{Hom}_Q(001,111)$ is 1-dimensional. But the morphism that spans this space factors through 011, so it lies in $\operatorname{rad}^2_Q(001,111)$. So there are no arrows from 001 to 111. The space $\operatorname{Hom}_Q(001,011)$ is also 1-dimensional, but the morphism that spans this space does not factor nontrivially. So $\operatorname{Hom}_Q(001,011) = \operatorname{rad}_Q(001,011)$ is 1-dimensional, and $\operatorname{rad}^2_Q(001,011) = 0$. We draw one arrow from 001 to 011. The space $\operatorname{Hom}_Q(011,110)$ is also 1-dimensional but any morphism from 011 to 110 factors through 111 (and through 010 as well).

Example 5.3.3. The AR quiver for the quiver

$$1 \longrightarrow 2 \longleftarrow 3 \longrightarrow 4$$

is

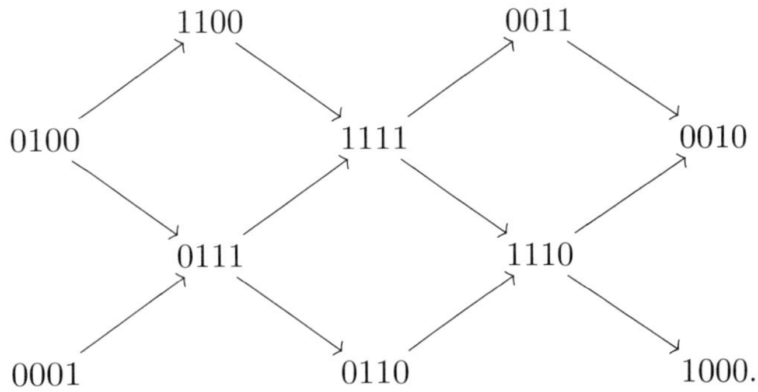

Example 5.3.4. Consider the algebra $A = \mathbb{C}Q/J$ where the quiver Q is

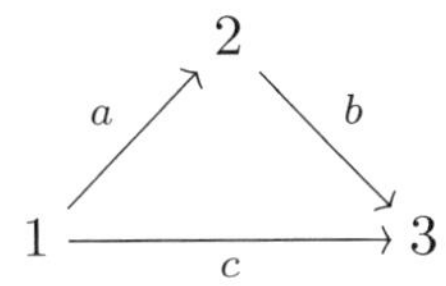

and J is generated by ba. The AR quiver is

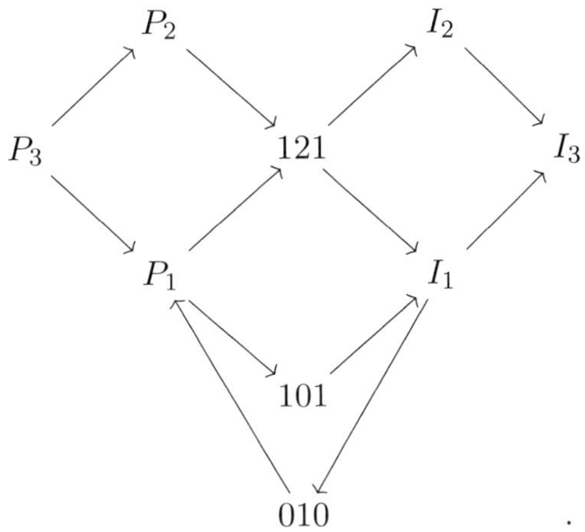

The dimension vectors of $P_1, P_2, P_3, I_1, I_2, I_3$ are 111 011, 001, 100, 110 and 111, respectively (P_1 and I_3 have the same dimension vector but are not isomorphic).

Exercises.

Exercise 5.3.1. *Verify the AR quiver in Example* 5.3.3.

Exercise 5.3.2. *Find the AR quiver for the quiver*

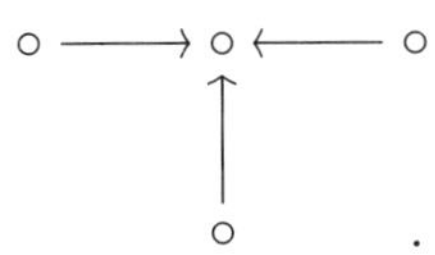

5.4. The Notion of an Almost Split Sequence

We will first introduce the notions of left/right minimal, and left/right almost split morphisms. These notions are closely related to the almost split sequence which we will define later.

Definition 5.4.1. Let L, M, N be A-modules.

(1) A homomorphism $f : L \to M$ is **left minimal** if for every $h \in \mathrm{End}_A(M)$, $hf = f$ implies h is an automorphism,

(2) A homomorphism $g : M \to N$ is **right minimal** if for every $k \in \mathrm{End}_A(M)$, $gk = g$ implies k is an automorphism,.

(3) A homomorphism $f : L \to M$ is **left almost split** if
   (a) f is not a section, and
   (b) for every homomorphism $u : L \to U$ which is not a section there exists a homomorphism $u' : M \to U$ such that the diagram

$$\begin{array}{ccc} L & \xrightarrow{f} & M \\ {\scriptstyle u}\downarrow & \swarrow{\scriptstyle u'} & \\ U & & \end{array}$$

   is commutative (i.e., $u'f = u$).

(4) A map $g : M \to N$ is **right almost split** if
   (a) g is not a retraction, and
   (b) for every homomorphism $v : V \to N$ which is not a retraction there exists a homomorphism $v' : V \to M$ such that the diagram

$$\begin{array}{ccc} & & V \\ & {\scriptstyle v'}\swarrow & \downarrow{\scriptstyle v} \\ M & \xrightarrow[g]{} & N \end{array}$$

   is commutative, i.e., $gv' = v$).

(5) A homomorphism $f : L \to M$ is **left minimal almost split** if it is both left minimal and left almost split.

(6) A homomorphism $g : M \to N$ is **right minimal almost split** if it is both right minimal and right almost split.

We notice that the "left" notions are dual to the "right" notions. In the sequel we will be thus proving only half of the statements. The following proposition shows that left/right almost split homomorphisms are unique in some sense.

Proposition 5.4.2.

(1) *If the homomorphisms $f : L \to M$ and $f' : L \to M'$ are both left minimal almost split, then there exists an isomorphism $h : M \to M'$ such that $f' = hf$.*

(2) *If the homomorphisms $g : M \to N$, $g' : M' \to N$ are both right minimal almost split, then there exists an isomorphism $k : M \to M'$ such that $g'k = g$.*

Proof. We prove the first part as the second part is dual. Since f and f' are almost split, there exist $h : M \to M'$ and $h' : M' \to M$ such that $f' = hf$ and $f = h'f'$. Hence $f = h'hf$ and $f' = hh'f'$. Since f and f' are left minimal, $h'h$ and hh' are automorphisms and we are done. □

Lemma 5.4.3.

(1) *If $f : L \to M$ is left almost split, then L is indecomposable.*

(2) *If $g : M \to N$ is right almost split, then N is indecomposable.*

Proof. We prove only (1) because of duality. Assume that $L = L_1 \oplus L_2$ with both L_1, L_2 nonzero, and let $p_i : L \to L_i$ be the projection ($i = 1, 2$). Then p_i is not a section. Thus there exist $u_i : M \to L_i$ such that $u_i f = p_i$:

$$\begin{array}{ccc} L & \xrightarrow{f} & M \\ {\scriptstyle p_i}\downarrow & \swarrow{\scriptstyle u_i} & \\ L_i & & \end{array} .$$

But then

$$u = \begin{pmatrix} u_1 \\ u_2 \end{pmatrix} : M \to L$$

satisfies $uf = 1_L$ which contradicts the fact that f is not a section. □

The next theorem makes a connection between the left/right almost split maps and irreducible morphisms.

Theorem 5.4.4.

(1) *If $f : L \to M$ is a left minimal almost split map, then it is irreducible. Moreover, a homomorphism $f' : L \to M'$ is irreducible if and only if $M' \neq 0$, and there exists a direct sum decomposition $M = M' \oplus M''$ and a homomorphism $f'' : L \to M''$ such that*

$$f := \begin{pmatrix} f' \\ f'' \end{pmatrix} .$$

(2) *If $g : M \to N$ is a right minimal almost split map, then it is irreducible. Moreover, a homomorphism $g' : M' \to N$ is irreducible if and only if $M' \neq 0$, and there exists a direct sum decomposition $M = M' \oplus M''$ and a homomorphism $fg' : M'' \to N$ such that*

$$g := \begin{pmatrix} g' & g'' \end{pmatrix}.$$

Proof. We prove only statement (1). Part (2) follows by duality. Let $f : L \to M$ be a left minimal almost split map. By definition, f is not a section. By Lemma 5.4.3, L is indecomposable. Since f is not an isomorphism, it is also not a retraction. Assume $f = f_1 f_2$ where $f_1 : X \to M$, $f_2 : L \to X$. We assume that f_2 is not a section and prove that f_1 is a retraction. Since f is left almost split, there exists $f_2' : M \to X$ such that $f_2 = f_2' f$. Hence $f = f_1 f_2 = f_1 f_2' f$. Because f is left minimal, $f_1 f_2'$ is an automorphism, so f_1 is a retraction. Thus f is irreducible.

Assume now $f' : L \to M'$ is irreducible. Then it is clear that $M' \neq 0$. Also, f' is not a section, so there exists a map $h : M \to M'$ such that $f' = hf$. Because f' is irreducible, and f is not a section, h is a retraction. So $hu = 1_{M'}$ for some $u : M' \to M$. Let $q : M \to M''$ be the cokernel of u. We may identify M with $M' \oplus M''$ via

$$\begin{pmatrix} h \\ q \end{pmatrix} : M \to M' \oplus M''.$$

With this identification we have

$$f = \begin{pmatrix} f' \\ f'' \end{pmatrix},$$

where $f'' = qf$.

To prove the reverse implication, assume that $M = M' \oplus M''$ and

$$f = \begin{pmatrix} f' \\ f'' \end{pmatrix}.$$

Since f is not a section, $f \in \operatorname{rad}_A(L, M)$ and therefore $f', f'' \in \operatorname{rad}_A$.

Assume that $f' = f_1 f_2$, $f_2 : L \to X$, $f_1 : X \to M'$. We assume that f_2 is not a section and prove that f_1 is a retraction. We have

$$f = \begin{pmatrix} f' \\ f'' \end{pmatrix} = \begin{pmatrix} f_1 & 0 \\ 0 & 1 \end{pmatrix} \begin{pmatrix} f_2 \\ f'' \end{pmatrix},$$

where $\begin{pmatrix} f_1 & 0 \\ 0 & 1 \end{pmatrix} : X \oplus M'' \to M' \oplus M''$, $\begin{pmatrix} f_2 \\ f'' \end{pmatrix} : L \to X \oplus M''$.

We have $f_2 \in \operatorname{rad}_A(L, X)$ because it is not a section. Because f_2, f'' are in the radical,

$$\begin{pmatrix} f_2 \\ f'' \end{pmatrix}$$

lies in the radical and is not a section. Because f is irreducible,

$$\begin{pmatrix} f_1 & 0 \\ 0 & 1 \end{pmatrix}$$

is a retraction. It follows that f_1 is a retraction. □

Definition 5.4.5. An exact sequence is called almost split if

$$0 \longrightarrow L \xrightarrow{f} M \xrightarrow{g} N \longrightarrow 0,$$

(1) f is left minimal almost split, and

(2) g is right minimal almost split.

Remark 5.4.6. The existence of almost split sequences is not obvious. However, it is clear from Lemma 5.4.3 that if the sequence above is almost split, then L and N are indecomposable. Since the sequence does not split, it is also clear that L cannot be injective and N cannot be projective. Finally, an almost split sequence is (by Proposition 5.4.2) determined uniquely up to isomorphism by each of its end terms.

The following lemma will be used in the proof of Theorem 5.4.8.

Lemma 5.4.7. *Let*

$$\begin{array}{ccccccccc} 0 & \longrightarrow & L & \xrightarrow{f} & M & \xrightarrow{g} & N & \longrightarrow & 0 \\ & & \downarrow u & & \downarrow v & & \downarrow w & & \\ 0 & \longrightarrow & L & \xrightarrow[f]{} & M & \xrightarrow[g]{} & N & \longrightarrow & 0 \end{array}$$

be a commutative diagram, with the rows exact and nonsplit.

(1) *If L is indecomposable and w is an automorphism, then u and hence v are automorphisms.*

(2) *If N is indecomposable and u is an automorphism, then w and hence v are automorphisms.*

Proof. As always, we just prove part (1). After composing with w^{-1} and taking the appropriate fibered product, we can assume that $w = 1_N$. If u is not an automorphism, it is nilpotent, so for some $m > 0$ we have $u^m = 0$. Then $v^m f = f u^m = 0$, so v^m factors through the cokernel N of f, so there exists $h : M \to N$ such that $v^m = hg$. Because $gv^m = g$, we get $ghg = g$ and consequently $gh = 1_N$ (because g is an epimorphism. This contradicts the fact that the sequence does not split. □

We continue with an important result giving equivalent conditions for an exact sequence to be almost split.

Theorem 5.4.8. *Let*

$$0 \longrightarrow L \xrightarrow{f} M \xrightarrow{g} N \longrightarrow 0$$

be a short exact sequence in A–mod. *The following conditions are equivalent.*

(1) *The given sequence is almost split.*

(2) L *is indecomposable and* g *is right almost split.*

(3) N *is indecomposable and* f *is left almost split.*

(4) *The homomorphism* f *is left minimal almost split.*

(5) *The homomorphism* g *is right minimal almost split.*

(6) L *and* N *are indecomposable and the maps* f *and* g *are irreducible.*

Proof. By definition of an almost split sequence, (1) implies (4) and (5). By Proposition 5.4.2. (1) implies (2) and (3). By Theorem 5.4.4 and Proposition 5.4.2 (1) implies (6) as well. We will show the following implications.

(a) (5) implies (2). Then dually, (4) implies (3).

(b) (2) implies (3). Then, dually, (3) implies (2).

(c) (2) and (3) imply (1).

(d) (6) implies (2).

By (b) and (c) the first three conditions are equivalent. Then (a) and the claims above show that the first five conditions are equivalent. Finally (d) and the implication (1)⇒(6) given above completes the proof.

(a) Assume that g is right minimal almost split. By Theorem 5.4.4, g is irreducible. Hence, by Corollary 5.2.7, $L = \ker(g)$ is indecomposable. This implies (2).

(b) To prove (2)⇒(3), by Lemma 5.4.3 it is enough to show that (2) implies that f is left almost split. Suppose that g is right almost split and L is indecomposable. Since g is not a retraction, f is not a section. Let $u : L \to U$ be such that $u'f \neq u$ for all $u' : M \to U$. We must prove that u is a section. There exists a commutative diagram with exact rows

$$\begin{array}{ccccccccc} 0 & \longrightarrow & L & \xrightarrow{f} & M & \xrightarrow{g} & N & \longrightarrow & 0 \\ & & \downarrow{\scriptstyle u} & & \downarrow{\scriptstyle v} & & \downarrow{\scriptstyle 1_N} & & \\ 0 & \longrightarrow & U & \xrightarrow[h]{} & V & \xrightarrow[k]{} & N & \longrightarrow & 0 \end{array}$$

where the left square

$$\begin{array}{ccc} L & \xrightarrow{f} & M \\ {\scriptstyle u}\downarrow & & \vdots\,{\scriptstyle v} \\ U & \underset{h}{\dashrightarrow} & V \end{array}$$

is a push-out diagram. Suppose that h is a section. Then there exists a homomorphism $w : V \to U$ such that $wh = 1_U$. If we define $z = wv$, then $zf = wvf = whu = u$. This contradicts our assumption on u. So h is not a section and k is not a retraction. Because g is right almost split, there exists a map $v' : V \to M$ such that $k = gv'$. Now we get a commutative diagram

$$\begin{array}{ccccccccc} 0 & \longrightarrow & L & \xrightarrow{f} & M & \xrightarrow{g} & N & \longrightarrow & 0 \\ & & \downarrow{\scriptstyle u'u} & & \downarrow{\scriptstyle v'v} & & \downarrow{\scriptstyle 1_N} & & \\ 0 & \longrightarrow & L & \xrightarrow[f]{} & M & \xrightarrow[g]{} & N & \longrightarrow & 0 \end{array}$$

where u' is derived from v' and 1_N by passing to the kernels. By Lemma 5.4.7 $u'u$ is an automorphism. Thus u is a section and we get (3).

(c) Assume that both (2) and (3) hold. We must prove that f and g are minimal. To prove that f is left minimal, consider $h : M \to M$ such that $f = hf$. We have a commutative diagram

$$\begin{array}{ccccccccc} 0 & \longrightarrow & L & \xrightarrow{f} & M & \xrightarrow{g} & N & \longrightarrow & 0 \\ & & \downarrow{\scriptstyle 1_L} & & \downarrow{\scriptstyle h} & & \downarrow & & \\ 0 & \longrightarrow & L & \xrightarrow[f]{} & M & \xrightarrow[g]{} & N & \longrightarrow & 0 \end{array}$$

with exact rows. By Lemma 5.4.7, h is an automorphism. So f is left minimal, and similarly g is right minimal.

(d) Assume (6). By hypothesis L is indecomposable and g is not a retraction. Assume that $v : V \to N$ is not a retraction. We may suppose that V is indecomposable (by replacing it by one of its indecomposable summands). Since f is irreducible, Lemma 5.2.6 gives either $v' : V \to M$ such that $v = gv'$ or an $h : M \to V$ such that $g = vh$. In the latter case, since g is irreducible and v is not a retraction, h must be a section. But since V is indecomposable, h has to be an isomorphism. Then $v' = h^{-1}$ satisfies $v = gv'$. □

The following two theorems give an exact description of the almost split sequences in terms of irreducible morphisms.

Theorem 5.4.9. *Suppose*

$$0 \longrightarrow L \xrightarrow{f} M \xrightarrow{g} N \longrightarrow 0$$

is an almost split sequence, and $M = M_1^{k_1} \oplus M_2^{k_2} \oplus \cdots \oplus M_r^{k_r}$*, where* $M_1, \ldots, M_r$ *are indecomposable and pairwise nonisomorphic. For any indecomposable representation* V *there exists an irreducible morphism to* N *if and only if* $V \cong M_i$ *for some* i*. If*

$$h = \begin{pmatrix} h_1 & h_2 & \cdots & h_{k_i} \end{pmatrix}$$

is the restriction of g *to* $M_i^{k_i}$*, then* $h_1, \ldots, h_{k_i}$ *form a basis of* $\operatorname{rad}_A(M_i, N)$ *modulo* $\operatorname{rad}_A^2(M_i, N)$*.*

Proof. Let V be an indecomposable representation and choose a basis $v_1, \ldots, v_l \in \operatorname{rad}_A(V, N)$ such that they form a basis modulo $\operatorname{rad}_A^2(V, N)$. Then

$$v = \begin{pmatrix} v_1 & v_2 & \cdots & v_l \end{pmatrix}$$

is irreducible by Lemma 5.2.4. So V^l is a direct summand of M. It follows that $V = M_i$ for some i and $\dim \operatorname{rad}_A(M_i, N)/\operatorname{rad}_A^2(M_i, N) = l \leq k_i$. The restriction of g to $M_i^{k_i} \to N$ is also irreducible, so $k_i \leq l$ by Lemma 5.2.4. $\square$

Theorem 5.4.10.

(1) *If* N *is an indecomposable finite dimensional* A*-module which is not projective, then there exists an almost split sequence*

$$0 \longrightarrow L \xrightarrow{f} M \xrightarrow{g} N \longrightarrow 0.$$

(2) *If* L *is an indecomposable finite dimensional* A*-module which is not injective, then there exists an almost split sequence*

$$0 \longrightarrow L \xrightarrow{f} M \xrightarrow{g} N \longrightarrow 0.$$

Proof. Let $M_1, \ldots, M_r$ be all the indecomposable modules for which there exists an irreducible morphism $M_i \to N$. Let $M = M_1^{k_1} \oplus \cdots \oplus M_r^{k_r}$ and $k_i = \dim \operatorname{rad}_A(M_i, N)/\operatorname{rad}_A^2(M_i, N)$. Define $g : M \to N$ such that, when

$$h = \begin{pmatrix} h_1 & \cdots & h_{k_i} \end{pmatrix}$$

is the restriction of g to $M_i^{k_i}$, then $h_1, \ldots, h_{k_i} \in \operatorname{rad}_A(M_i, N)$ form a basis modulo $\operatorname{rad}_A^2(M_i, N)$. Suppose that $v : V \to N$ is not a retraction. Decompose $V = V_1 \oplus \cdots \oplus V_s$ and write

$$v = \begin{pmatrix} v_1 \\ \vdots \\ v_s \end{pmatrix}$$

Now $v_i \in \operatorname{rad}_A(V_i, N)$. Since $\operatorname{Hom}_A(V_i, N)$ is spanned by special morphisms by Proposition 5.2.9 , we can write $v_i = \sum_{j=1}^m w_j z_j$ where w_j is an irreducible morphism between two indecomposable representations. It follows that w_j factors through M by the construction of M, and hence v_i factors through M. We conclude that V factors through M. This shows that $g : M \to N$ is right almost split. Suppose that $u : M \to M$ satisfies $gu = g$. Suppose that u is not an isomorphism. Let $\mathfrak{m} = \operatorname{rad}_A(M, M)$. Then $M/\mathfrak{m}M \cong \mathbb{C}^{k_1+\cdots+k_r}$ and u induces The morphism g induces an isomorphism

$$\overline{g} : M/\mathfrak{m}M \to \operatorname{rad}_A(M, N)/\operatorname{rad}_A^2(M, N).$$

The morphism $u : M \to M$ induces a linear map $\overline{u} : M/\mathfrak{m}M \to M/\mathfrak{m}M$. We have $\overline{gu} = \overline{g}$ and $\overline{u}$ is the identity. It follows that u is an isomorphism. We conclude that g is right minimal.

Therefore, g is right minimal almost split and the sequence is an almost split sequence.

$\square$

The almost split sequences can be read off from the AR quiver as follows. If N is an indecomposable module that is not a projective, and $h_i : M_i \to N$, $1 \leq i \leq k$ are all the incoming arrows at N in the AR quiver, then there exists an almost split sequence

$$0 \longrightarrow L \longrightarrow M_1 \oplus \cdots \oplus M_k \longrightarrow N \longrightarrow 0,$$

where L is another indecomposable module. The dual statement of course is also true.

Example 5.4.11. Consider the example 5.3.3:

$$\circ \longrightarrow \circ \longleftarrow \circ \longrightarrow \circ$$

with AR quiver

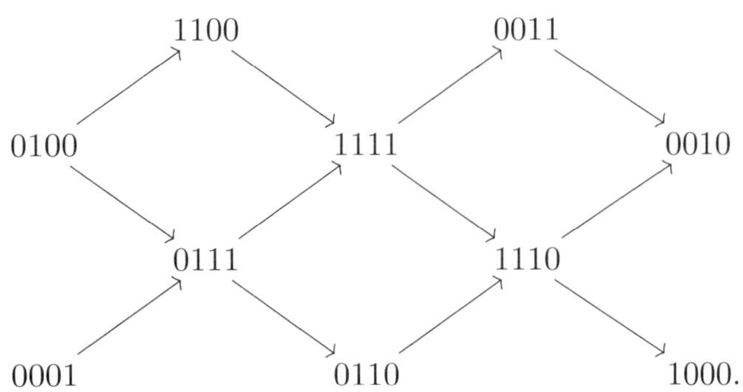

The almost split sequences are

$$\begin{aligned}
&0 \to 1100 \to 1111 \to 0011 \to 0\\
&0 \to 0100 \to 1100 \oplus 01111 \to 1111 \to 0\\
&0 \to 1111 \to 0011 \oplus 1110 \to 0011 \to 0\\
&0 \to 0111 \to 0110 \oplus 1111 \to 1110 \to 0\\
&0 \to 0111 \to 1111 \to 0110 \to 0\\
&0 \to 0110 \to 1110 \to 1000 \to 0.
\end{aligned}$$

Example 5.4.12. Consider the quiver

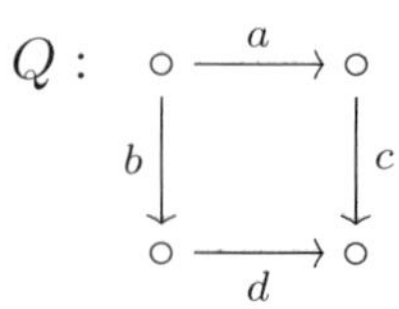

together with the relation given $ca - db$, and let $A = \mathbb{C}Q/(ca - db)$. The AR quiver is

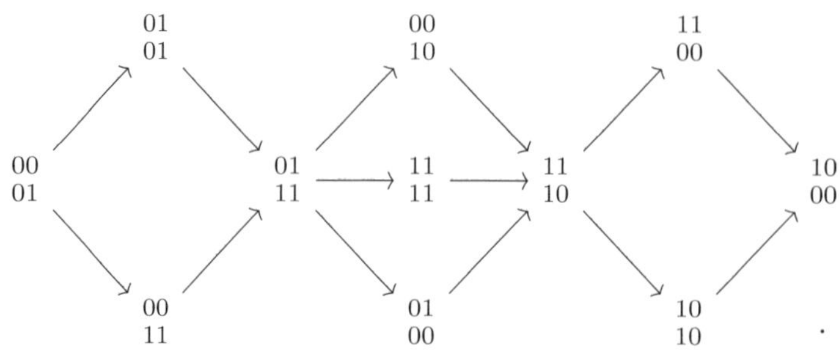

The almost split sequence that ends with $\begin{smallmatrix}11\\10\end{smallmatrix}$ is

$$0 \to \begin{smallmatrix}01\\11\end{smallmatrix} \to \begin{smallmatrix}00\\10\end{smallmatrix} \oplus \begin{smallmatrix}11\\11\end{smallmatrix} \oplus \begin{smallmatrix}01\\00\end{smallmatrix} \to \begin{smallmatrix}11\\10\end{smallmatrix} \to 0.$$

Exercises.

Exercise 5.4.1. *Write down all almost split sequences in Example* 5.4.12.

Exercise 5.4.2. *Find all almost split sequences in Example* 5.3.4.

5.5. Bibliographical Remarks

This material of this chapter is due to Auslander and Reiten [**4**], [**5**] who introduced the fundamental notion of almost split sequences.

The technique of almost split sequences and of constructing the Auslander-Reiten quivers (known as knitting) proved to be very important in analyzing the structure of algebras of finite representation type and of the tame type. In the hereditary case it gives the second part of the proof of Gabriel theorem, by exhibiting all indecomposable representations in the Dynkin cases. It also provides the functorial definition of Coxeter functor, which was introduced in [**8**].

However, the significance of this technique goes much further, since the Auslander-Reiten quiver encodes many properties of representations in a finite representation type and tame cases. It is still unknown, however, what is the general shape of the Auslander-Reiten quiver of an algebra of finite representation type.

For wild algebras the Auslander-Reiten quiver has many components and it is not clear how they interact. To understand the situation better one needs more geometric methods. This will be the subject of the chapter on semi-invariants.

In writing this chapter we followed the approach of Assem, Simson and Skowroński [**38**]. This book is an excellent source and we recommend it to the reader.

Chapter 6

Auslander-Reiten Theory

The Coxeter functor described in Chapter 4 turns out to have a simple homological description as a dual transpose. This definition generalizes for arbitrary finite dimensional algebra to the Auslander-Reiten translation. In this way one can understand not only the indecomposable representations but all the maps between them. The Auslander-Reiten quiver assembles all information about the indecomposables and the irreducible maps in a compact way. For quivers of finite representation type the information is basically complete as we can read off the dimensions of Hom and Ext spaces from the Auslander-Reiten quiver.

In this chapter we work over a hereditary algebra $A = \mathbb{C}Q$ where Q is acyclic and connected, except for Sections 6.1 and 6.2, where A may be an arbitrary finite dimensional algebra.

6.1. Injective Envelopes and Projective Covers

The Auslander-Reiten translation is defined using minimal presentations. For this, we need the notion of projective covers and injective envelopes.

Definition 6.1.1. An injective module homomorphism $q : V \hookrightarrow I$ is called the **injective envelope** of V if I is an injective module, and for any homomorphism $f : I \to N$, fq injective implies that f is injective.

Example 6.1.2. Consider the equioriented $\mathbf{A}_n$ quiver from Example 4.5.2:

$$1 \xrightarrow{a_1} 2 \xrightarrow{a_2} \cdots \xrightarrow{a_{n-1}} n$$

Let $A = \mathbb{C}Q$. Consider the embedding of the representation $E_{n,n}$ into $E_{1,n}$. The representation $E_{1,n}$ is injective, and all its nonzero submodules are isomorphic to $E_{i,n}$ for $1 \leq i \leq n$ so they contain $E_{n,n}$. This means our embedding is essential.

The dual notion of an injective envelope is as follows.

Definition 6.1.3. A surjective module homomorphism $p : P \twoheadrightarrow V$ is called the **projective cover** if P is a projective module and if for any homomorphism $g : M \to P$, pg surjective implies that g surjective.

Lemma 6.1.4. *Suppose that $p : P \twoheadrightarrow V$ is a surjective module homomorphism and P is projective. Then p is the projective cover if and only if the induced map $\overline{p} : P/\mathfrak{m}P \to V/\mathfrak{m}V$ is an isomorphism.*

Proof. Assume that $\overline{p} : P/\mathfrak{m}P \to V/\mathfrak{m}V$ is an isomorphism. Suppose that $pg : M \to V$ is surjective for some homomorphism $g : M \to P$. Then g induces a linear map $\overline{g} : M/\mathfrak{m}M \to P/\mathfrak{m}P$ such that $\overline{pg}$ is surjective. Since $\overline{p}$ is an isomorphism, $\overline{g}$ is surjective. It follows that g is surjective by Corollary 3.2.5.

Conversely, suppose that $\overline{p}$ is not an isomorphism. Because p is surjective, we still have that $\overline{p}$ is surjective. So $\overline{p}$ is not injective. We can choose a submodule M' containing $\mathfrak{m}M$ such that $M' \neq M$ and $\overline{p}(M'/\mathfrak{m}M) = P/\mathfrak{m}P$. By Corollary 3.2.5, pg is surjective where $g : M' \hookrightarrow M$ is the inclusion. Since g is not surjective, p is not a projective cover.

□

Example 6.1.5. Assume that Q has no oriented cycles and let $x \in Q_0$. Let $A = \mathbb{C}Q/J$. The natural surjection from the indecomposable projective representation P_x onto the simple representation S_x is a projective cover, because $P_x/\mathfrak{m}P_x \cong S_x/\mathfrak{m}S_x \cong S_x$.

The injective envelopes and projective covers are unique up to isomorphism.

Proposition 6.1.6.

(1) *Let V be an A-module and let $q : V \to I$, $q' : V \to I'$ be two injective envelopes. Then there exists an isomorphism $f : I \to I'$ such that $fq = q'$.*

(2) *Let V be an A-module and let $p : P \to V$, $p' : P' \to V$ be two projective covers. Then there exists an isomorphism $g : P \to P'$ such that $p'g = p$.*

Proof. The two parts are dual to each other. Let us prove (1). Consider the diagram

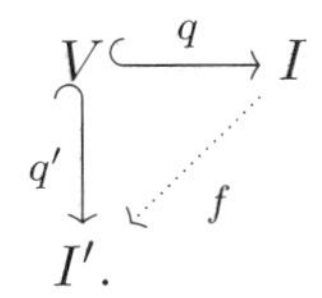

By the definition of injective modules, the diagram can be filled by a map $f : I \to I'$ such that $fq = q'$. Similarly, there exists a homomorphism $f' : I' \to I$ such that $f'q' = q$. Since $q = f'q'$ and $q' = fq$ are injective, the morphisms f' and f are injective as well. The homomorphism $f'f : I \to I$ is injective and I is finite dimensional, so $f'f$ is an isomorphism. The homomorphism ff' is an isomorphism as well for the same reason, so f and f' are isomorphisms. □

Proposition 6.1.7. *Injective envelopes and projective covers exist in* A–mod.

Proof. A module P is projective if an only if $\mathrm{Hom}_A(P, -)$ is an exact functor. A module homomorphism $\varphi : M \to N$ is surjective if and only if it is an epimorphism. This shows that the definitions of a projective cover and an injective envelope (Definitions 6.1.3 and 6.1.1) can be expressed entirely in categorical terms. Since A is Morita equivalent to an algebra of the form $\mathbb{C}Q/J$ (Corollary 3.4.12) where J is an admissible ideal, it suffices to prove the proposition in the case where $A = \mathbb{C}Q/J$. For this, we refer to Exercise 6.1.1. □

We recall some basic definitions from noncommutative algebra. Let V be an A-module. The **socle** of a module V is

$$\mathrm{soc}(V) := \{x \in V \mid \mathfrak{m}x = 0\}.$$

This is the biggest semi-simple submodule in V. Dually, the **top** of a module V is $\mathrm{top}(V) := V/\mathfrak{m}V$. This is the biggest semi-simple quotient of V.

Exercises.

Exercise 6.1.1. *Suppose that* $A = \mathbb{C}Q/J$ *is a quiver with relations. Let* $R \subseteq A$ *be the subalgebra generated by* $e_x, x \in Q_0$. *Suppose that* V *is a finite dimensional* A*-module. Let* $u : V/\mathfrak{m}V \to V$ *be an* R*-module homomorphism that is a section to* $V \to V/\mathfrak{m}V$. *There is a unique* A*-module homomorphism*

$$A \otimes_R V/\mathfrak{m}V \to V$$

defined by $a \otimes v = au(v)$. *Show that this is a projective cover. Note that if the semi-simple module* $V/\mathfrak{m}V$ *decomposes as* $V/\mathfrak{m}V = \bigoplus_{x \in Q_0} S_x^{k_x}$, *then* $A \otimes_R V = \bigoplus_{x \in Q_0} P_x^{k_x}$. *Similarly, show that injective envelopes exist.*

Exercise 6.1.2. *Formulate the dual statement to Lemma* 6.1.4

6.2. The Transpose Functor

In this section we construct the transpose functor from minimal presentations of modules. The Auslander-Reiten Translation is constructed from the transpose by composing it with the duality functor, and will be studied for hereditary algebras in the next section.

Suppose that A is an arbitrary finite dimensional algebra, and that V is an A-module.

Lemma 6.2.1. *If P is a projective left A-module, then $\widetilde{P} := \mathrm{Hom}_A(P, A)$ is a projective right A-module.*

Proof. Since P is projective, it is a direct summand of A^l for some $l > 0$. But then $\widetilde{P} := \mathrm{Hom}_A(P, A)$ is a direct summand of $\mathrm{Hom}_A(A^l, A) \cong A^l$. So $\widetilde{P}$ is projective. □

A **minimal presentation** of V is an exact sequence

$$P_1 \longrightarrow P_0 \longrightarrow V \longrightarrow 0 \tag{6.1}$$

such that P_0 is the projective cover of V, and P_1 is the projective cover of the kernel of the map $P_0 \to V$. If we apply $\mathrm{Hom}_A(-, A)$ to the exact sequence (6.1), we get an exact sequence of right A-modules

$$0 \longrightarrow \mathrm{Hom}_A(V, A) \longrightarrow \widetilde{P_0} \longrightarrow \widetilde{P_1}. \tag{6.2}$$

Definition 6.2.2. The **transpose** $\mathrm{Tr}(V)$ is the cokernel of the right hand side map in (6.2). This gives an exact sequence of right A-modules

$$0 \longrightarrow \mathrm{Hom}_A(V, A) \longrightarrow \widetilde{P_0} \longrightarrow \widetilde{P_1} \longrightarrow \mathrm{Tr}(V) \longrightarrow 0\ .$$

If V is projective, then $P_1 = 0$, and $\mathrm{Tr}(V) = 0$. Note that $\mathrm{Tr}(V)$ is a right A-module. For a morphism $\phi : V \to V'$ we can also define $\mathrm{Tr}(\phi)$ (see Exercise 6.2.2). This makes Tr into a functor from A–mod to mod–A (see Exercise 6.2.3).

Lemma 6.2.3. *If P is a projective A-module, then we have an isomorphism*

$$P \cong \mathrm{Hom}_{A^{\mathrm{op}}}(\widetilde{P}, A)$$

of left A-modules.

Proof. We have a natural transformation

$$\gamma :\ \mathrm{id} \longrightarrow \mathrm{Hom}_{A^{\mathrm{op}}}(\mathrm{Hom}_A(-, A), A)$$

defined as follows. The homomorphism

$$\gamma(M): \ M \longrightarrow \mathrm{Hom}_{A^{\mathrm{op}}}(\mathrm{Hom}_A(M, A), A)$$

maps $m \in M$ to the function $f \mapsto f(m)$. It is trivial to see that $\gamma(A)$ is an isomorphism. Since P is a direct summand of A^l for some $l > 0$, we can apply Lemma 3.4.2 to show that

$$\gamma(P): \ P \longrightarrow \mathrm{Hom}_{A^{\mathrm{op}}}(\widetilde{P}, A)$$

is an isomorphism. □

In the next section, we will study the transpose and translation functors for hereditary algebras.

Exercises.

Exercise 6.2.1. *Show that the minimal presentation of V is unique: If $P_1 \to P_0 \to V \to 0$ and $P_1' \to P_0' \to V \to 0$ are minimal presentations, then there exists a commutative diagram:*

$$\begin{array}{ccccccc} P_1 & \longrightarrow & P_0 & \longrightarrow & V & \longrightarrow & 0 \\ \cong\big\downarrow & & \cong\big\downarrow & & \big\downarrow 1_V & & \\ P_1' & \longrightarrow & P_0' & \longrightarrow & V & \longrightarrow & 0 \end{array}$$

where the vertical maps are isomorphisms. A similar result is true for minimal resolutions.

Exercise 6.2.2. *Suppose that $\varphi: V \to V'$ is a morphism of A-modules and that $P_1 \to P_0 \to V \to 0$ and $P_1' \to P_0' \to V' \to 0$ are minimal presentations. Show that there exists a commutative diagram:*

$$\begin{array}{ccccccc} P_1 & \longrightarrow & P_0 & \longrightarrow & V & \longrightarrow & 0 \\ \big\downarrow & & \big\downarrow & & \big\downarrow \phi & & \\ P_1' & \longrightarrow & P_0' & \longrightarrow & V' & \longrightarrow & 0. \end{array}$$

Use this to define $\mathrm{Tr}(\phi)$ *(and show that this is well-defined).*

Exercise 6.2.3. *Show that* Tr *is a functor from A–*mod *to* mod*–A.*

Exercise 6.2.4. *Let $P_x = Pe_x$ be the indecomposable projective representation corresponding to vertex x. Show that $\widetilde{P}_x := \mathrm{Hom}_A(P_x, A) \cong e_x A$ and $\mathrm{Hom}_A(A, \widetilde{P}_x) \cong P_x$ Use this to give another proof that Lemma* 6.2.3.

Exercise 6.2.5. *If I_x is the indecomposable injective module corresponding to vertex x. Show that $I_x \cong \widetilde{P}_x^\star$. (Use Exercise* 6.2.4.*)*

6.3. The Translation Functor for Quivers

For the remainder of this chapter, we take $A = \mathbb{C}Q$ where Q is acyclic and connected.

There is a way of defining the transpose that does not depend on the choice of a minimal presentation.

Proposition 6.3.1. *Let $A = \mathbb{C}Q$ and let V be a representation without projective direct summands. Then*

$$\mathrm{Tr}(V) = \mathrm{Ext}_Q(V, A).$$

Proof. Since A is hereditary, our minimal presentation is in fact a projective resolution

$$0 \longrightarrow P_1 \longrightarrow P_0 \longrightarrow V \longrightarrow 0\ .$$

Therefore the cokernel of the induced map $\mathrm{Hom}_Q(P_0, A) \to \mathrm{Hom}_Q(P_1, A)$ is by definition equal to $\mathrm{Ext}_Q(V, A)$. □

Lemma 6.3.2. *If P is a projective module, and V does not have any projective summands, then* $\mathrm{Hom}_A(V, P) = 0$.

Proof. Assume that P is projective and that V does not have any projective summands. Suppose that $\varphi : V \to P$ is an A-module homomorphism and let P' be its image. Then P' is a submodule of P. Since $\mathbb{C}Q$ is hereditary, P' is projective as well. The induced surjective map $V \twoheadrightarrow P'$ splits because P' is projective. So P' is a direct summand of V. Since V does not have any projective summands, we have $P' = 0$ and so $\varphi = 0$. □

We can also define the transpose $\mathrm{Tr} : \text{mod–}A \to A\text{–mod}$ by going through left-right equivalences:

$$\text{mod–}A \cong A^{\mathrm{op}}\text{–mod} \xrightarrow{\quad \mathrm{Tr} \quad} \text{mod–}A^{\mathrm{op}} \cong A\text{–mod}.$$

Proposition 6.3.3. *If V does not have any projective summands, then* $\mathrm{Tr}(\mathrm{Tr}(V))$ *is isomorpic to V.*

Proof. Suppose that V does not have any projective summands. Let

$$0 \longrightarrow P_1 \longrightarrow P_0 \longrightarrow V \longrightarrow 0$$

be the minimal resolution of V. We have an exact sequence

$$0 = \mathrm{Hom}_A(V, A) \longrightarrow \widetilde{P}_0 \longrightarrow \widetilde{P}_1 \longrightarrow \mathrm{Tr}(V) \longrightarrow 0$$

and $\widetilde{P}_0$ and $\widetilde{P}_1$ are projective A^{op}-modules. So by definition, $\mathrm{Tr}(\mathrm{Tr}(V))$ is the cokernel of the map

$$P_0 \cong \mathrm{Hom}_{A^{\mathrm{op}}}(\widetilde{P}_0, A) \longrightarrow P_1 \cong \mathrm{Hom}_{A^{\mathrm{op}}}(\widetilde{P}_1, A).$$

which is equal to V. □

The transpose has the drawback that it switches left modules to right modules and vice versa. Thus it is natural to consider a functor

$$\tau^+ : A\text{–mod} \xrightarrow{\quad \mathrm{Tr} \quad} \text{mod–}A \cong A^{\mathrm{op}}\text{–mod} \xrightarrow{\quad D \quad} A\text{–mod},$$

where D is the **duality functor** sending the representation V to V^*. The functor τ^+ is a covariant functor vanishing on projective modules.

Similarly, we can define a functor

$$\tau^- : A\text{–mod} \xrightarrow{\quad D \quad} A^{\mathrm{op}}\text{–mod} \xrightarrow{\quad \mathrm{Tr} \quad} \text{mod–}A^{\mathrm{op}} \cong A\text{–mod},$$

which is a covariant functor vanishing on injective modules. We call the functors $\tau^+ = D \circ \mathrm{Tr}$, $\tau^- = \mathrm{Tr} \circ D$ the **Auslander-Reiten translation functors**.

Proposition 6.3.4.

(1) *If V does not have any projective summands, then*

$$\tau^-(\tau^+(V)) \cong V.$$

(2) *If V does not have any injective summands, then*

$$\tau^+(\tau^-(V)) \cong V.$$

Proof. If V does not have any projective summands, then

$$\tau^-(\tau^+(V)) = \mathrm{Tr} \circ D \circ D \circ \mathrm{Tr}(V) = \mathrm{Tr}(\mathrm{Tr}(V)) \cong V.$$

If V does not have any injective summands, then $V^\star$ does not have any projective summands, so

$$\tau^+(\tau^-(V)) = D \circ \mathrm{Tr} \circ \mathrm{Tr} \circ D(V) = (\mathrm{Tr}(\mathrm{Tr}(V^\star)))^\star \cong (V^\star)^\star \cong V.$$

□

Exercises.

Exercise 6.3.1. *Show that for every module V, $\mathrm{Tr}(V)$ does not have any projective summands. Also show that $\tau^-(V)$ does not have any projective summands, and that $\tau^+(V)$ does not have any injective summands.*

6.4. Auslander-Reiten Duality

The translation functor not only gives a duality between Hom and Ext spaces, but is also closely related to the notion of almost split sequences. We will explore this in this section.

Theorem 6.4.1 (Auslander-Reiten duality).

(1) *Assume that V has no projective summands. Then we have natural isomorphisms of vector spaces:*

$$\begin{aligned} \mathrm{Hom}_Q(U, \tau^+ V) &\cong \mathrm{Ext}_Q(V, U)^*, \\ \mathrm{Ext}_Q(U, \tau^+ V) &\cong \mathrm{Hom}_Q(V, U)^*. \end{aligned}$$

(2) *Assume that V has no injective summands. Then we have natural isomorphisms of vector spaces:*

$$\begin{aligned} \mathrm{Hom}_Q(\tau^- V, U) &\cong \mathrm{Ext}_Q(U, V)^*, \\ \mathrm{Ext}_Q(\tau^- V, U) &\cong \mathrm{Hom}_Q(U, V)^*. \end{aligned}$$

Proof. We prove part (1), since the proof of (2) is dual. Obviously it is enough to prove the result for U, V indecomposable. Assume V is indecomposable, nonprojective. Consider the canonical projective resolution of V from Proposition 2.3.4:

$$0 \longrightarrow \bigoplus_{a \in Q_1} V(ta) \otimes P_{ha} \xrightarrow{d^V} \bigoplus_{x \in Q_0} V(x) \otimes P_x \longrightarrow V \longrightarrow 0.$$

Let us apply the functor $\mathrm{Hom}_Q(-, A)$ to this resolution. We get an exact sequence of Q^{op}-representations

$$0 \longrightarrow \bigoplus_{x \in Q_0} V(x)^* \otimes \widetilde{P}_x \longrightarrow \bigoplus_{a \in Q_1} V(ta)^* \otimes \widetilde{P}_{ha} \longrightarrow \mathrm{Ext}_Q(V, A) \longrightarrow 0$$

because $\mathrm{Hom}_Q(V, A) = 0$ by Lemma 6.3.2. The above sequence is a sequence of A^{op}-modules. Apply the duality D. We get an exact sequence (using Exercise 6.2.5)

$$0 \longrightarrow \tau^+ V \longrightarrow \bigoplus_{a \in Q_1} V(ta) \otimes I_{ha} \longrightarrow \bigoplus_{x \in Q_0} V(x) \otimes I_x \longrightarrow 0$$

of A-modules. This gives an injective resolution of $\tau^+ V$. Thus we can calculate the spaces $\mathrm{Hom}_Q(U, \tau^+ V)$, $\mathrm{Ext}_Q(U, \tau^+ V)$ by applying the functor $\mathrm{Hom}_Q(U, -)$ to the above sequence and calculating homology spaces. This means the spaces $\mathrm{Hom}_Q(U, \tau^+ V)$ and $\mathrm{Ext}_Q(U, \tau^+ V)$ are, respectively, the kernel and cokernel of the linear map

$$\bigoplus_{a \in Q_1} \mathrm{Hom}_Q(U, V(ta) \otimes I_{ta}) \longrightarrow \bigoplus_{x \in Q_0} \mathrm{Hom}_Q(U, V(x) \otimes I_x)$$

which can be rewritten as

$$\bigoplus_{a \in Q_1} V(ta) \otimes U(ha)^* \longrightarrow \bigoplus_{x \in Q_0} V(x) \otimes U(x)^*. \tag{6.3}$$

On the other hand we can calculate the spaces $\mathrm{Hom}_Q(V,U)$, $\mathrm{Ext}_Q(V,U)$ using the projective resolution of V we started with, applying the functor $\mathrm{Hom}_Q(-,U)$ and calculating homology. We get these spaces as the kernel and cokernel of the linear map

$$\bigoplus_{x\in Q_0} \mathrm{Hom}(V(x),U(x)) \longrightarrow \bigoplus_{a\in Q_1} \mathrm{Hom}(V(ta),U(ha))\ .$$

Identifying $\mathrm{Hom}(X,Y)$ with $X^*\otimes Y$ we see that the terms in this linear map are

$$\bigoplus_{x\in Q_0} V(x)^*\otimes U(x) \longrightarrow \bigoplus_{a\in Q_1} V(ta)^*\otimes U(ha). \tag{6.4}$$

We see that the terms of the maps (6.3) and (6.4) are dual to each other. Tracing the differentials we can easily check that these linear maps are dual to each other, which proves the theorem. □

Remark 6.4.2. Tracing through the definitions we can see that the Auslander-Reiten duality is also the duality of $(\mathrm{End}_Q(U),\mathrm{End}_Q(V))$-bimodule structures on the vector spaces $\mathrm{Hom}_Q(U,\tau^+V)$, $\mathrm{Ext}_Q(V,U)^*$ and on $\mathrm{Ext}_Q(U,\tau^+V)$, $\mathrm{Hom}_Q(V,U)^*$.

Under mild assumptions, we can now show that the translation functor preserves the Hom and Ext spaces.

Proposition 6.4.3.

(1) *If U and V do not have any projective summands, then we have an isomorphism of* $(\mathrm{End}_Q(U),\mathrm{End}_Q(V))$ *bimodules:*

$$\begin{aligned}\mathrm{Hom}_Q(U,V) &\cong \mathrm{Hom}_Q(\tau^+U,\tau^+V),\\ \mathrm{Ext}_Q(U,V) &\cong \mathrm{Ext}_Q(\tau^+U,\tau^+V).\end{aligned}$$

(2) *If U and V do not have any injective summands, then we have an isomorphism of* $(\mathrm{End}_Q(U),\mathrm{End}_Q(V))$ *bimodules:*

$$\begin{aligned}\mathrm{Hom}_Q(U,V) &\cong \mathrm{Hom}_Q(\tau^-U,\tau^-V),\\ \mathrm{Ext}_Q(U,V) &\cong \mathrm{Ext}_Q(\tau^-U,\tau^-V).\end{aligned}$$

Proof. We use Theorem 6.4.1. For (1) we have

$$\mathrm{Hom}_Q(\tau^+U,\tau^+V)\cong \mathrm{Ext}_Q(V,\tau^+U)^\star \cong \mathrm{Hom}_Q(U,V)$$

and

$$\mathrm{Ext}_Q(\tau^+U,\tau^+V)\cong \mathrm{Ext}_Q(V,\tau^+U)^\star \cong \mathrm{Ext}_Q(U,V).$$

The proof of (2) is similar. □

Proposition 6.4.4. *If N is not projective, then there exists an almost split exact sequence*

$$\xi : 0 \longrightarrow \tau^+N \longrightarrow M \longrightarrow N \longrightarrow 0\ .$$

Proof. Let $\mathfrak{m}$ be the radical of $\mathrm{Hom}_Q(N,N)$. Since N is indecomposable, we have $\mathrm{Hom}_Q(N,N)/\mathfrak{m} \cong \mathbb{C}$. Let $\xi \in \mathrm{Hom}_Q(N,N)^\star$ be the quotient map

$$\xi : \mathrm{Hom}_Q(N,N) \longrightarrow \mathrm{Hom}_Q(N,N)/\mathfrak{m} \cong \mathbb{C}\,.$$

Because of AR-duality, we have $\mathrm{Ext}_Q(N,\tau^+N) \cong \mathrm{Hom}_Q(N,N)^\star$ we may view ξ as an extension

$$\xi : 0 \longrightarrow \tau^+N \longrightarrow M \longrightarrow N \longrightarrow 0.$$

We claim that ξ is right almost split. Suppose that $f : V \to N$ lies in the radical. Let $\eta \in \mathrm{Hom}_Q(N,V)^\star$ be the map defined by $g \mapsto \xi(fg)$. The exact sequence η is obtained from ξ as a pull-back

$$\begin{array}{ccccccccc}
\xi : 0 & \longrightarrow & \tau^+N & \longrightarrow & M & \longrightarrow & N & \longrightarrow & 0 \\
 & & {\scriptstyle =}\big\uparrow & & \big\uparrow & & \big\uparrow{\scriptstyle f} & & \\
\eta : 0 & \longrightarrow & \tau^+N & \longrightarrow & W & \longrightarrow & V & \longrightarrow & 0.
\end{array}$$

For every $g \in \mathrm{Hom}_Q(N,V)$, fg lies in $\mathrm{rad}_A(N,N) = \mathfrak{m}$ and $\xi(fg) = 0$. So $\eta = 0$ and the exact sequence on the bottom is split. This implies that ξ is a right almost split sequence.

Since τ^+N is indecomposable, the sequence ξ is almost split by Theorem 5.4.8. □

Corollary 6.4.5. *If*

$$0 \longrightarrow L \longrightarrow M \xrightarrow{\;g\;} N \longrightarrow 0$$

is an almost split exact sequence, then $L = \tau^+N$ and $N = \tau^-L$.

Proof. Since the sequence is almost split, N is indecomposable and not projective. There exists an almost split exact sequence:

$$0 \longrightarrow \tau^+N \longrightarrow M' \xrightarrow{\;g'\;} N \longrightarrow 0.$$

The morphisms g and g' are right minimal almost split, so there exists a morphism $k : M \to M'$ such that $g'k = g$ by Proposition 5.4.2. By diagram chasing, there exists a isomorphism $h : L \to \tau^+N$ such that the diagram

$$\begin{array}{ccccccccc}
0 & \longrightarrow & L & \longrightarrow & M & \xrightarrow{\;g\;} & N & \longrightarrow & 0 \\
 & & {\scriptstyle \cong}\big\downarrow & & \big\downarrow{\scriptstyle k} & & \big\downarrow{\scriptstyle =} & & \\
0 & \longrightarrow & \tau^+N & \longrightarrow & M' & \xrightarrow[g']{} & N & \longrightarrow & 0
\end{array}$$

commutes. □

Exercises.

Exercise 6.4.1. *Take the equioriented quiver of type* $\mathbf{A}_n$*, or the quiver from Example* 4.5.3*. Check that functors* τ^+*,* τ^- *are the same as the functors* C^+*,* C^- *defined in Chapter* 4*.*

Exercise 6.4.2. *Let* Q *be a Dynkin quiver. Prove that if* V *and* W *are indecomposable modules, then either* $\operatorname{Hom}_Q(V, W) = 0$ *or* $\operatorname{Ext}_Q(V, W) = 0$*. Deduce that the values of* Hom *and* Ext *can be obtained from the values of Euler form* $\langle \underline{\dim}\, V, \underline{\dim}\, W \rangle$*.*

6.5. Coxeter Functors Revisited

In this section we show that the translation functors τ^+, τ^- defined in Section 6.2 coincide with the Coxeter functors C^+, C^- defined in Chapter 4.

We start by recovering the reflection functors C_x^+, C_x^- in the language of modules. Suppose that x is a sink. We assume that x is not an isolated vertex. The projective module P_x is isomorphic to the simple module S_x, but it is not injective.

Consider the representation

$$T_x := \Big(\bigoplus_{y \neq x} P_y \Big) \oplus \tau^- P_x.$$

Proposition 6.5.1. *We have an isomorphism*

$$\operatorname{Hom}_Q(T_x, T_x) \cong \mathbb{C}\sigma_x(Q)^{\mathrm{op}}.$$

The reflection functor

$$C_x^+ : \operatorname{Rep}(Q) \to \operatorname{Rep}(\sigma_x Q)$$

defined in Chapter 4 *is isomorphic to the functor* $\operatorname{Hom}_Q(T_x, -)$*.*

Proof. Let us describe $\tau^- P_x$ explicitly. By definition τ^- is equal to the composition $\operatorname{Tr} \circ D$. We have $\operatorname{Tr} \circ D(S_x) = \operatorname{Tr}(S_x^\star)$ where $S_x^\star$ is the simple representation over Q^{op}. In order to calculate $\operatorname{Tr}(S_x^\star)$ we have to describe its projective resolution which is

$$0 \longrightarrow \bigoplus_{\substack{a \in Q_1 \\ ha = x}} \widetilde{P}_{ta} \longrightarrow \widetilde{P}_x \longrightarrow S_x^\star \longrightarrow 0. \tag{6.5}$$

Applying the functor $\operatorname{Hom}_{Q^{\mathrm{op}}}(-, \mathbb{C}Q^{\mathrm{op}})$ to this resolution, and identifying the cokernel of the resulting map with $\tau^- P_x$, we get the projective resolution of $\tau^- P_x$:

$$0 \longrightarrow P_x \longrightarrow \bigoplus_{\substack{a \in Q_1 \\ ha = x}} P_{ta} \longrightarrow \tau^- P_x \longrightarrow 0. \tag{6.6}$$

We have natural isomorphisms

$$\operatorname{Hom}_Q(P_y, P_z) \cong e_y\mathbb{C}Qe_z = e_y\mathbb{C}\sigma_x(Q)e_z,$$

$$\operatorname{Hom}_Q(P_y, \tau^- P_x) = \bigoplus_{\substack{a\in Q_1\\ ha=x}} e_y\mathbb{C}Qe_{ta} \cong e_y\mathbb{C}\sigma_x(Q)e_x,$$

$$\operatorname{Hom}_Q(\tau^- P_x, P_y) \cong \operatorname{Ext}_Q(P_y, P_x)^* = 0 = e_x\mathbb{C}\sigma_x(Q)e_y,$$

$$\operatorname{Hom}_Q(\tau^- P_x, \tau^- P_x) \cong \operatorname{Hom}_Q(P_x, P_x) \cong \mathbb{C} \cong e_x\mathbb{C}\sigma_x(Q)e_x$$

for $y, z \neq x$. The first isomorphism follows from Proposition 2.2.2 and the description of P_x, the second from the resolution (6.6), the third from the Auslander-Reiten duality, Theorem 6.4.1 (2), and the fourth isomorphism follows from Proposition 6.4.3. The components of $\operatorname{Hom}_Q(T_x, T_x)$ are the same as the components of $\mathbb{C}\sigma_x(Q)$. The reader may verify, that the isomorphism of vector spaces is in fact an isomorphism of algebras between $\operatorname{Hom}_Q(T_x, T_x)$ and $\mathbb{C}\sigma_x(Q)^{\mathrm{op}}$.

Let us calculate $\operatorname{Hom}_Q(T_x, V)$. Assume that V is indecomposable. If V is isomorphic to the simple representation $S_x \cong P_x$, then we have

$$\begin{aligned}\operatorname{Hom}_Q(T_x, S_x) = \operatorname{Hom}_Q(\tau^- P_x, S_x) &\cong \operatorname{Ext}_Q(S_x, P_x)\\ &= \operatorname{Ext}_Q(P_x, P_x) = 0 = C_x^+(S_x).\end{aligned}$$

For $y \neq x$ we have $\operatorname{Hom}_Q(P_y, V) = V(y)$, so $\operatorname{Hom}_Q(P_y, V)$ is isomorphic to $C_x^+(V)(y)$. We can calculate $\operatorname{Hom}_Q(\tau^- P_x, V)$ by applying $\operatorname{Hom}_Q(-, V)$ to the resolution (6.6):

$$0 \longrightarrow \operatorname{Hom}_Q(\tau^- P_x, V) \longrightarrow \bigoplus_{\substack{a\in Q_1\\ ha=x}} V(ta) \longrightarrow V(x).$$

We see that $\operatorname{Hom}_Q(\tau^- P_x, V)$ is isomorphic to $C_x^+(V)(x)$. We leave to the reader to check that the above isomorphism respects the module structures. □

We generalize this result in the following way. Without loss of generality, we assume that $Q_0 = \{1, 2, \ldots, n\}$ and that $ta < ha$ for all arrows $a \in Q_1$. So the sequence $1, 2, 3, \ldots, n$ is an admissible sequence of sinks. For $k \in \{1, 2, \ldots, n\}$ we define a module

$$U_k = C_1^- \cdots C_k^-(\mathbb{C}\sigma_k \cdots \sigma_1(Q)).$$

Proposition 6.5.2.

(1) *We have*

$$U_k = \tau^-(P_1 \oplus \ldots \oplus P_k) \oplus P_{k+1} \oplus \cdots \oplus P_n.$$

(2) *We have an isomorphism of algebras*

$$\operatorname{Hom}_Q(U_k, U_k) = \mathbb{C}\sigma_k \dots \sigma_1(Q)^{\mathrm{op}}.$$

(3) *The partial Coxeter functor* $C_k^+ \dots C_1^+$ *can be identified with* $\operatorname{Hom}_Q(U_k, -)$.

Proof. We prove parts (2) and (3) of the proposition by induction on k. The case $k = 1$ follows from Proposition 6.5.1. Suppose that $k > 1$. Let U_k' be the representation $C_2^- \cdots C_k^-(\mathbb{C}\sigma_k \cdots \sigma_1(Q))$ of the quiver $\sigma_1(Q)$. By the induction hypothesis, we have

$$\operatorname{Hom}_{\sigma_1(Q)}(U_k', U_k') \cong \mathbb{C}\sigma_k \cdots \sigma_1(Q)^{\mathrm{op}}.$$

Suppose that the simple representation S_1 of $\sigma_1(Q)$ appears in U_k' as a direct summand. Then $C_k^+ \cdots C_2^+(S_1)$ is an injective direct summand of $\mathbb{C}\sigma_k \cdots \sigma_1(Q)$ by Exercise 4.3.5 of Chapter 4. Because the module $\mathbb{C}\sigma_k \cdots \sigma_1(Q)$ is a direct sum of projective representations, this means that there must be a representation that is injective and projective. This is not possible because $n \geq 2$ and Q is connected. So U_k' does not have the simple representation S_1 as a direct summand. It follows that

$$\begin{aligned}\operatorname{Hom}_Q(U_k, U_k) = \operatorname{Hom}_Q(C_1^-(U_k'), C_1^-(U_k')) \cong \\ \cong \operatorname{Hom}_{\sigma_1(Q)}(U_k', U_k') \cong \mathbb{C}\sigma_k \cdots \sigma_1(Q)^{\mathrm{op}}.\end{aligned}$$

By the induction hypothesis, we get

$$\operatorname{Hom}_{\sigma_1(Q)}(U_k', V) \cong C_k^+ \cdots C_2^+(V)$$

for every representation V of $\sigma_1(Q)$. For a representation V of Q we have

$$\begin{aligned}\operatorname{Hom}_Q(U_k, V) = \operatorname{Hom}_Q(C_1^-(U_k'), V) = \operatorname{Hom}_{\sigma_1(Q)}(U_k', C_1^+(V)) \\ \cong C_k^+ \cdots C_1^+(V)\end{aligned}$$

by Exercise 4.3.6 of Chapter 4.

Now we show part (1). Let $P_1', \dots, P_n'$ be the projectives for the quiver $\sigma_k \cdots \sigma_1 Q$. If $j > k$, then we have

$$C_1^- \cdots C_k^- P_j' \cong P_j$$

by Exercise 4.3.5 of Chapter 4. We also have

$$\begin{aligned}\operatorname{Hom}_Q(C_1^- \cdots C_k^- P_k', V) &\cong \operatorname{Hom}_{\sigma_{k-1}\cdots\sigma_1 Q}(C_k^- P_k', C_{k-1}^+ \cdots C_1^+ V) \\ \cong \operatorname{Hom}_{\sigma_{k-1}\cdots\sigma_1 Q}(\tau^- P_k'', C_{k-1}^+ \cdots C_1^+ V) &\cong \operatorname{Ext}_{\sigma_{k-1}\cdots\sigma_1 Q}(C_{k-1}^+ \cdots C_1^+ V, P_k')^\star \\ \cong \operatorname{Ext}_Q(V, C_1^+ \cdots C_{k-1}^+ P_k')^\star &= \operatorname{Ext}_Q(V, P_k)^\star = \operatorname{Hom}_Q(\tau^- P_k, V),\end{aligned}$$

where P_k'' is the projective of the quiver $\sigma_{k-1} \cdots \sigma_1(Q)$ corresponding to vertex k. These isomorphisms are functorial in V. It follows that

$$C_1^- \cdots C_k^- P_k \cong' \tau^- P_k.$$

If $j < k$, then we have

$$C_1^- \cdots C_k^- P_j' = C_1 - \cdots C_j^- P_j''' = \tau^- P_j,$$

where P_j''' is the projective for the quiver $\sigma_1 \cdots \sigma_j Q$. □

In particular, taking $k = n$ we get that $\operatorname{Hom}_Q(U_n, V)$ is isomorphic to $C^+(V)$. On the other hand $U_n = \tau^-(\bigoplus_{x \in Q_0} P_x) = \tau^-(\mathbb{C}Q)$. Thus using 6.4.1 (2) we have, for any indecomposable V,

$$C^+(V) = \operatorname{Hom}_Q(U_n, V) = \operatorname{Hom}_Q(\tau^-(\mathbb{C}Q), V)$$

$$= \operatorname{Ext}_Q(V, \mathbb{C}Q)^* = D \circ \operatorname{Tr}(V) = \tau^+(V).$$

Let us just formulate the corresponding results for the functors C^-.

Suppose that $Q_0 = \{1, 2, \ldots, n\}$ and that $ha > ta$ for all $a \in Q_1$. Then $1, 2, \ldots, n$ is an admissible sequence of sources. Consider the representation

$$Z_k = \tau^+(I_1 \oplus \cdots \oplus I_k) \oplus I_{k+1} \oplus \cdots \oplus I_n.$$

Proposition 6.5.3. *We have an isomorphism of algebras*

$$\operatorname{Hom}_Q(Z_k, Z_k) = \mathbb{C}\sigma_k \ldots \sigma_1(Q)^{\mathrm{op}}.$$

The partial Coxeter functor $C_k^- \ldots C_1^-$ *can be identified with* $D \circ \operatorname{Hom}_Q(-, Z_k)$.

In particular, taking $k = n$ we get that $\operatorname{Hom}_Q(V, Z_n)^\star$ is isomorphic to $C^-(V)$. On the other hand

$$Z_n = \tau^+\Big(\bigoplus_{x \in Q_0} I_x\Big) = \tau^+((\mathbb{C}Q)^\star),$$

where we use the isomorphism of left $\mathbb{C}Q$-modules $(\mathbb{C}Q)^\star \cong \bigoplus_{x \in Q_0} I_x$. Thus using 6.4.1 (1) we have, for any indecomposable V,

$$\begin{aligned} C^-(V) &= \operatorname{Hom}_Q(V, U_n)^* = \operatorname{Hom}_Q(V, \tau^+(\mathbb{C}Q)^\star)^* \\ &= \operatorname{Ext}_Q((\mathbb{C}Q)^\star, V) = \operatorname{Ext}_{Q^{\mathrm{op}}}(V^\star, (\mathbb{C}Q)^\star) = \operatorname{Tr} \circ D(V) = \tau^-(V). \end{aligned}$$

We will write τ^{-i} for the i-th power $(\tau^-)^i$ and τ^i or τ^{+i} for $(\tau^+)^i$.

Exercises.

Exercise 6.5.1. *Deduce Proposition* 6.5.3 *from Proposition* 6.5.2 *by using duality.*

Exercise 6.5.2. *Prove that* $(\underline{\dim}\, \tau^+(V))_x = \dim \operatorname{Ext}_Q(V, P_x)$. *Similarly,* $(\underline{\dim}\, \tau^-(W))_x = \dim \operatorname{Ext}_Q(Q_x, W)$.

Exercise 6.5.3. *Let Q be an acyclic quiver and let x be a sink. Denote Q' the quiver Q restricted to $Q_0 \setminus \{x\}$. Let* $\operatorname{res} : \operatorname{Rep}(Q) \to \operatorname{Rep}(Q')$ *be a restriction functor sending V to $V|_{Q'}$. Prove that there exists a left adjoint functor* $\operatorname{ind} : \operatorname{Rep}(Q') \to \operatorname{Rep}(Q)$ *such that we have functorial isomorphisms*

$$\operatorname{Hom}_Q(\operatorname{ind}(V), W) = \operatorname{Hom}_{Q'}(V, \operatorname{res}(W)).$$

Interpret $\operatorname{ind}(V)$ *in terms of projective resolutions. Prove that $\tau(V)_x = 0$ if and only if there exists $W \in \operatorname{Rep}(Q')$ such that $V = \operatorname{ind}(W)$.*

6.6. The Auslander-Reiten Quiver for Hereditary Algebras

Recall the definition of the Auslander-Reiten quiver (Definition 5.3.1). The vertices of the Auslander-Reiten quiver are the indecomposable representations and for two indecomposable representations V and W we draw $\dim \operatorname{rad}_Q(V, W)/\operatorname{rad}^2_Q(V, W)$ arrows from V to W. We give an explicit construction for the AR quiver when Q is of finite representation type.

Lemma 6.6.1. *For every $x \in Q_0$ we have an almost split sequence:*

$$0 \longrightarrow P_x \longrightarrow \bigoplus_{\substack{a \\ ha=x}} P_{ta} \oplus \bigoplus_{\substack{b \\ tb=x}} \tau^-(P_{hb}) \longrightarrow \tau^-(P_x) \longrightarrow 0\,. \tag{6.7}$$

Proof. Suppose that V is an indecomposable representation and that $\varphi : P_x \to V$ is an irreducible morphism. If V is projective, then there exists an arrow a with $ta = x$ and $V \cong P_{ha}$. Up to a scalar, the morphism φ is the unique map $P_x \to P_{ha}$ that sends $e_x \in P_x(x)$ to $a \in P_{ha}(x)$ (see Exercise 6.6.1).

If V is not projective, then there exists an almost split sequence

$$0 \longrightarrow \tau^+(V) \longrightarrow M \longrightarrow V \longrightarrow 0$$

such that P_x is a direct summand of M. There exists an irreducible map $\tau^+(V) \to P_x$. Irreducible maps are either injective or surjective, and the map $\tau^+(V) \to P_x$ clearly cannot be surjective. So $\tau^+(V) \to P_x$ is injective and $\tau^+(V)$ is projective. Again by Exercise 6.6.1, there exists an arrow b with $tb = x$ and $\tau^+(V) \cong P_{hb}$. So we have $V = \tau^-\tau^+(V) = \tau^-(P_{hb})$. □

Let Q be a quiver of finite representation type. We define a graph $\operatorname{AR}(Q)$ as follows:

(1) The objects in $\operatorname{AR}(Q)$ are all the nonzero Auslander-Reiten translates $\tau^{-i}(P_x)$ where $i \geq 0$ and $x \in Q_0$.

(2) Suppose that $a \in Q_1$ and that $\tau^{-i}(P_{ha})$ and $\tau^{-i}(P_{ta})$ are both nonzero. The we draw an arrow $\tau^{-i}(P_{ha}) \to \tau^{-i}(P_{ta})$ in $\mathrm{AR}(Q)$ and the map is obtained by applying τ^{-i} to the inclusion $P_{ha} \hookrightarrow P_{ta}$.

(3) Suppose that $b \in Q_1$ and that $\tau^{-i}(P_{tb})$ and $\tau^{-i-1}(P_{hb})$ are both nonzero. Then we draw an arrow $\tau^{-i}(P_{tb}) \to \tau^{-i-1}(P_{hb})$ and the map is obtained by applying τ^{-i} to the inclusion $P_{tb} \to \tau^{-}(P_{hb})$ which appears in the almost split sequence (6.7) with $x = tb$.

Proposition 6.6.2. *The graph* $\mathrm{AR}(Q)$ *is the Auslander-Reiten quiver for* Q.

Example 6.6.3. Consider the quiver

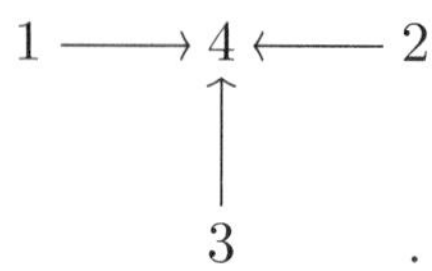

The almost split sequences with a projective representation on the left are

$$0 \longrightarrow P_4 \longrightarrow P_1 \oplus P_2 \oplus P_3 \longrightarrow \tau^{-1}(P_4) \longrightarrow 0$$

and

$$0 \longrightarrow P_j \longrightarrow \tau^{-1}(P_4) \longrightarrow \tau^{-1}(P_j) \longrightarrow 0$$

for $j = 1, 2, 3$. The indecomposable representations are $\tau^{-i}P_j$ with $j = 1, 2, 3, 4$ and $i = 0, 1, 2$. All other Auslander-Reiten translations of projective indecomposable representations are 0. The irreducible maps are $\tau^{-i}(P_4) \to \tau^{-i}(P_j)$ for $j = 1, 2, 3$ and $i = 0, 1, 2$ and $\tau^{-i}(P_j) \to \tau^{-i-1}(P_4)$ for $j = 1, 2, 3$ and $i = 0, 1$. So the AR quiver is

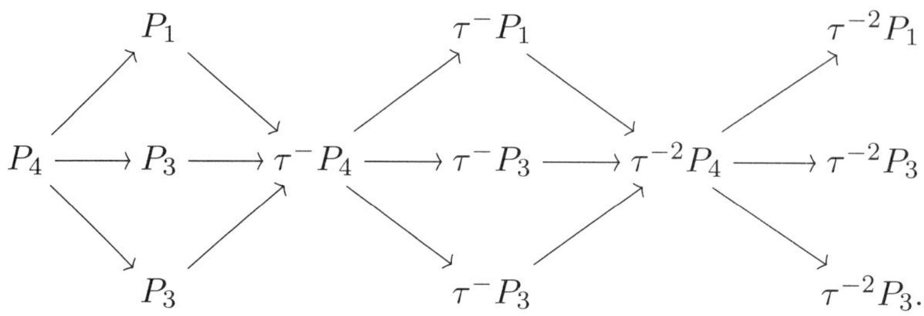

Proof of Proposition 6.6.2. It is clear that all the arrows in $\mathrm{AR}(Q)$ are irreducible. We need to show that every irreducible map between the indecomposable objects in $\mathrm{Rep}(Q)$ is an arrow in $\mathrm{AR}(Q)$ (up to a nonzero scalar). Suppose that $\varphi : V \to W$ is an irreducible map between 2 indecomposable representations. For some i, we have that $\tau^i(V)$ is projective. If $\tau^i(W) = 0$, then there exists a $j < i$ such that $\tau^j(W)$ is projective. The morphism $\tau^j(V) \to \tau^j(W)$ is irreducible. This map cannot be surjective, because $\tau_j(W)$ is projective and not a direct summand of $\tau^j(V)$. So $\tau^j(V)$

is a subrepresentation of $\tau^j(W)$ and therefore also projective. But then we have $\tau^i(V) = \tau^{i-j}(\tau^j(V)) = 0$. So we can assume that $\tau^j(W)$ is nonzero. Then $\tau^j(W)$ is either isomorphic to P_{ta} where a is an arrow with $ha = x$, or isomorphic to $\tau^-(P_{hb})$ where b is an arrow with $tb = x$. So φ is, up to a scalar, equal to one of the arrows in AR(Q). □

Remark 6.6.4. Suppose that Q is a quiver of finite representation type. Then Q does not have multiple arrows, so $\mathrm{Hom}_Q(P_{ha}, P_{ta})$ is 1-dimensional for all $a \in Q_1$. Also, we have $\mathrm{Hom}_Q(P_{tb}, \tau^- P_{hb}) = \tau^- P_{hb}(tb)$ is 1-dimensional (see Exercise 6.6.2). It follows that whenever $V \to W$ is an arrow in AR(Q), then $\mathrm{Hom}_Q(V, W)$ is 1-dimensional. We have $\mathrm{rad}^2_Q(V, W) = 0$ and there is a unique irreducible map $V \to W$, up to a nonzero scalar.

Remark 6.6.5. From the Auslander-Reiten quiver we can read off easily the dimensions of the spaces $\mathrm{Hom}_Q(U, V)$ and $\mathrm{Ext}_Q(U, V)$ for indecomposable representations U and V. Each homomorphism between U and V is a linear combination of paths in AR(Q). The only relations are the relations given by the composition zero in each of the almost split sequences we see in $AR(Q)$. The dimension of the $\mathrm{Ext}_Q(U, V)$ reduces by Auslander-Reiten duality to the dimension of $\mathrm{Hom}_Q(V, \tau^+(U))$ (if U is not projective) or to the dimension of $\mathrm{Hom}_Q(\tau^- V, U)$ (if V is not injective). If U is projective or V is injective, we obviously have $\mathrm{Ext}_Q(U, V) = 0$.

Note that an interesting consequence of the construction of the Auslander-Reiten quiver is also that the category Rep(Q) for Q Dynkin is **directed**, i.e., there can be no *cycles*, i.e., nonzero homomorphisms $V = V_1 \to V_2 \to \dots \to V_m = V$ with $V_1, \dots, V_m$ indecomposable.

Example 6.6.6. Let Q be the quiver

$$1 \longrightarrow 2 \longrightarrow 4 \longleftarrow 3.$$

Applying τ^- to $P_1 = 1110$ gives the injective 0001. Applying τ^- to $P_2 = 0110$ gives the injective 1111. Repeatedly apllying τ^- to $P_3 = 0010$ gives 0111, 1100, and repeatedly applying τ^- to $P_4 = 0011$ gives 0100 and 1000.

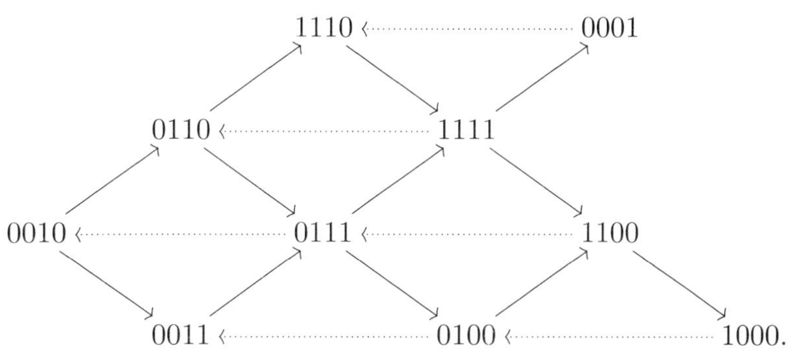

The Auslander-Reiten translation τ^+ is drawn as a dotted line.

Exercises.

Exercise 6.6.1. *Suppose that Q is an acyclic quiver and that P_x and P_y are indecomposable projective representations corresponding to $x, y \in Q_0$. To an arrow $a \in Q_1$ with $ha = x$ and $ta = y$, we can associate a morphism $P_x \to P_y$ that sends a path $p \in P_x(z) = [x, z]$ to $pa \in P_y(z) = [y, z]$. Show that* $\operatorname{rad}_Q(P_x, P_y)/\operatorname{rad}_Q^2(P_x, P_y)$ *is spanned by all arrows a with $ha = x$ and $ta = y$. In particular, the number of arrows $P_x \to P_y$ in the Auslander-Reiten quiver is equal to the number of arrows from y to x in Q.*

Exercise 6.6.2. *Suppose that Q is an acyclic quiver with no multiple arrows between vertices. Then we have $\tau^{-}(P_{hb})(tb)$ is one-dimensional for every arrow $b \in Q_1$.*

Exercise 6.6.3. *Give the Auslander-Reiten quiver of the quiver*

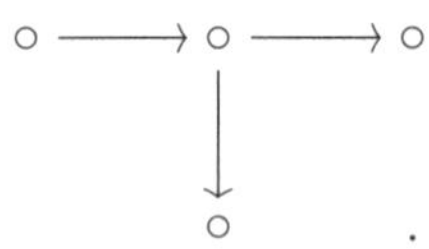

Also, indicate how the Auslander translation acts.

Exercise 6.6.4. *Prove the statements from Remark* 6.6.5.

6.7. The Preprojective Algebra

In this section we assume that Q is acyclic, but not of finite representation type. We will construct a preprojective algebra that contains the path algebra of Q. It has the interesting property that, as a module over $\mathbb{C}Q$, it is a direct sum of the projective representations and all their translates.

We can follow the construction of $\mathrm{AR}(Q)$ in the previous section, but we will see that this no longer gives the whole Auslander-Reiten quiver. A representation is called **preprojective** if it is of the form $\tau^{-i}(P_x)$ where $i \geq 0$ and P_x is indecomposable.

Lemma 6.7.1. *Not every indecomposable representation is preprojective.*

Proof. Suppose that every indecomposable representation is preprojective. Then every indecomposable injective I_y is of the form $\tau^{-i(y)}(P_{\gamma(y)})$ for some $\gamma(y) \in Q_1$, and $i(y) \in \mathbb{N}$. It is clear that $\gamma : Q_1 \to Q_1$ is injective, so it must be surjective as well. So for every projective indecomposable representation P_x there exists an i such that $\tau^{-i}(P_x)$ is injective, and $\tau^{-i-1}(P_x) = 0$. But then there are only finitely many preprojective representations. But since there are infinitely many indecomposable representations, not all of them can be preprojective. $\square$

Let Q be an acyclic quiver. We define the quiver $\bar{Q} = (\bar{Q}_0, \bar{Q}_1, h, t)$ as follows. For every arrow $a \in Q_1$ we construct a new arrow $a^\star$ with $ha^\star = ta$ and $ta^\star = ha$ and $a^{\star\star} = a$. Let $Q_1 = \{a^\star \mid a \in Q_1\}$. We define $\bar{Q}_0 = Q_0$, $\bar{Q}_1 = Q_1 \coprod Q_1^*$ (disjoint union). Thus we double the arrows of the quiver by adding to each arrow a dual arrow in reverse direction.

We introduce a grading on $\mathbb{C}\bar{Q}$ by setting the degree of paths e_x and arrows $a \in Q_1$ to be zero, and defining the degree of dual arrows $a^* \in Q_1^*$ to be one. The degree of a path $p = q_1 \dots q_r$ with $q_i \in \bar{Q}_1$ equals $\sum_{i=1}^r \deg(q_i)$. We identify the path algebra $\mathbb{C}Q$ with the part of $\mathbb{C}\bar{Q}_0$ of degree zero.

Let $u, v : Q_1 \to \mathbb{C}$ be two arbitrary functions. We will only assume that for any $a \in Q_1$ we have $u(xa) \neq 0 \neq v(a)$. For each $x \in Q_0$ we define

$$m_x = \sum_{\substack{a \in Q_1 \\ ta = x}} u(a) a^* a - \sum_{\substack{a \in Q_1 \\ ha = x}} v(a) a a^*.$$

Let $\bar{J}$ be the two-sided ideal in $\mathbb{C}\bar{Q}$ generated by the elements m_x, $x \in Q_0$.

Definition 6.7.2. The (u, v)-**preprojective algebra** $\mathcal{P}(Q)_{u,v}$ is the factor algebra $\mathbb{C}\bar{Q}/\bar{J}$.

All elements m_x are homogeneous of degree 1 in $\mathbb{C}\bar{Q}$. Thus the algebra $\mathcal{P}(Q)_{u,v}$ is a homogeneous algebra, and its zero graded part is identified with $\mathbb{C}Q$.

For $x \in Q_0$, $d \in \mathbb{N}$ we define $\mathcal{W}_d \subseteq \mathbb{C}\bar{Q}$ to be the span of all paths p of degree d. We have

$$\mathbb{C}\bar{Q} = \bigoplus_{d \in \mathbb{N}} \mathcal{W}_d.$$

We also define $\mathcal{V}_d = (\mathcal{W}_d / \mathcal{W}_d \cap \bar{J})$ so that

$$\mathcal{P}_{u,v}(Q) = \bigoplus_{d \in \mathbb{N}} \mathcal{V}_d.$$

Theorem 6.7.3 (Kleiner). *Let Q be an acyclic quiver. For every $d \in \mathbb{N}$ and $x \in Q_0$ we have an exact sequence*

$$\mathcal{V}_d e_x \longrightarrow \bigoplus_{\substack{a \in Q_1 \\ ta = x}} \mathcal{V}_{d+1} e_{ha} \oplus \bigoplus_{\substack{b \in Q_1 \\ hb = x}} \mathcal{V}_d e_{tb} \longrightarrow \mathcal{V}_{d+1} e_x \to 0 \,.$$

For $a \in Q_1$ with $ta = x$, the map $\mathcal{V}_d e_x \to \mathcal{V}_{d+1} e_{ha}$ is given by left multiplication with $a^\star$, and the map $\mathcal{V}_{d+1} e_{ha} \to \mathcal{V}_{d+1} e_x$ is given by left multiplication by $u(a)a$. For $b \in Q_1$ with $hb = x$, the map $\mathcal{V}_d e_x \to \mathcal{V}_d e_{tb}$ is given by left multiplication with b, and the map $\mathcal{V}_d e_{tb} \to \mathcal{V}_{d+1} e_x$ is given by left multiplication with $-v(a)a^\star$.

The comparison with reflection functors is given as follows.

Theorem 6.7.4 (Kleiner). *Let $x \in Q_0$ be a source. We define functions u', v' on $(\sigma_x Q)_1$ by setting $u'(\sigma_x a) = v(a)$, $v'(\sigma_x a) = u(a)$ for the arrows a with $ta = x$. Let $\mathcal{V}'_d$ be the degree d part of the preprojective algebra $\mathcal{P}(\sigma_x(Q))_{u',v'}$. Let C_x^- be the reflection functor corresponding to x. We have*

$$C_x^-(\mathcal{V}_d e_y) = \begin{cases} \mathcal{V}'_d e_y, & \text{if } y \neq x, \\ \mathcal{V}'_{d+1} e_y, & \text{if } y = x. \end{cases}$$

Corollary 6.7.5. *Let Q be an acyclic quiver.*

(1) *$\mathcal{V}_0 e_x = P_x$ for all $x \in Q_0$.*

(2) *$\mathcal{V}_d e_x = \tau^{-d} P_x$ for all $x \in Q_0$.*

(3) *If $\mathcal{V}_d e_x \neq 0$, then the left morphism in the exact sequence in Theorem* 6.7.3 *is radical and left almost split in* $\mathbb{C}Q$–mod.

(4) *If $\mathcal{V}_{d+1} e_x \neq 0$, then the sequence in Theorem* 6.7.3 *is an almost split short exact sequence.*

Finally, we mention the dual construction. The representations of the form $\tau^i I_x$ where I_x is an indecomposable injective module are called **preinjective**. By passing to the opposite quiver we can construct the part of the Auslander-Reiten quiver corresponding to the preinjective representations.

6.8. Bibliographical Remarks

The results in this chapter are mostly due to Auslander and Reiten. The Auslander-Reiten duality was introduced in [**6**].

We again follow the approach from the book [**38**] except we restrict Auslander-Reiten duality to hereditary case. We also prove that the Coxeter functor defined via reflection functors in [**8**] and via the Auslander-Reiten transform are the same. There is in fact a sign difference between morphisms defined in two ways, but we do not get into this subtlety.

We also include the section on preprojective algebra and its connection with preprojective modules. We learned this material from the paper of Mark Kleiner [**59**].

The deformed preprojective algebras will be important in the proof of Kac theorem in Chapter 8.

Chapter 7

Extended Dynkin Quivers

Beyond the algebras of finite representation type there exists a class of algebras (so-called tame algebras) for which the indecomposable representations can be described very explicitly and they occur as isolated points or one-parameter families (akin to the Jordan form of an endomorphism or a Kronecker classification of two linear maps). For hereditary algebras there is again the connection to Lie theory; the tame path algebras $\mathbb{C}Q$ correspond to extended Dynkin quivers and to affine Kac-Moody Lie algebras. The one-parameter families of indecomposable modules occur in the dimension vectors $n\alpha$ where α is the only primitive isotropic root. The other indecomposable occur in dimension vectors that correspond to real roots.

In this chapter we do not develop this theory fully as there are excellent notes of Dlab and Ringel [**34**] available. We start with the representations of Kronecker quiver which is exactly the classification of pairs of linear maps. Then we describe the Auslander-Reiten quiver of a Kronecker quiver and give examples of these notions for other extended Dynkin quivers. We pay particular attention to real Schur roots which are the dimension vectors of so-called regular representations. For each Dynkin type and a particular orientation of Q we describe explicitly the dimension vectors of simple regular modules.

We need these results for the discussion of rings of semi-invariants of tame quivers to be carried out in Chapter 10.

7.1. Representations of the Kronecker Quiver

In this section we classify the indecomposable representations of the Kronecker quiver

$$\theta(2):\ 1 \underset{b}{\overset{a}{\rightrightarrows}} 2.$$

A representation V of Q consists of two vector spaces $V(1), V(2)$ and a pair of linear maps $V(a), V(b) : V(1) \to V(2)$. Each representation V of Q has dimension vector

$$\underline{\dim}\, V = (\dim V(1), \dim V(2)) \in \mathbb{Z}^2.$$

Let

$$J_\lambda^n = \begin{pmatrix} \lambda & 1 & & & \\ & \lambda & 1 & & \\ & & \ddots & \ddots & \\ & & & \lambda & 1 \\ & & & & \lambda \end{pmatrix}$$

be the $n \times n$ Jordan block with eigenvalue λ. We define a family $V_{x,y}^n$, $x, y \in \mathbb{C}$ of representations of dimension (n, n) by

$$V_{x,y}^n(1) = V_{x,y}^n(2) = \mathbb{C}^n$$

and

$$V_{x,y}^n(a) = J_x^n, \quad V_{x,y}^n(b) = J_y^n.$$

If $(x, y) \neq (0, 0)$, then the representation $V_{x,y}^n$ is indecomposable (see Exercise 7.1.1). The representations $V_{x,y}^n$ and $V_{x',y'}^n$ are isomorphic if and only if there exists a nonzero $\lambda \in \mathbb{C}$ such that $(x', y') = \lambda(x, y)$ (see Exercise 7.1.2). We will write $V_{[x:y]}^n$ instead of $V_{x,y}^n$, since the isomorphism class of $V_{[x:y]}^n$ only depends on the projective coordinates $[x : y]$ of (x, y). Now we have a family of indecomposable representations of dimension (n, n) parametrized by $[x : y] \in \mathbb{P}^1$.

We define a representation W^n of dimension $(n, n+1)$ by $W^n(1) = \mathbb{C}^n$, $W^n(2) = \mathbb{C}^{n+1}$,

$$W^n(a) = \begin{pmatrix} 1 & 0 & \cdots & 0 \\ 0 & 1 & \cdots & 0 \\ \vdots & \vdots & \ddots & \vdots \\ 0 & 0 & \cdots & 1 \\ 0 & 0 & \cdots & 0 \end{pmatrix}, \quad \text{and } W^n(b) = \begin{pmatrix} 0 & 0 & \cdots & 0 \\ 1 & 0 & \cdots & 0 \\ 0 & 1 & \cdots & 0 \\ \vdots & \vdots & \ddots & \vdots \\ 0 & 0 & \cdots & 1 \end{pmatrix}. \tag{7.1}$$

We also define a representation Z^n of dimension $(n+1,n)$ by $Z^n(1)=\mathbb{C}^{n+1}$, $Z^n(2)=\mathbb{C}^n$,
(7.2)

$$Z^n(a)=\begin{pmatrix}1 & 0 & \cdots & 0 & 0\\ 0 & 1 & \cdots & 0 & 0\\ \vdots & \vdots & \ddots & \vdots & \vdots\\ 0 & 0 & \cdots & 1 & 0\end{pmatrix},\qquad \text{and } W^n(b)=\begin{pmatrix}0 & 1 & 0 & \cdots & 0\\ 0 & 0 & 1 & \cdots & 0\\ \vdots & \vdots & \vdots & \ddots & \vdots\\ 0 & 0 & 0 & \cdots & 1\end{pmatrix}.$$

Theorem 7.1.1 (Kronecker). *The indecomposable representations of the quiver $\theta(2)$ are, up to isomorphism, the representations listed below:*

(1) *the representations W^n, $n\geq 0$,*
(2) *the representations Z^n, $n\geq 0$,*
(3) *the representations $V^n_{[x:y]}$, $n\geq 1$, $[x:y]\in\mathbb{P}^1$.*

Proof. Suppose that U is an indecomposable representation of dimension (m,n) of $\theta(2)$. By induction on the total dimension $m+n$, we will show that U must be one of the representations listed in the theorem. If $m=0$ or $n=0$, then U must be simple and isomorphic to $W^0=S_2$ or $Z^0=S_1$. We assume that $m>0$ and $n>0$.

Suppose that $n>m$. Then C_2^+U is a representation of $\sigma_2\theta(2)$ of dimension $(m,2m-n)$. If we switch the labels of the vertices, we get an indecomposable representation U' of dimension $(2m-n,m)$ of $\theta(2)$. The total dimension of this representation is $(2m-n)+m=(m+n)-2(n-m)<m+n$, so by induction we can assume that U' is one of the listed indecomposables. Since $2m-n<m$, the only possibility is $U'\cong W^{m-1}$. It follows that $U\cong W^m$ by Exercise 7.1.3. Similarly, if $n<m$, then we get $U\cong Z^n$.

Suppose that $m=n$. Suppose that $U(a)$ is invertible. After base changes in $U(1)$ and $U(2)$ we may assume that $U(a)$ is the identity matrix. After the same change of coordinates in $U(1)$ and $U(1)$ we may also assume that $U(b)$ is the Jordan block J^n_y. The representation U is isomorphic to $V^n_{[1:y]}$ by Exercise 7.1.4. Similarly, if $U(b)$ is invertible then U is isomorphic to $V^n_{[x:1]}$ for some x.

Suppose that $U(a)$ and $U(b)$ are both singular. We can choose a basis such that both matrices are upper triangular by Lemma 7.1.3 below. Suppose that $U(a)+tU(b)$ is invertible for some t. After a change of coordinates we may assume that $U(a)+tU(b)$ is equal to the identity, and that $U(b)$ is a Jordan block. But then either $U(a)$ or $U(b)$ must be invertible which is a case already dealt with. Suppose that $U(a)+tU(b)$ is singular for all t. We may assume that $U(a)$ and $U(b)$ are upper triangular. If $U(a)_{i,i}=U(b)_{i,i}=0$ for some i, then there exists an i-dimensional subspace $A\subseteq U(1)$ that is mapped to a subspace of dimension $i-1$. By applying reflection functors we obtain an indecomposable representation of dimension (n,n) with the subrepresentation S_1 of dimension $(1,0)$. But this is a contradiction because S_1 is injective and therefore a direct summand. $\square$

Lemma 7.1.2. *Every representation of dimension (n,n) has a subrepresentation of dimension $(1,1)$.*

Proof. Suppose that U is a representation of dimension (n,n) and that $U(a)$ is invertible. Let v be an eigenvector of $U(a)^{-1}U(b)$, say $U(a)^{-1}U(b)v = \lambda v$. Then we have $U(b)v = \lambda U(a)v$. This shows that U has a subrepresentation of dimension $(1,1)$. If $U(a)$ is not invertible, we can choose a sequence of invertible matrices $A_1, A_2, \ldots$ that converges to $U(a)$. For every i there exists a vector v_i of length 1 and a complex number λ_i such that $A_i v_i = \lambda_i U(a) v_i$. By changing to a subsequence, we may assume that $v_1, v_2, \ldots$ and $\lambda_1, \lambda_2, \ldots$ converge. Let v and λ be their limits, respectively. By continuity, we have that

$$A(b)v = \lim_{i\to\infty} A_i v_i = \lim_{i\to\infty} \lambda_i U(a) v_i = \lambda U(a) v.$$

So we conclude again that U has a subrepresentation of dimension $(1,1)$. $\square$

Lemma 7.1.3. *If U is a representation of dimension (n,n), then we can choose bases in $U(1)$ and $U(2)$ such that $U(a)$ and $U(b)$ are upper triangular matrices.*

Proof. By the previous lemma, we can choose a subrepresentation $U^{(1)} \subseteq U$ of dimension $(1,1)$. Next we can choose a subrepresentation of $U/U^{(1)}$ of dimension $(1,1)$ this yields a subrepresentation of U of dimension $(2,2)$. We obtain a flag $U^{(1)} \subset U^{(2)} \subset \cdots \subset U^{(n)} = U$ where $U^{(i)}$ has dimension (i,i). Choose bases $e_1, \ldots, e_n$ of $U(1)$ and $f_1, \ldots, f_n$ of $U(2)$ such that $e_1, \ldots, e_i$ spans $U^{(i)}(1)$ and $f_1, \ldots, f_i$ spans $U^{(i)}(2)$ for all i. Now $U(a)$ and $U(b)$ have the desired form. $\square$

Exercises.

Exercise 7.1.1. *Show that $V^n_{x,y}$ is indecomposable when $(x,y) \neq 0$. What is $\operatorname{Hom}_Q(V^n_{x,y}, V^n_{x,y})$?*

Exercise 7.1.2. *Show that $V^n_{x,y}$ and $V^n_{x',y'}$ are isomorphic if and only if there exists a nonzero $\lambda \in \mathbb{C}$ such that $(x',y') = \lambda(x,y)$.*

Exercise 7.1.3. *Show that $C_1^+ W^n$ is isomorphic to W^{n+1} after relabeling of the vertices of $\sigma_1\theta(2)$ so that we can identify $\sigma_1\theta(2)$ with $\theta(2)$. Similarly, show that $C_2^+ Z^n$ is isomorphic to Z^{n+1} (after relabeling). Use this to show that the representations W^n, Z^n, $n \geq 0$ are all indecomposable.*

Exercise 7.1.4. *Define a representation U by $U(1) = 1_n$ and $U(2) = J^n_\lambda$. Show that U is isomorphic to $V^n_{[1:\lambda]}$.*

7.2. The Auslander-Reiten Quiver of the Kronecker Quiver

We define the **defect** of a dimension vector $\alpha = (\alpha_1, \alpha_2)$ by $\partial(\alpha) = \alpha_1 - \alpha_2$. The three classes $\{W^n\}$, $\{V^n_{[x:y]}\}$ and $\{Z^n\}$ of indecomposable representations correspond to the cases $\partial(\alpha) < 0$, $\partial(\alpha) = 0$ and $\partial(\alpha) > 0$, respectively.

Consider first the representations of negative defect. The representations W^0 and W^1 are just indecomposable projective representations. The Auslander-Reiten transform τ^- takes W^n to W^{n+2}. We also have irreducible embeddings

$$f_a^n, f_b^n : W^n \to W^{n+1}$$

defined as follows. Choose bases $\{u_1, \ldots, u_{n+1}\}$ of $W^n(1)$ and $\{v_1, \ldots, v_{n+2}\}$ of $W^n(2)$ in which W^n has matrix representation given by (7.1). Then f_n^a is an inclusion of W^n onto subrepresentation spanned by $u_1, \ldots, u_n$ and $v_1, \ldots, v_{n+1}$, and f_n^b is an inclusion of W^n onto a subrepresentation spanned by $u_2, \ldots, u_{n+1}, v_2, \ldots, v_{n+2}$.

Thus we get the component of Auslander-Reiten quiver which can be written as follows:

$$W^0 \underset{f_b^0}{\overset{f_a^0}{\rightrightarrows}} W^1 \underset{f_b^1}{\overset{f_a^1}{\rightrightarrows}} W^2 \underset{f_b^2}{\overset{f_a^2}{\rightrightarrows}} \cdots.$$

The important point is that every indecomposable with negative defect is obtained from indecomposable projective representation by repeated application of τ^-. These are the preprojective representations that we encountered in Section 6.7.

Similarly we get the component of representations with $\partial(\alpha) > 0$. We have two epimorphisms, $g_a^n, g_b^n : Z^{n+1} \to Z^n$, and we get the Auslander-Reiten component

$$\cdots \underset{g_b^2}{\overset{g_a^2}{\rightrightarrows}} Z^2 \underset{g_b^1}{\overset{g_a^1}{\rightrightarrows}} Z^1 \underset{g_b^0}{\overset{g_a^0}{\rightrightarrows}} Z^0.$$

The representations Z^0 and Z^1 are injective, and $\tau^+ Z^n = Z^{n+2}$ for all n. Every indecomposable with positive defect is obtained from an injective indecomposable representation by repeated application of τ^+. These are the preinjective representations.

Finally, the indecomposable representations with the defect equal to zero have the property that they are fixed by τ^+. They form a tube:

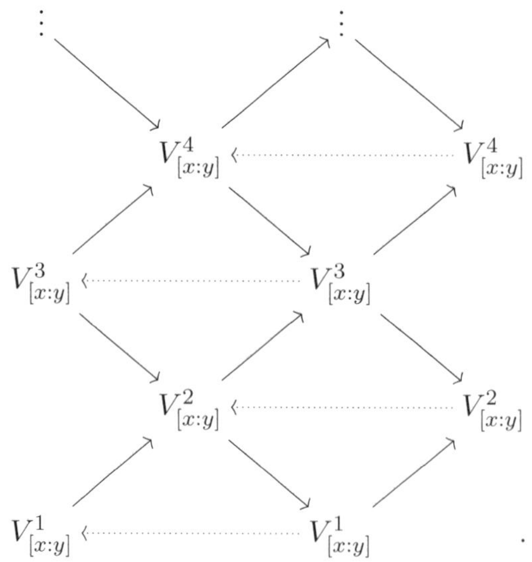

In the diagram above we identify X with $\tau^+(X)$ so that we get an infinite tube.

7.3. AR Quivers for other Extended Dynkin Types

The classification of indecomposable representations of extended Dynkin quivers is similar to the classification for the Kronecker quiver. There are some complications but the general features are the same. An excellent reference for this theory is the memoir [**34**]. We will just give examples and outline the proofs. Our goal is to make sure the reader can calculate examples.

We recall some facts from Chapter 4. The quadratic form B_Q for extended Dynkin quiver is semi-positive definite. Let $(-,-)$ be the associated Cartan form. The kernel of the form $(-,-)$ is one-dimensional. Choose a dimension vector α as in Lemma 4.2.1. This dimension vector is called the **primitive isotropic root**. We have

$$(\alpha, \beta) = \langle \alpha, \beta \rangle + \langle \beta, \alpha \rangle = 0$$

for every dimension vector β. For a dimension vector β, define its **defect** by

$$\partial(\beta) = \langle \alpha, \beta \rangle = -\langle \beta, \alpha \rangle.$$

The **defect** of a representation V is $\partial(\underline{\dim}(V))$. The defect can be negative, positive or equal to 0. The structure of representations of positive and negative defect is quite transparent.

Theorem 7.3.1. *Let Q be an extended Dynkin quiver.*

(1) *Every indecomposable representation of negative defect is preprojective, i.e., is of the type $\tau^{-d}P_x$, where $d \geq 0$, $x \in Q_0$. These representations are pairwise nonisomorphic. The component of the Auslander-Reiten graph of preprojective representations is given by Theorem* 6.7.3.

(2) *Every indecomposable representation of positive defect is preinjective, i.e. of the type $\tau^{d}I_x$ where $d \geq 0$, $x \in Q_0$. All such representations are nonisomorphic. The component of the Auslander-Reiten graph of preinjective representations is given by reversing the arrows in the graph of the preprojective component of the path algebra opposite quiver Q^{op} in Theorem* 6.7.3

Proof. The proof is similar to the proof of Gabriel theorem using Coxeter functors. The main point is that if β is a vector of negative defect, then $c^k\beta$ is not nonnegative for some power k of the Coxeter element c. Similarly, if β has positive defect, then for some $k > 0$ the vector $c^{-k}\beta$ is not nonnegative. □

The main problem is the classification of the indecomposable **regular representations**, i.e., indecomposable representations of defect zero. The results are based on the classification of roots of the quadratic form B_Q. The regular representations turn out to be periodic with respect to the Coxeter functor C.

We define a representation to be **regular** if all of its direct summands are regular. The subcategory $\mathrm{Reg}(Q)$ of regular representations is an abelian category, closed under extensions. They are of two kinds of simple regular representations: nonhomogeneous and homogeneous.

The nonhomogeneous indecomposable regular representations occur in (two for $\widehat{\mathbf{A}}_n$, three for all other extended Dynkin quivers) collections of dimension vectors that are periodic with respect to c. Each such family is generated by the set of simple regular representations. We will denote three families of simple regular representations by $\{E_i\}_{0\leq i\leq l-1}$, $\{E'_j\}_{0\leq j\leq l'-1}$, $\{E''_k\}_{0\leq k\leq l''-1}$.

The three families are orthogonal to each other, that is, the spaces $\mathrm{Hom}_Q(U,V)$ and $\mathrm{Ext}_Q(U,V)$ are zero if U and V are from different families.

For such a family, say $E_0, E_1 = C(E_0), \ldots, E_{l-1} = C^{l-1}(E_0), E_l = C^l(E_0) = E_0$ we have the following properties.

Proposition 7.3.2.

(1) $\operatorname{Hom}_Q(E_i, E_j) = 0$ *unless* $i = j$.

(2) $\operatorname{Ext}_Q(E_i, E_j) = \begin{cases} \mathbb{C} \text{ if } j \equiv i + 1 (\operatorname{mod} l), \\ 0 \text{ otherwise.} \end{cases}$

(3) $\underline{\dim}(E_0) + \cdots + \underline{\dim}(E_{l-1}) = \alpha$.

The family $\{E_i\}_{0 \leq i \leq l-1}$ generates **the nonhomogeneous tube** of period l, i.e., the connected component of the Auslander-Reiten quiver that looks like the infinite tube (folded after l horizontal steps)

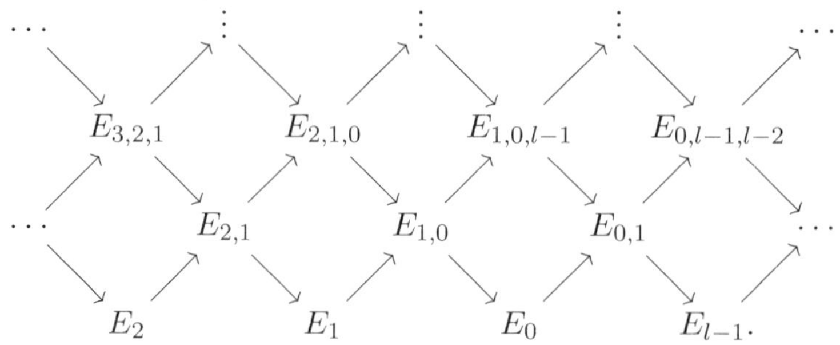

All the triangles at the bottom and the squares in the middle are almost split sequences.

The homogeneous indecomposable representations occur in dimensions $m\alpha$ for $m \in \mathbb{N}$. Each of these representations has a filtration whose associated graded object is a direct sum of m copies of the same representation of dimension α. Each such family is **a homogeneous tube**. In dimension vector α there is a one-parameter family of indecomposable representations V_t where $t \in \mathbb{P}^1(\mathbb{C}) \setminus \{0, 1, \infty\}$ (resp. $\mathbb{P}^1(\mathbb{C}) \setminus \{0, \infty\}$ for type $\widehat{\mathbf{A}}_n$). This family can be extended to $\mathbb{P}^1$ family, by including three representations of dimension α from components corresponding to 3 (resp. 2) nonhomogeneous tubes. There are several such extensions, adding different representations from nonhomogeneous tubes. The homogeneous indecomposable regular representations are explicitly described in [**34**] for particular orientations of the extended Dynkin quivers. For other orientations we can use the reflection functors.

Example 7.3.3. Consider the quiver Q below of type $\widehat{A}_2$:

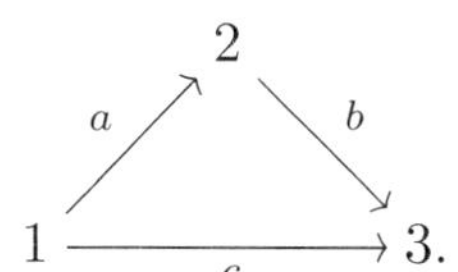

Consider the family

$V^1_{[x:y]}$:

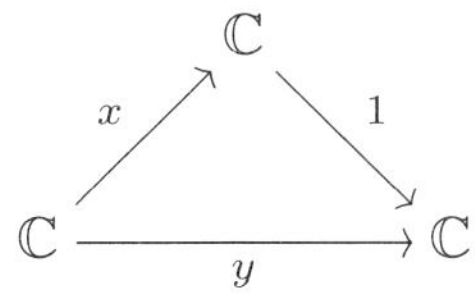

of representations parametrized by $[x : y] \in \mathbb{P}^1(\mathbb{C})$.

Instead, we could have chosen

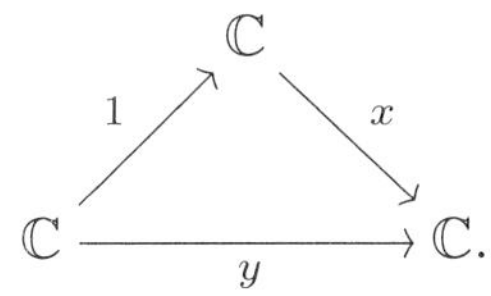

Two families coincide for all $[x : y] \in \mathbb{P}^1(\mathbb{C}) \setminus \{[1 : 0]\}$. Each of the families contains exactly one representation from both nonhomogeneous tubes.

If one finds the dimension vectors of simple regular representations for a given extended Dynkin quiver, one can get those for other orientations via the reflection functors. We exhibit the class of the distinguished orientations from which one can get to an arbitrary one by applying the reflection functors. For each of these orientations we exhibit the simple regular modules by their dimension vectors. Indeed, these dimension vectors are the Schur roots, i.e., there is one indecomposable representation in such dimension and its isomorphism class forms an open orbit in $\mathrm{Rep}(Q, \alpha)$.

The orientations we use are as follows.

(1) Type $\widehat{\mathbf{A}}_n$. Consider the quiver

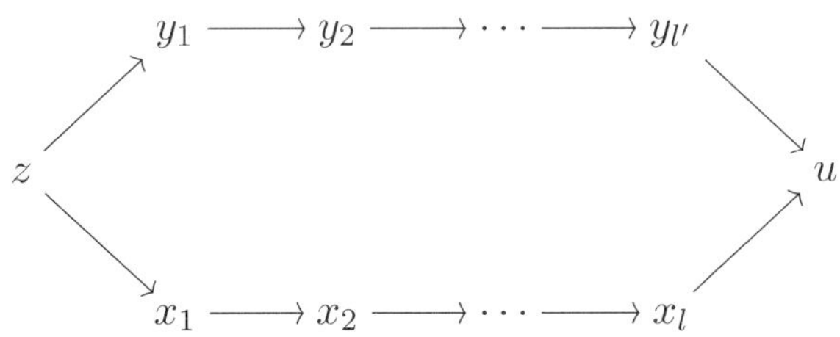

with $l + l' = n - 1$.

(a) The first family is

$$E_0 : 1 \begin{matrix} 1 & 1 & \cdots & 1 & 1 \\ & & & & \\ 0 & 0 & \cdots & 0 & 0 \end{matrix} 1$$

and $E_i = S_{x_i}$ for $1 \leq i \leq l$.

(b) The second family is

$$E_0' = 1 \begin{array}{ccccc} 0 & 0 & \cdots & 0 & 0 \\ & & & & \\ 1 & 1 & \cdots & 1 & 1 \end{array} 1.$$

and $E_i' = S_{y_i}$ for $1 \leq i \leq l'$

$$E_j' = S_{y_j}, \text{ for } 1 \leq j \leq u.$$

(2) Type $\widehat{\mathbf{D}}_n$. Consider the quiver

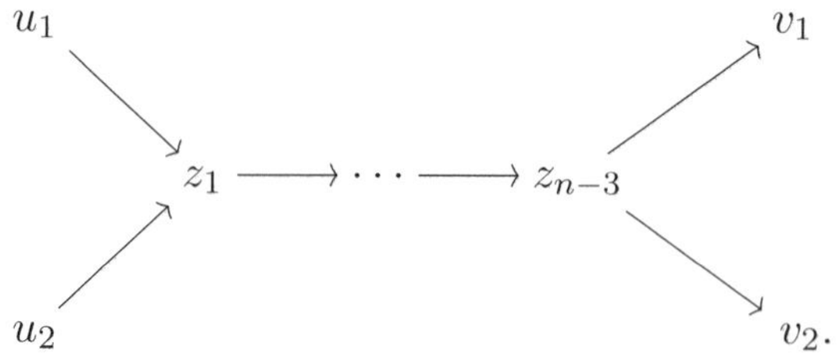

(a) The first family is

$$E_0 : \begin{array}{ccccccc} 1 & & & & & & 1 \\ & 1 & 1 & \ldots & 1 & 1 & \\ 1 & & & & & & 1 \end{array},$$

and $E_i = S_{z_i}$ if $1 \leq i \leq n-3$.

(b) The second family is

$$E_0' : \begin{array}{ccccccc} 0 & & & & & & 0 \\ & 1 & 1 & \ldots & 1 & 1 & \\ 1 & & & & & & 1 \end{array},$$

$$E_1' : \begin{array}{ccccccc} 1 & & & & & & 1 \\ & 1 & 1 & \ldots & 1 & 1 & \\ 0 & & & & & & 0 \end{array}.$$

(c) The third family is

$$E_0'' : \begin{array}{ccccccc} 0 & & & & & & 1 \\ & 1 & 1 & \ldots & 1 & 1 & \\ 1 & & & & & & 0 \end{array},$$

$$E_1'' : \begin{array}{ccccccc} 1 & & & & & & 0 \\ & 1 & 1 & \ldots & 1 & 1 & \\ 0 & & & & & & 1 \end{array}.$$

(3) Type $\widehat{\mathrm{E}}_6$.

$$\begin{array}{ccccccccc} & & & & \circ & & & & \\ & & & & \downarrow & & & & \\ & & & & \circ & & & & \\ & & & & \downarrow & & & & \\ \circ & \longrightarrow & \circ & \longrightarrow & \circ & \longleftarrow & \circ & \longleftarrow & \circ. \end{array}$$

(a) The first family is

$$E_0 = \begin{array}{ccccc} & & 0 & & \\ & & 1 & & \\ 0 & 1 & 1 & 1 & 0 \end{array}, \quad E_1 = \begin{array}{ccccc} & & 1 & & \\ & & 1 & & \\ 1 & 1 & 2 & 1 & 1 \end{array}.$$

(b) The second family is

$$E'_0 = \begin{array}{ccccc} & & 0 & & \\ & & 0 & & \\ 0 & 1 & 1 & 1 & 1 \end{array}, \quad E'_1 = \begin{array}{ccccc} & & 1 & & \\ & & 1 & & \\ 0 & 0 & 1 & 1 & 0 \end{array}, \quad E'_2 = \begin{array}{ccccc} & & 0 & & \\ & & 1 & & \\ 1 & 1 & 1 & 0 & 0 \end{array}.$$

(c) The third family is

$$E''_0 = \begin{array}{ccccc} & & 0 & & \\ & & 1 & & \\ 0 & 0 & 1 & 1 & 1 \end{array}, \quad E''_1 = \begin{array}{ccccc} & & 1 & & \\ & & 1 & & \\ 0 & 1 & 1 & 0 & 0 \end{array}, \quad E''_2 = \begin{array}{ccccc} & & 0 & & \\ & & 0 & & \\ 1 & 1 & 1 & 1 & 0 \end{array}.$$

(4) Type $\widehat{\mathrm{E}}_7$.

$$\begin{array}{ccccccccccccc} & & & & & & \circ & & & & & & \\ & & & & & & \downarrow & & & & & & \\ \circ & \longrightarrow & \circ & \longrightarrow & \circ & \longrightarrow & \circ & \longleftarrow & \circ & \longleftarrow & \circ & \longleftarrow & \circ \end{array}$$

(a) The first family is

$$E_0 = \begin{array}{ccccccc} & & & 1 & & & \\ 0 & 0 & 1 & 1 & 1 & 0 & 0 \end{array}, \quad E_1 = \begin{array}{ccccccc} & & & 1 & & & \\ 1 & 1 & 1 & 2 & 1 & 1 & 1 \end{array},$$

$$E_2 = \begin{array}{ccccccc} & & & 0 & & & \\ 0 & 1 & 1 & 1 & 1 & 1 & 0 \end{array}.$$

(b) The second family is

$$E'_0 = \begin{array}{ccccccc} & & & 1 & & & \\ 0 & 0 & 0 & 1 & 1 & 1 & 0 \end{array}, \quad E'_1 = \begin{array}{ccccccc} & & & 0 & & & \\ 1 & 1 & 1 & 1 & 1 & 0 & 0 \end{array},$$

$$E'_2 = \begin{array}{ccccccc} & & & 1 & & & \\ 0 & 1 & 1 & 1 & 0 & 0 & 0 \end{array}, \quad E'_3 = \begin{array}{ccccccc} & & & 0 & & & \\ 0 & 0 & 1 & 1 & 1 & 1 & 1 \end{array}.$$

(c) The third family is

$$E_0'' = \begin{smallmatrix} & & & 1 & & & \\ 0 & 1 & 1 & 2 & 2 & 1 & 1 \end{smallmatrix}, \quad E_1'' = \begin{smallmatrix} & & & 1 & & & \\ 1 & 1 & 2 & 2 & 1 & 1 & 0 \end{smallmatrix}.$$

(5) Type $\widehat{\mathrm{E}}_8$

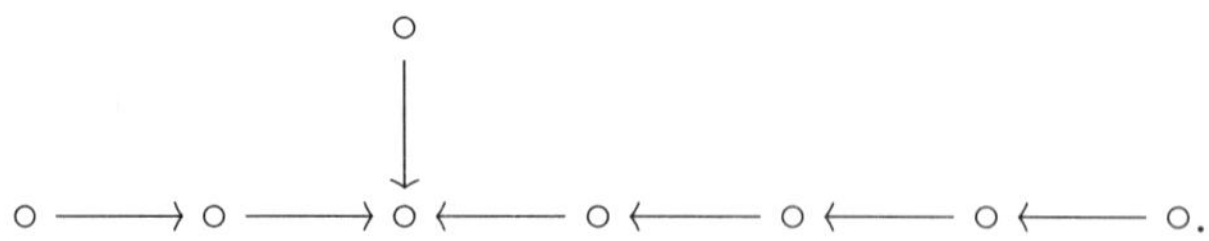

(a) The first family is

$$E_0 = \begin{smallmatrix} & & 1 & & & & & \\ 0 & 1 & 1 & 1 & 0 & 0 & 0 & 0 \end{smallmatrix}, \quad E_1 = \begin{smallmatrix} & & 1 & & & & & \\ 1 & 1 & 2 & 1 & 1 & 1 & 1 & 1 \end{smallmatrix},$$

$$E_2 = \begin{smallmatrix} & & 0 & & & & & \\ 0 & 1 & 1 & 1 & 1 & 1 & 1 & 0 \end{smallmatrix}, \quad E_3 = \begin{smallmatrix} & & 1 & & & & & \\ 0 & 0 & 1 & 1 & 1 & 1 & 0 & 0 \end{smallmatrix},$$

$$E_4 = \begin{smallmatrix} & & 0 & & & & & \\ 1 & 1 & 1 & 1 & 1 & 0 & 0 & 0 \end{smallmatrix}.$$

(b) The second family is

$$E_0' = \begin{smallmatrix} & & 1 & & & & & \\ 1 & 2 & 2 & 1 & 1 & 1 & 0 & 0 \end{smallmatrix}, \quad E_1' = \begin{smallmatrix} & & 1 & & & & & \\ 0 & 1 & 2 & 2 & 2 & 1 & 1 & 1 \end{smallmatrix},$$

$$E_2' = \begin{smallmatrix} & & 1 & & & & & \\ 1 & 1 & 2 & 2 & 1 & 1 & 1 & 0 \end{smallmatrix}.$$

(c) The third family is

$$E_0'' = \begin{smallmatrix} & & 2 & & & & & \\ 1 & 2 & 3 & 2 & 2 & 1 & 1 & 0 \end{smallmatrix}, \quad E_1'' = \begin{smallmatrix} & & 1 & & & & & \\ 1 & 2 & 3 & 3 & 2 & 2 & 1 & 1 \end{smallmatrix}.$$

Exercises.

Exercise 7.3.1. *Let Q be an extended Dynkin quiver and let $\{E_0, \ldots, E_m\}$ be the family of simple regular modules periodic under τ^+, i.e., $\tau^+ E_i = E_{i-1}$ for $i = 1, \ldots, m$, $\tau^+ E_0 = E_m$. Prove that* $\dim \operatorname{Hom}_Q(E_i, E_j) = \delta_{i,j}$, $\dim \operatorname{Ext}_Q(E_i, E_j) = \delta_{i+1,j}$.

Exercise 7.3.2. *Let Q be an extended Dynkin quiver. Let $V, W \in \operatorname{Rep}(Q)$ be indecomposable modules.*

(1) $\operatorname{Hom}_Q(V, W) = 0$ *if V is preinjective and W is preprojective or regular,*

(2) $\operatorname{Hom}_Q(V, W) = 0$ *if W is preprojective and V is preinjective or regular.*

Exercise 7.3.3. *Define a module $V \in \operatorname{Rep}(Q)$ for an extended Dynkin quiver Q to be regular if every indecomposable summand is regular.*

(1) *Prove that every submodule of V has only preprojective and regular summands.*

(2) *Prove that every factor module of V has only preinjective and regular summands.*

(3) *Let $f : V \to W$ be a homomorphism of regular modules. Prove that the kernel and the cokernel of f are regular.*

(4) *If a module V has a filtration*

$$0 = V_0 \subset V_1 \subset \cdots \subset V_{m-1} \subset V_m = V$$

with each factor V_i/V_{i-1} regular, then V is regular.

7.4. Bibliographical Remarks

The description of indecomposable representations of extended Dynkin quivers is due to Nazarova [**66**] and Donovan and Freislich [**37**]. We did not fully develop this theory here because the notes by Dlab and Ringel [**34**] are beautifully written and not difficult to read so we recommend them to the reader. We discuss enough of the theory needed for the discussion of semi-invariants. The case of extended Dynkin quivers and their isotropic roots is the first when the ring of semi-invariants is not a polynomial ring and in that case one can see the direct link between the semi-invariants and regular modules. We will describe this more precisely in the chapter on semi-invariants.

Chapter 8

Kac's Theorem

In 1980 Victor Kac proved a beautiful result generalizing Gabriel's theorem to arbitrary quivers. The general statement says that for any quiver Q and a dimension vector α there exists an indecomposable representation of dimension vector α if and only if α is a root of the corresponding infinite roots system (so it is a root of the corresponding Kac-Moody Lie algebra).

We give a proof of this result, following the ideas of Crawley-Boevey and Van den Bergh. First one proves the result for indivisible dimension vectors using the deformed preprojective algebras and their representations. Then one uses counting techniques over a finite field to show first that the existence of indecomposables and the existence of infinitely many indecomposables in a dimension vector α does not depend on the orientation of arrows of the quiver. This allows us to prove a general result using the Weyl group.

One should point out that the original proof of Kac also used finite fields and Weyl conjectures to prove the result. So far there is no proof of Kac's theorem over $\mathbb{C}$ that does not use the fields of positive characteristic.

As far as we know this is the first proof of Kac's theorem in book form.

8.1. Deformed Preprojective Algebras

In this section we will use deformed preprojective algebras to prove that it is independent of the orientation whether there exists an indecomposable representation of a given dimension vector α.

Suppose that $Q = (Q_0, Q_1, h, t)$ is a quiver. We define the double quiver $\overline{Q}$ as in Section 6.7 of Chapter 6. For every arrow $a \in Q_1$ we define a dual arrow $a^\star$ with $h(a^\star) = ta$ and $t(a^\star) = ha$. Set $Q_1^\star = \{a^\star \mid a \in Q_1\}$ and define $\overline{Q}_1 = Q_1 \cup Q_1^\star$. The double of the quiver Q is $\overline{Q} = (Q_0, \overline{Q}_1, h, t)$, and the opposite quiver is $Q^\star = Q^{\mathrm{op}} = (Q_0, Q_1^\star, h, t)$. For $\lambda \in \mathbb{C}^{Q_0}$, we define the **deformed preprojective algebra** by

$$\Pi_\lambda = \mathbb{C}\overline{Q}/(r_\lambda),$$

where

$$r_\lambda = \sum_{a \in Q_1} (a^\star a - a a^\star) - \lambda \in \mathbb{C}\overline{Q}.$$

Here $\lambda \in \mathbb{C}^{Q_0}$ has been identified with $\sum_{x \in Q_0} \lambda(x) e_x \in \mathbb{C}\overline{Q}$. Note that $e_y r_\lambda e_x = 0$ for $x \neq y$ and

$$e_x r_\lambda e_x = \sum_{\substack{a \in Q_1 \\ ta = x}} a^\star a - \sum_{\substack{a \in Q_1 \\ ha = x}} a a^\star - \lambda(x) e_x.$$

The algebra Π_0 is the preprojective algebra introduced in Section 6.7 of Chapter 6.

We may view $\mathrm{Rep}_\alpha(\Pi_\lambda)$ as a subset of $\mathrm{Rep}_\alpha(Q) \oplus \mathrm{Rep}_\alpha(Q^\star)$ consisting of all pairs (V, W) such that

$$\sum_{\substack{a \in Q_1 \\ ta = x}} W(a^\star)V(a) - \sum_{\substack{a \in Q_1 \\ ha = x}} V(a)W(a^\star) = \lambda(x) 1_{\alpha(x)}$$

for all $x \in Q_0$.

Define the **moment map**

$$\mu_\alpha : \mathrm{Rep}_\alpha(Q) \oplus \mathrm{Rep}_\alpha(Q^\star) \to \bigoplus_{x \in Q_0} \mathrm{End}(\mathbb{C}^{\alpha(x)})$$

by

$$\mu_\alpha(V, W) = \sum_{\substack{a \in Q_1 \\ ta = x}} W(a^\star)V(a) - \sum_{\substack{a \in Q_1 \\ ha = x}} V(a)W(a^\star).$$

Then we have

$$\mu_\alpha^{-1}(\lambda) = \mathrm{Rep}_\alpha(\Pi_\lambda),$$

where $\lambda \in \mathbb{C}^{Q_0}$ has been identified with $(\lambda(x) 1_{\alpha(x)}, x \in Q_0) \in \mathrm{End}(\mathbb{C}^{\alpha(x)})$.

Remark 8.1.1. If

$$\mu_\alpha(V, W) = (A(x), x \in Q_0),$$

then we have

$$\sum_{x \in Q_0} \mathrm{Tr}(A(x)) = \sum_{a \in Q_1} \mathrm{Tr}(W(a^\star)V(a) - V(a)W(a^\star)) = 0.$$

For $\lambda \in \mathbb{C}^{Q_0}$ and $\alpha \in \mathbb{N}^{Q_0}$ we define

$$\lambda(\alpha) = \lambda \cdot \alpha = \sum_{x \in Q_0} \lambda(x)\alpha(x).$$

Lemma 8.1.2. *If $\lambda(\alpha) \neq 0$, then we get $\operatorname{Rep}_\alpha(\Pi_\lambda) = \emptyset$.*

Proof. Suppose that

$$(V, W) \in \operatorname{Rep}_\alpha(\Pi_\lambda) = \mu_\alpha^{-1}(\lambda).$$

Then it follows from Remark 8.1.1 that

$$0 = \sum_{x \in Q_0} \operatorname{Tr}(\lambda(x) 1_{\alpha(x)}) = \sum_{x \in Q_0} \lambda(x)\alpha(x) = \lambda(\alpha).$$

□

Suppose that $V \in \operatorname{Rep}_\alpha(Q)$. Then we have an exact sequence

$$\begin{array}{c} 0 \\ \downarrow \\ \operatorname{Hom}_Q(V, V) \\ \downarrow \\ \bigoplus_{x \in Q_0} \operatorname{End}(\mathbb{C}^{\alpha(x)}) \\ \downarrow \\ \bigoplus_{a \in Q_1} \operatorname{Hom}(V(ta), V(ha)) \\ \downarrow \\ \operatorname{Ext}^1_Q(V, V) \\ \downarrow \\ 0 \end{array},$$

where $\mathrm{End}(\mathbb{C}^{\alpha(x)}) \cong \mathrm{Hom}_{\mathbb{C}}(V(x), V(x))$. Dualizing gives an exact sequence

$$
\begin{array}{c}
0\\
\downarrow\\
\mathrm{Ext}^1_Q(V,V)^\star\\
\downarrow\\
\mathrm{Rep}_\alpha(Q^\star)\\
\downarrow{\scriptstyle \mu_\alpha^V}\\
\bigoplus_{x\in Q_0}\mathrm{End}(\mathbb{C}^{\alpha(x)})\\
\downarrow{\scriptstyle \gamma}\\
\mathrm{Hom}_Q(V,V)^\star\\
\downarrow\\
0
\end{array} \quad . \tag{8.1}
$$

Here we have identified

$$
\begin{aligned}
\mathrm{Rep}_\alpha(Q^\star) &= \bigoplus_{a^\star\in Q_1^\star} \mathrm{Hom}_Q(\mathbb{C}^{t(a^\star)}, \mathbb{C}^{h(a^\star)})\\
&= \bigoplus_{a\in Q_1} \mathrm{Hom}_Q(\mathbb{C}^{h(a)}, \mathbb{C}^{t(a)}) = \bigoplus_{a\in Q_1} \mathrm{Hom}_Q(V(ta), V(ha))^\star
\end{aligned}
$$

and $\mathrm{End}(\mathbb{C}^{\alpha(x)})$ with $\mathrm{End}(\mathbb{C}^{\alpha(x)})^\star$ using the symmetric bilinear pairing on $\mathrm{End}(\mathbb{C}^{\alpha(x)})$ given by

$$(A,B) = \sum_{x\in Q_0} \mathrm{Tr}(A(x)B(x)).$$

The map $\mathrm{Rep}_\alpha(Q^\star) \to \bigoplus_{x\in Q_0}\mathrm{End}(\mathbb{C}^{\alpha(x)})$ is equal to

$$\mu_\alpha^V = \mu_\alpha(V,-) : \mathrm{Rep}_\alpha(Q^\star) \to \bigoplus_{x\in Q_0}\mathrm{End}(\mathbb{C}^{\alpha(x)}).$$

Theorem 8.1.3. *We have*

$$(\mu_\alpha^V)^{-1}(\lambda) \neq \emptyset$$

if and only if for every indecomposable summand Y of V we have

$$\lambda(\underline{\dim}\ Y) = 0.$$

Proof. If $(\mu_\alpha^V)^{-1}(\lambda)$ is nonempty, then we have $\gamma(\lambda) = \lambda(\alpha) = 0$ because of the exactness of (8.1). Let $p : V \to V$ be the projection onto Y. Then we have

$$0 = \gamma(\lambda)(p) = \sum_{x \in Q_0} \lambda(x) \operatorname{Tr}(p(x)) = \lambda(\underline{\dim}\ Y).$$

Conversely, suppose that V is indecomposable and $\lambda(\alpha) = 0$. Note that $\operatorname{End}_Q(V)$ is spanned by the identity and nilpotent endomorphisms. We have

$$\gamma(\lambda)(1_V) = \lambda(\alpha) = 0,$$

and for every nilpotent endomorphism $f \in \operatorname{End}_Q(V)$ we have

$$\gamma(\lambda)(f) = \sum_{x \in Q_0} \operatorname{Tr}(\lambda(x) f(x)) = 0$$

because $f(x)$ is nilpotent. We have $\gamma(\lambda) = 0$, and conclude that $(\mu_\alpha^V)^{-1}(\lambda) \neq \emptyset$ because of the exactness of (8.1).

If $V = V_1 \oplus \cdots \oplus V_r$ with V_i indecomposable of dimension α_i and $\lambda(\underline{\dim}\, V_i) = 0$ for all i, then there exist $W_1, \ldots, W_r$ such that $(V_i, W_i) \in \operatorname{Rep}_{\alpha_i}(\Pi_\lambda)$. We have $(V, W) \in \operatorname{Rep}_\alpha(\Pi_\lambda) = \mu_\alpha^{-1}(\lambda)$, where $W = W_1 \oplus \cdots \oplus W_r$. □

A dimension vector α is called **indivisible** if α cannot be written in the form $\alpha = k\alpha'$ where α' is a dimension vector and k is an integer ≥ 2. A dimension vector α is indivisible when the greatest common divisor of all $\alpha(x)$, $x \in Q_0$ is equal to 1.

Corollary 8.1.4. *Suppose that α is an indivisible dimension vector. Whether there exists an indecomposable representation of Q of dimension α is independent of the orientation of Q.*

Proof. Suppose that the quivers Q and Q' are the same up to orientation. Then the double quivers $\overline{Q}, \overline{Q}'$ are the same. Choose $\lambda \in \mathbb{C}^{Q_0}$ such that for all $\beta \in \mathbb{Z}^{Q_0}$ we have $\lambda(\beta) = 0$ if and only if β is a rational multiple of α (see Exercise 8.1.1). If V is an indecomposable representation of Q, then V can be lifted to a representation $Z = (V, W)$ of $\overline{Q}$ by Theorem 8.1.3. Restricting Z to Q' gives a representation V' of Q'. For every indecomposable summand Y of V' we have $\lambda(\underline{\dim}\ Y) = 0$, so $\underline{\dim}\ Y$ is a rational multiple of α. Since α is indivisible, we must have $\underline{\dim}\, Y = 0$ or $\underline{\dim}\, Y = \alpha$. This shows that V' is indecomposable. □

We will see later that the condition that α is indivisible can be omitted.

Exercises.

Exercise 8.1.1. *Suppose that $\alpha \in \mathbb{Q}^n \subseteq \mathbb{C}^n$. Show that there exists a vector $\lambda \in \mathbb{C}^n$ such that for all $\beta \in \mathbb{Q}^n$ we have: $\lambda(\beta) = 0$ if and only if $\beta = t\alpha$ for some $t \in \mathbb{Q}$. (Hint: Choose $\delta^{(1)}, \ldots, \delta^{(n-1)} \in \mathbb{Q}^n$ such that*

$$\{\beta \in \mathbb{Q}^n \mid \delta^{(1)}(\beta) = \cdots = \delta^{(n-1)}(\beta) = 0\}$$

is the $\mathbb{Q}$-span of α. Then take $\lambda = t_1\delta^{(1)} + \cdots + t_{n-1}\delta^{(n-1)}$, where $t_1, \ldots, t_{n-1} \in \mathbb{C}$ are chosen to be linearly independent over $\mathbb{Q}$.)

8.2. Reflections

For $\alpha \in \mathbb{Z}^{Q_0}$ we defined the reflection

$$\sigma_x(\alpha) = \alpha - (\alpha, \epsilon_x)\epsilon_x.$$

Dually, for $\lambda \in \mathbb{C}^{Q_0}$, we define $(\sigma_x^\star \lambda)(y) = \lambda(y) - (\epsilon_x, \epsilon_y)\lambda(x)$. In particular we get $\sigma_x^\star(\lambda)(x) = -\lambda(x)$. We have

$$\lambda(\sigma_x(\alpha)) = \sigma_x^\star(\lambda)(\alpha).$$

The **Weyl group** $\mathcal{W}$ is the group generated by σ_x, $x \in Q_0$. The length $\ell(w)$ of an element $w \in \mathcal{W}$ is the smallest integer r such that we can write

$$w = \sigma_{x_r} \cdots \sigma_{x_2}\sigma_{x_1}.$$

Lemma 8.2.1. *Let α be an indivisible dimension vector. Suppose that there exists an indecomposable representation V of Q of dimension α. Then either $\alpha = \epsilon_x$ and V is isomorphic to the simple representation S_x, or $\sigma_x(\alpha) \in \mathbb{N}^{Q_0}$ and there exists an indecomposable representation of dimension $\sigma_x(\alpha)$.*

Proof. Let Q' be a quiver such that $\overline{Q} = \overline{Q'}$ and such that x is a sink of Q'. There exists an indecomposable representation V' of Q' of dimension α by Corollary 8.1.4. If $C_x^+(V') = 0$, then we have $V' = S_x$ by Theorem 4.3.9, so $\alpha = \epsilon_x$ and $V = S_x$. If $C_x^+(V') \neq 0$, then $C_x^+(V')$ is an indecomposable representation of $\sigma_x Q'$ of dimension $\sigma_x(\alpha)$ by Theorem 4.3.9. So there also exists an indecomposable representation of Q of dimension $\sigma_x(\alpha)$ by Corollary 8.1.4. □

Theorem 8.2.2. *Suppose $\alpha \in \mathbb{N}^{Q_0}$ is indivisible, and $\beta = w(\alpha)$ for some $w \in \mathcal{W}$. Then there exists an indecomposable representation of dimension α if and only if there exists an indecomposable representation of dimension β or $-\beta$.*

Proof. Let us write

$$w = \sigma_{x_r} \cdots \sigma_{x_1}.$$

By induction on r we will show that there exists an indecomposable representation of dimension β or $-\beta$. The case $r = 0$ is clear. By the induction hypothesis we may assume that there exist an indecomposable representation of dimension γ where $\gamma = \pm\sigma_{x_{r-1}} \cdots \sigma_{x_1}(\alpha)$ and $\beta = \pm\sigma_{x_r}(\gamma)$. We apply Lemma 8.2.1. In the first case $\sigma_{x_r}(\gamma) \notin \mathbb{N}^{Q_0}$ and $\gamma = \epsilon_{x_r}$. Then $\beta = \pm\epsilon_{x_r}$ so there exists an indecomposable representation of dimension β or $-\beta$, namely the simple representation S_{x_r}; or in the second case $\sigma_{x_r}(\gamma) \in \mathbb{N}^{Q_0}$ and there exists an indecomposable representation of dimension $\sigma_{x_r}(\gamma) = \pm\beta$. □

We can also reflect representations of the deformed preprojective algebra (but we do not need it for the proof of Kac's theorem). Suppose that $Z = (V, W)$ is a finite dimensional indecomposable Π_λ module, and $x \in Q_0$. Let $a_1, \dots, a_t \in \overline{Q}_1$ be the incoming arrows and $a_1^\star, \dots, a_t^\star$ be the outgoing arrows, and let $y_i = ta_i = ha_i^\star$. Define

$$\mathrm{in}_x = \begin{pmatrix} a_1 & a_2 & \cdots & a_t \end{pmatrix}$$

and

$$\mathrm{out}_x = \begin{pmatrix} \delta_1 a_1^\star \\ \delta_2 a_2^\star \\ \vdots \\ \delta_t a_t^\star \end{pmatrix},$$

where $\delta_i = 1$ if $a_i \in Q_1$ and $\delta_i = -1$ if $a_i \in Q_1^\star$. We have $Z(\mathrm{in}_x)Z(\mathrm{out}_x) = \lambda(x)1_{Z(x)}$. Assume that $\lambda(x) \neq 0$.

We have

$$\bigoplus_{i=1}^{t} Z(y_i) = \mathrm{im}\, Z(\mathrm{out}_x) \oplus \ker Z(\mathrm{in}_x)$$

and $Z(x)$ is isomorphic to $\mathrm{im}\, Z(\mathrm{out}_x)$.

We now define a representation $\sigma_x Z$ over $\overline{Q}$ as follows:

Define $\sigma_x Z(x) = \ker Z(\mathrm{in}_x)$, and $\sigma_x Z(y) = Z(y)$ for $y \neq x$. Define

$$\sigma_x Z(\mathrm{out}_x) : \sigma_x Z(x) \longrightarrow \bigoplus_{i=1}^{t} \sigma_x Z(y_i) = \bigoplus_{i=1}^{t} Z(y_i)$$

as the inclusion, and

$$\sigma_x Z(\mathrm{in}_x) : \bigoplus_{i=1}^{t} Z(y_i) = \bigoplus_{i=1}^{t} \sigma_x Z(y_i) \longrightarrow \sigma_x Z(x)$$

as the projection multiplied with the scalar $-\lambda(x)$. This defines $\sigma_x Z(a_i)$ and $\sigma_x Z(a_i^\star)$ for $i = 1, 2, \dots, t$. For any arrow $b \in \overline{Q}_1 \setminus \{a_1, \dots, a_t, a_1^\star, \dots, a_t^\star\}$ we define $\sigma_x Z(b) = Z(b)$.

Lemma 8.2.3. *The representation $\sigma_x Z$ is a $\Pi_{\sigma_x^\star(\lambda)}$-module.*

Proof. See Exercise 8.2.3 □

There is a canonical way to define σ_x for morphisms (see Exercise 8.2.2).

Lemma 8.2.4.

(1) σ_x *is a functor from* Π_λ–mod *to* $\Pi_{\sigma_x^\star(\lambda)}$–mod.

(2) σ_x^2 *is naturally equivalent to the identity functor. In particular,* σ_x *is an equivalence of categories.*

Proof. See Exercise 8.2.2. □

Exercises.

Exercise 8.2.1. *Prove Lemma* 8.2.3

Exercise 8.2.2. *Show that* σ_x *can be defined for morphisms as well and prove Lemma* 8.2.4.

8.3. Root Systems

In this section we define the roots sytems associated to Q (in fact it will only depend on the underlying graph $\Gamma(Q)$). These are exactly the root systems of the corresponding Kac-Moody Lie algebra. The structure is a bit more complicated than in the finite type case. There are two types of roots, namely the real roots (which are similar to the roots in the finite type case) and imaginary roots. We will show Kac's theorem in a special situation: If $\alpha \in \mathbb{N}^{Q_0}$ is indivisible, then it is the dimension vector of an indecomposable if and only if it is a positive root.

Define the set of **real roots** by

$$\Phi_{\mathrm{re}} = \bigcup_{x \in Q_0} \mathcal{W}\epsilon_x.$$

Define $\Phi_{\mathrm{re}}^+ = \Phi_{\mathrm{re}} \cap \mathbb{N}^{Q_0}$ and $\Phi_{\mathrm{re}}^- = \Phi_{\mathrm{re}}^+ \cap -\mathbb{N}^{Q_0}$. Because $-\epsilon_x = \sigma_x(\epsilon_x)$ we have $\Phi_{\mathrm{re}} = -\Phi_{\mathrm{re}}$ and $\Phi_{\mathrm{re}}^- = -\Phi_{\mathrm{re}}^+$.

Lemma 8.3.1. *For every* $\alpha \in \Phi_{\mathrm{re}}^+$ *there exists an indecomposable representation of dimension* α *and* $\Phi_{\mathrm{re}} = \Phi_{\mathrm{re}}^+ \cup \Phi_{\mathrm{re}}^-$.

Proof. Suppose that $\alpha \in \Phi_{\mathrm{re}}$. Then we can write $\alpha = w(\epsilon_x)$ for some $w \in \mathcal{W}$ and $x \in Q_0$. From Theorem 8.2.2 it follows that there exists an indecomposable representation of dimension α or $-\alpha$. In particular, if $\alpha \in \Phi_{\mathrm{re}}^+ \cup \Phi_{\mathrm{re}}^-$ and if $\alpha \in \Phi_{\mathrm{re}}^+$, then there exists an indecomposable representation of dimension α. □

For a dimension vector α we define $\mathrm{sub}(\alpha)$ as the set of all nonzero dimension vectors β for which $\alpha - \beta$ is also a nonzero dimension vector.

Lemma 8.3.2. *Suppose that α is a dimension vector such that $\alpha(x) > 0$ for all $x \in Q_0$, and $(\alpha, \varepsilon) \leq 0$ for all $x \in Q_0$. If $\beta \in \mathrm{sub}(\alpha)$, then we have $(\beta, \alpha - \beta) \leq 0$. Moreover, if $(\beta, \alpha - \beta) = 0$, then β is proportional to α and $(\alpha, \epsilon_x) = 0$ for all x.*

Proof. Without loss of generality we may assume that the support of α is Q_0. We have

$$
\begin{aligned}
(8.2)\quad 0 \geq \frac{1}{2}\sum_{x,y}\left(\frac{\beta(x)}{\alpha(x)} - \frac{\beta(y)}{\alpha(y)}\right)^2 \alpha(x)\alpha(y)(\epsilon_x, \epsilon_y) \\
= \sum_{x,y}\frac{\beta(x)^2\alpha(y)}{\alpha(x)}(\epsilon_x, \epsilon_y) - \sum_{x,y}\beta(x)\beta(y)(\epsilon_x, \epsilon_y) \\
= \sum_{x}\frac{\beta(x)^2}{\alpha(x)}(\epsilon_x, \alpha) - (\beta, \beta).
\end{aligned}
$$

Replacing β by $\alpha - \beta$ in (8.2) and then adding that to (8.2) again gives

$$
\begin{aligned}
(8.3)\quad 0 \geq \sum_{x}\frac{\beta(x)^2 + (\alpha(x) - \beta(x))^2}{\alpha(x)}(\epsilon_x, \alpha) - (\beta, \beta) - (\alpha - \beta, \alpha - \beta) \\
= \sum_{x}\alpha(x)(\epsilon_x, \alpha) - 2\sum_{x}\frac{\beta(x)(\alpha(x) - \beta(x))}{\alpha(x)}(\epsilon_x, \alpha) - (\beta, \beta) \\
- (\alpha - \beta, \alpha - \beta) \geq (\alpha, \alpha) - (\beta, \beta) - (\alpha - \beta, \alpha - \beta) = 2(\alpha - \beta, \beta)
\end{aligned}
$$

because $(\epsilon_x, \alpha) \leq 0$ for all $x \in Q_0$.

Suppose that $(\beta, \alpha - \beta) = 0$. Then we must have equality in (8.2), so β is proportional to α. We also have equality in (8.3), so $(\epsilon_x, \alpha) = 0$ for all $x \in Q_0$. □

Define $K \subseteq \mathbb{N}^{Q_0}$ as the set of all nonzero dimension vectors α such that the support of α is connected and $(\alpha, \epsilon_x) \leq 0$ for all $x \in Q_0$.

Lemma 8.3.3. *If $\alpha \in K$, then there exist infinitely many indecomposable representation of dimension α.*

Proof. Without loss of generality we may assume that the support of α is Q_0, i.e., $\alpha(x) > 0$ for all $x \in Q_0$. So Q is a connected quiver.

Suppose that $\beta \in \mathrm{sub}(\alpha)$. We can view $\mathrm{Rep}_\beta(Q) \oplus \mathrm{Rep}_{\alpha-\beta}(Q)$ as a subspace of $\mathrm{Rep}_\alpha(Q)$. Consider the morphism

$$\psi_\beta : \mathrm{GL}_\alpha \times \mathrm{Rep}_\beta(Q) \oplus \mathrm{Rep}_{\alpha-\beta}(Q) \to \mathrm{Rep}_\alpha(Q)$$

defined by $(A, V, W) \mapsto A \cdot (V \oplus W)$. Let Z_β be the Zariski closure of the image of ψ. The tangent space of GL_α at the identity 1_α is naturally isomorphic to $\bigoplus_{x \in Q_0} \mathrm{End}(\mathbb{C}^{\alpha(x)})$ and the tangent spaces of $\mathrm{Rep}_\alpha(Q)$ can be identified with $\operatorname{Rep}_\alpha(Q)$ itself. The tangent map $(d\psi_\beta)_{(1,V,W)}$ at $(1,V,W)$ is given by

$$\Big((A(x), x \in Q_0), V', W'\Big) \\ \mapsto \left(\begin{pmatrix} V'(a) & 0 \\ 0 & W'(a) \end{pmatrix} + A(ha)\begin{pmatrix} V(a) & 0 \\ 0 & W(a) \end{pmatrix} \right. \\ \left. - \begin{pmatrix} V(a) & 0 \\ 0 & W(a) \end{pmatrix} A(ta),\ a \in Q_1\right).$$

If the $(A(x), x \in Q_0)$ has the block form

$$\begin{pmatrix} B & 0 \\ 0 & C \end{pmatrix},$$

then the image lies in $\operatorname{Rep}_\beta(Q) \oplus \operatorname{Rep}_{\alpha-\beta}(Q)$. So the dimension of the image of the tangent map is at most

$$\dim \operatorname{Rep}_\beta(Q) + \dim \operatorname{Rep}_{\alpha-\beta}(Q) + \sum_{x \in Q_0} (\alpha(x)^2 - \beta(x)^2 - (\alpha(x) - \beta(x))^2) \\ = -\langle \beta, \beta\rangle - \langle \alpha-\beta, \alpha-\beta\rangle + \langle \alpha, \alpha\rangle + \dim \operatorname{Rep}_\alpha(Q) = \dim \operatorname{Rep}_\alpha(Q) + (\beta, \alpha-\beta).$$

So the tangent map induces a linear map

$$\bigoplus_{x \in Q_0} \operatorname{End}(\mathbb{C}^{\alpha(x)})/(\operatorname{End}(\mathbb{C}^{\beta(x)}) \oplus \operatorname{End}(\mathbb{C}^{\alpha(x)-\beta(x)})) \\ \to \operatorname{Rep}_\alpha(Q)/(\operatorname{Rep}_\beta(Q) \oplus \operatorname{Rep}_{\alpha-\beta}(Q)).$$

The codimension of the image is at least

$$\dim \operatorname{Rep}_\alpha(Q) - \dim \operatorname{Rep}_\beta(Q) - \dim \operatorname{Rep}_{\alpha-\beta}(Q) \\ - \sum_{x \in Q_0} (\alpha(x)^2 - \beta(x)^2 - (\alpha-\beta)^2(x)) \\ = -\langle \alpha, \alpha\rangle + \langle \beta, \beta\rangle + \langle \alpha-\beta, \alpha-\beta\rangle = -(\beta, \alpha-\beta).$$

We have $(\beta, \alpha - \beta) \geq 0$ for all $\beta \in \operatorname{sub}(\alpha)$ by Lemma 8.3.2.

Suppose that $(\beta, \alpha-\beta) = 0$ for some $\beta \in \operatorname{sub}(\alpha)$. Then β is proportional to α and $(\alpha, \epsilon_x) = 0$ for all $x \in Q_0$ by Lemma 8.3.2. In particular we have that $(\alpha, \alpha) = 0$. If γ is any dimension vector, then there exist a positive integer m such that $m\alpha - \gamma$ is a dimension vector. So we have

$$-m(\gamma, \gamma) = (\gamma, m\alpha - \gamma) \leq 0.$$

This shows that $(\gamma, \gamma) \geq 0$ for all dimension vectors. So Q must be a extended Dynkin Quiver. In this case we already know that there exists an indecomposable representation of dimension α.

Suppose that $(\beta, \alpha-\beta) > 0$ for all $\beta \in \operatorname{sub}(\alpha)$. Then Z_β has codimension ≥ 1 for all $\beta \in \operatorname{sub}(\alpha)$. Any representation in $\operatorname{Rep}_\alpha(Q) \setminus \bigcup_{\beta \in \operatorname{sub}(\alpha)} Z_\beta$ is indecomposable. $\square$

Define

$$\Phi_{\mathrm{im}}^+ = \mathcal{W}K, \quad \Phi_{\mathrm{im}}^- = -\Phi_{\mathrm{im}}^+, \quad \Phi_{\mathrm{im}} = \Phi_{\mathrm{im}}^+ \cup \Phi_{\mathrm{im}}^-.$$

and finally $\Phi = \Phi_{\mathrm{re}} \cup \Phi_{\mathrm{im}}$. Note that $(\alpha, \alpha) = (\epsilon_x, \epsilon_x) = 2$ for all $\alpha \in \Phi_{\mathrm{re}}$. For $\alpha \in K$ we have $(\alpha, \alpha) \leq 0$, so $(\alpha, \alpha) \leq 0$ for all $\alpha \in \Phi_{\mathrm{im}}$. This shows that $\Phi_{\mathrm{re}} \cap \Phi_{\mathrm{im}} = \emptyset$.

Lemma 8.3.4. *We have $\Phi_{\mathrm{im}}^+ \subseteq \mathbb{N}^{Q_0}$. If $\alpha \in \Phi_{\mathrm{im}}^+$ is indivisible, then there exists an indecomposable representation of dimension α.*

Proof. Suppose that $\alpha \in K$, $\beta = w(\alpha) \in \Phi_{\mathrm{im}}$ and $w = \sigma_{x_r} \cdots \sigma_{x_1}$. Assume that α is indivisible. There exist an indecomposable representation of dimension α. By induction on r we prove that $\beta \in \mathbb{N}^{Q_0}$ and that there exists an indecomposable representation of dimension β. The case $r = 0$ is clear. Assume that have already proven that there exists an indecomposable representation of dimension $\gamma = \sigma_{x_{r-1}} \cdots \sigma_{x_1}(\alpha)$. If $\beta = \sigma_{x_r}(\gamma) \notin \mathbb{N}^{Q_0}$, then $\gamma = \epsilon_{x_r}$ by Lemma 8.2.1. But then $\beta \in \Phi_{\mathrm{re}} \cap \Phi_{\mathrm{im}} = \emptyset$. Contradiction. Therefore, $\beta \in \mathbb{N}^{Q_0}$ and there exists an indecomposable representation of dimension β. If α is not indivisible, then we can write $\alpha = s\alpha'$ and we have that $w(\alpha') \in \mathbb{N}^{Q_0}$. So it follows that $\beta = w(\alpha) = w(s\alpha') = sw(\alpha') \in \mathbb{N}^{Q_0}$. $\square$

Again, the lemma above is true, even if α is not indivisible.

We define

$$\Phi_+ = \Phi \cap \mathbb{N}^{Q_0} = \Phi_{\mathrm{re}}^+ \cup \Phi_{\mathrm{im}}^+$$

and $\Phi_- = -\Phi_+$. Then $\Phi = \Phi^+ \cup \Phi^-$.

Theorem 8.3.5. *If there exists an indecomposable representation of dimension α, then $\alpha \in \Phi_+$.*

Proof. We prove this statement by induction on $|\alpha| = \sum_{x \in Q_0} \alpha(x)$. The case $|\alpha| = 1$ is clear.

Suppose now that $|\alpha| > 1$ and that there exists an indecomposable representation of dimension α. Assume that $(\alpha, \epsilon_x) > 0$ for some $x \in Q_0$. Then

$$|s_x\alpha| = |\alpha| - (\alpha, \epsilon_x) < |\alpha|.$$

There exists an indecomposable representation of dimension $s_x\alpha$ by Lemma 8.2.1. Hence, $s_x\alpha \in \Phi_+$. Suppose that $(\alpha, \epsilon_x) \leq 0$ for all $x \in Q_0$. The support of α is connected, otherwise every representation of dimension α decomposes. So $\alpha \in K$ and there exists an indecomposable representation of dimension α by Lemma 8.3.3. $\square$

We have proven a weak version of Kac's theorem:

Corollary 8.3.6. *Suppose that α is indivisible. Then there exists an indecomposable representation of dimension α if and only if $\alpha \in \Phi^+$.*

The condition that α is indivisible is not necessary. But to prove this, we have to work over finite fields.

8.4. Quiver Representations over Finite Fields

In this section, we will sketch a proof of Kac's theorem, using the representation theory of quivers over finite fields.

A representation V of $Q = (Q_0, Q_1, h, t)$ over the field $\mathbb{F}$ is the set of finite dimensional $\mathbb{F}$-vector spaces

$$\{V(x) \mid x \in Q_0\}$$

together with a set of $\mathbb{F}$-linear maps

$$\{V(a) : V(ta) \to V(ha) \mid a \in Q_1\}.$$

The notions of morphisms and direct sums are similar to the case where we work over $\mathbb{C}$. The Krull-Remak-Schmidt theorem is true for representations over the field $\mathbb{F}$ (see Exercises). Let $\operatorname{Rep}_\alpha(Q, \mathbb{F}) = \prod_{a \in Q_1} \operatorname{Mat}_{ha,ta}(\mathbb{F})$ and $\operatorname{GL}_\alpha(\mathbb{F}) = \prod_{x \in Q_0} \operatorname{GL}_{\alpha(x)}(\mathbb{F})$. Then $\operatorname{GL}_\alpha(\mathbb{F})$ acts on $\operatorname{Rep}_\alpha(Q, \mathbb{F})$ by base change, and the orbits are exactly the isomorphism classes of α-dimensional representations over the field $\mathbb{F}$.

Suppose that V is a representation of Q over the field $\mathbb{F}$, and $\mathbb{F}'$ is a field extension of $\mathbb{F}$. We define a representation $\mathbb{F}' \otimes_{\mathbb{F}} V$ of Q over $\mathbb{F}'$ by

$$(\mathbb{F}' \otimes_{\mathbb{F}} V)(x) := \mathbb{F}' \otimes_{\mathbb{F}} V(x)$$

for all $x \in Q_0$ and

$$(\mathbb{F}' \otimes_{\mathbb{F}} V)(a) = 1 \otimes_{\mathbb{F}} V(a) : \mathbb{F}' \otimes_{\mathbb{F}} V(ta) \to \mathbb{F}' \otimes_{\mathbb{F}} V(ha).$$

Lemma 8.4.1. *If $\mathbb{F}'$ is a finite field extension of $\mathbb{F}$ and $\mathbb{F}' \otimes_{\mathbb{F}} V$ and $\mathbb{F}' \otimes_{\mathbb{F}} W$ are isomorphic, then V and W are isomorphic.*

Proof. Suppose that V, W are α-dimensional representations of Q, and let d be the degree of the field extension. We may view $\mathbb{F}' \otimes_{\mathbb{F}} V$ and $\mathbb{F}' \otimes_{\mathbb{F}} W$ as representations of Q over $\mathbb{F}$. As representations over $\mathbb{F}$, they are isomorphic to V^d and W^d, respectively. An isomorphism $\phi : \mathbb{F}' \otimes_{\mathbb{F}} V \to F' \otimes_{\mathbb{F}} W$ over $\mathbb{F}'$ is $\mathbb{F}$-linear, so V^d and W^d are isomorphic representations over $\mathbb{F}$. By the Krull-Remak-Schmidt theorem, we have that V is isomorphic to W. □

We say that V is **absolutely indecomposable** if $\overline{\mathbb{F}} \otimes_{\mathbb{F}} V$ is indecomposable as a representation over the field $\overline{\mathbb{F}}$, the algebraic closure of $\mathbb{F}$.

Let $\mathbb{F}_q$ be the (unique) field with q elements. The map $x \mapsto x^q$ defines an **Frobenius automorphism** of $\mathbb{F}_q$. We will denote it by Φ.

The proof of Kac's theorem depends on counting isomorphism classes of (absolutely) indecomposable representations of a given dimension vector and we want to show that this does not depend on the orientation of Q. This counting problem can be reduced to counting the number of isomorphism classes of a given dimension. To show that this does not depend on the orientation of the quiver, we need the following three lemmas.

Lemma 8.4.2. *Suppose that V is an n-dimensional $\mathbb{F}_q$-vector space, and G is a finite group acting on V by linear transformations and denote the set of G-orbits on V by V/G. Then we have $|V/G| = |V^\star/G|$.*

Proof. The cardinality of the set

$$Y = \{(g, v) \in G \times V \mid g \cdot v = v\}$$

is

$$\sum_{v \in V} |G_v| = |G| \sum_{v \in V} \frac{1}{|G \cdot v|} = |G| \cdot |V/G|$$

because if we sum $1/|G \cdot w|$ for $w \in G \cdot v$, then we get 1.

Consider the set

$$X = \{(g, v, f) \in G \times V \times V^\star \mid f(g \cdot v) = f(v)\}.$$

We have

$$\begin{aligned}
|X| &= \sum_{(g,v) \in Y} |\{f \in V^\star \mid f(v) = f(g \cdot v)\}| \\
&\quad + \sum_{(g,v) \notin Y} |\{f \in V^\star \mid f(v) = f(g \cdot v\}| \\
&= |G| \cdot |V/G| \cdot q^n + |G|(q^n - |V/G|) \cdot q^{n-1} \\
&= |G|(|V/G|(q^n - q^{n-1}) + q^{2n-1}).
\end{aligned}$$

Since the role of V and $V^\star$ in X is symmetric, we also have

$$|X| = |G|(|V^\star/G|(q^n - q^{n-1}) + q^{2n-1}).$$

It follows that $|V/G| = |V^\star/G|$. □

Lemma 8.4.3. *If G acts linearly on the $\mathbb{F}_q$-vector spaces V and W, then $|(V \oplus W)/G| = |V^\star \oplus W)/G|$.*

Proof. We have

$$|(V \oplus W)/G| = \frac{1}{|G|} \sum_{w \in W} \sum_{v \in V} \frac{1}{|G_v \cap G_w|} = \sum_{w \in W} |V/G_w|.$$

Similarly,

$$|(V^\star \oplus W)/G| = \sum_{w \in W} |V^\star/G_w| = \sum_{w\in W} |V/G_w| = |(V \oplus W)/G|$$

by Lemma 8.4.2. □

Lemma 8.4.4. *The number of α-dimensional representations (up to isomorphism) of Q over $\mathbb{F}_q$ does not depend on the orientation of Q'.*

Proof. The number of α-dimensional representations of Q over $\mathbb{F}_q$ is the number of $\mathrm{GL}_\alpha(\mathbb{F}_q)$ orbits in $\mathrm{Rep}_\alpha(Q, \mathbb{F}_q)$. Reversing the arrow α means that we replace the summand $\mathrm{Mat}_{ha,ta}(\mathbb{F}_q)$ by the dual representation. Now the lemma follows from Lemma 8.4.3. □

Lemma 8.4.5. *The number of indecomposable representations of dimension α does not depend on the orientation of Q.*

Proof. We will use generating functions to prove this. Let $t_x, x \in Q_0$ be formal variables. For a dimension vector α we use the notation $t^\alpha = \prod_{x\in Q_0} t_x^\alpha$. Let $r_\alpha(q)$ be the number of α-dimensional representations over $\mathbb{F}_q$. Define the generating function

$$R = \sum_{\alpha \in \mathbb{N}^{Q_0}} r_\alpha(q) t^\alpha.$$

Let $i_\alpha(q)$ be the number of indecomposable representations of dimension α. We have

$$R = \prod_{\alpha\in\mathbb{N}^{Q_0}} \frac{1}{(1-t^\alpha)^{i_\alpha(q)}}.$$

It follows that the numbers $i_\alpha(q)$, $\alpha \in \mathbb{N}^{Q_0}$ are completely determined by the numbers $r_\alpha(q)$, $\alpha \in \mathbb{N}^{Q_0}$ and the latter do not depend on the orientation. □

Lemma 8.4.6. *Suppose that V is an indecomposable representation over $\mathbb{F}_q$ and d is a positive integer. If W is an indecomposable summand of $\mathbb{F}_{q^d} \otimes_{\mathbb{F}} V$, then*

$$\mathbb{F}_{q^d} \otimes_{\mathbb{F}} V = W \oplus \Phi(W) \oplus \cdots \oplus \Phi^{e-1}(W),$$

where e is the smallest positive integer such that W and $\Phi^e(W)$ are isomorphic.

Proof. Let W be an indecomposable summand of $V' = \mathbb{F}_{q^d} \otimes_{\mathbb{F}} V$, and let $Z = W \oplus \Phi(W) \oplus \cdots \oplus \Phi^{e-1}(W)$. Note that e divides d, the order of Φ.

If U is a representation of Q over $\mathbb{F}_{q^d}$, we may view it as a representation over $\mathbb{F}_q$. Then $\mathbb{F}_{q^d} \otimes_{\mathbb{F}_q} U$ is again a representation over $\mathbb{F}_q$. We have an isomorphism of $\mathbb{F}_{q^d}$-spaces

$$\mathbb{F}_{q^d} \otimes_{\mathbb{F}_q} U(x) \to U(x)^d$$

defined by

$$\lambda \otimes v \mapsto \lambda \begin{pmatrix} v \\ \Phi(v) \\ \vdots \\ \Phi^{d-1}(v) \end{pmatrix}.$$

For every arrow a we have a commutative diagram

$$\begin{array}{ccc} \mathbb{F}_{q^d} \otimes_{\mathbb{F}_q} U(ta) & \longrightarrow & U(ta)^d \\ {\scriptstyle 1\otimes U(a)} \downarrow & & \downarrow {\scriptstyle \widetilde{U}(a)} \\ \mathbb{F}_{q^d} \otimes_{\mathbb{F}_q} U(ha) & \longrightarrow & U(ha)^d, \end{array}$$

where

$$\widetilde{U}(a) = \begin{pmatrix} U(a) & & & \\ & \Phi(U(a)) & & \\ & & \ddots & \\ & & & \Phi^{d-1}(U(a)) \end{pmatrix}.$$

So we have

$$\widetilde{U} = U \oplus \Phi(U) \oplus \cdots \oplus \Phi^{d-1}(U) = Z^{d/e}.$$

Since W is a direct summand of V', we get that $\widetilde{W}$ is a direct summand of $\widetilde{V'} = (V')^d$, so $\widetilde{W} \cong (V')^s$ for some s. It follows that $Z^{d/e} = (V')^s$ and since W has no multiple summands, we have $V' = Z^t$ for some positive integer t. We claim that $t = 1$.

The endomorphism ring $\operatorname{End}_Q(V)$ has a maximal ideal $\mathfrak{m}$ which is nilpotent, and $\operatorname{End}_Q(V)/\mathfrak{m}$ is a skew-field (see Exercise 8.4.2). It is in fact a field, because every finite skew-field is a field. We have $\operatorname{End}_Q(V') = \operatorname{End}_Q(V) \otimes_{\mathbb{F}_q} \mathbb{F}_{q^d}$. The ideal $\mathfrak{m}' = \mathfrak{m} \otimes \mathbb{F}_{q^d}$ is nilpotent and the quotient ring $R := \operatorname{End}_Q(V')/\mathfrak{m}'$ is isomorphic to $\operatorname{End}_Q(V)/\mathfrak{m} \otimes_{\mathbb{F}_q} \mathbb{F}_{q^d}$. Since $\mathbb{F}_{q^d}/\mathbb{F}_q$ is separable, the ring R does not have any nilpotents. So all nilpotent elements in $\operatorname{End}_Q(V')$ are contained in $\mathfrak{m}'$. If $t > 1$, then $\operatorname{End}_Q(V') = \operatorname{End}_Q(Z) \otimes_{\mathbb{F}_{q^d}} \operatorname{Mat}_{t,t}(\mathbb{F}_{q^d})$. In the ring $\operatorname{End}_Q(V')$ the ideal generated by all nilpotent elements is the whole ring. This implies that $\mathfrak{m}' = \operatorname{End}_Q(V')$. Contradiction. □

Lemma 8.4.7. *The number of absolutely indecomposable representations of dimension α does not depend on the orientation of Q.*

Proof. Let $a_\alpha(q)$ be the number of absolutely indecomposable representations over $\mathbb{F}_q$. The number of absolutely indecomposable representations of dimension α over $\mathbb{F}_{q^k}$ that cannot be defined over any proper subfield of $\mathbb{F}_{q^k}$

containing $\mathbb{F}_q$ is

$$b_{k,\alpha}(q) = \sum_{l|k} a_\alpha(q^l)\mu(k/l),$$

where μ is the Möbius function. We have

$$i_\alpha(q) = \sum_{k|\alpha} b_{k,\alpha/k}(q)/k = \sum_{l|k|\alpha} a_{\alpha/k}(q^l)\mu(k/l)/k.$$

By induction on $|\alpha|$ we see that $a_\alpha(q)$ is determined by the numbers $i_{k,\beta}(q)$. □

Proposition 8.4.8. *For every algebraically closed field $\mathbb{F}$, the subset of $\operatorname{Rep}_\alpha(Q,\mathbb{F})$ of all absolutely indecomposable representations is denoted by $A_\alpha(Q,\mathbb{F})$.*

(1) *The subset of $A_\alpha(Q,\mathbb{F})$ is a constructible set for any algebraically closed field $\mathbb{F}$.*

(2) *The set $A_\alpha(Q,\mathbb{C})$ contains k elements if and only if $A_\alpha(Q,\overline{\mathbb{F}}_p)$ contains k elements for infinitely many primes p.*

(3) *The set $A_\alpha(Q,\mathbb{C})$ is infinite if and only if $A_\alpha(Q,\overline{\mathbb{F}}_p)$ is infinite for infinitely many primes p.*

Lemma 8.4.9. *If $\alpha \neq \epsilon_x$ and $a_\alpha(q) \neq 0$, then $a_\alpha(q) = a_{\sigma(\alpha)}(q)$.*

Proof. The proof is the same as the proof of Lemma 8.2.1, except using Lemma 8.4.7 instead of Corollary 8.1.4. □

Theorem 8.4.10. *Suppose that $\alpha \in \mathbb{N}^{Q_0}$ and $\beta = w(\alpha)$ for some $w \in \mathcal{W}$. Then we have $a_\beta(q) = a_\alpha(q)$ if $\beta \in \mathbb{N}^{Q_0}$ or $a_{-\beta}(q) = a_\alpha(q)$ if $-\beta \in \mathbb{N}^{Q_0}$.*

Proof. The proof goes like the proof of Theorem 8.2.2, except for using Lemma 8.4.9 instead of Lemma 8.2.1. □

Theorem 8.4.11 (Kac).

(1) *There exists an indecomposable representation of dimension α if and only if $\alpha \in \Phi^+$.*

(2) *If $\alpha \in \Phi^+_{\mathrm{re}}$, then there exists exactly* 1 *indecomposable representation of dimension α.*

(3) *If $\alpha \in \Phi^+_{\mathrm{im}}$, then there are infinitely many indecomposable representations of dimension α.*

Proof. (2) Suppose that $\alpha \in \Phi^+_{\mathrm{re}}$. We can write $\alpha = w(\epsilon_x)$. Since $I_{\epsilon_x}(Q,\overline{\mathbb{F}}_p)$ contains 1 element, it follows that $I_\alpha(Q,\overline{\mathbb{F}}_p)$ contains 1 element for infinitely many primes by Proposition 8.4.8. So there is a unique indecomposable representation of dimension α over $\mathbb{C}$ by Proposition 8.4.8.

(3) Suppose that $\alpha \in \Phi^+_{\rm im}$. Then $\alpha = w(\beta)$ for some $\beta \in K$. The set $A_\beta(Q, \mathbb{C})$ is infinite by Lemma 8.3.3 and Theorem 8.4.10. So $A_\beta(Q, \overline{\mathbb{F}}_p)$ is inifinite for infinitely many primes p. Therefore $A_\alpha(Q, \overline{\mathbb{F}}_p)$ is infinite by Lemma 8.3.3. Hence $A_\alpha(Q, \mathbb{C})$ is infinite by Proposition 8.4.8.

(2) and (3) imply one direction of (1) and the other direction is Theorem 8.3.5. □

Exercises.

Exercise 8.4.1. *Suppose that $p(t) \in \mathbb{F}[t]$ is an irreducible polynomial, and V is a representation of a quiver Q over a field $\mathbb{F}$, and $\phi : V \to V$ is an endomorphism. Define*

$$V_{p(t)}(x) = \{v \in V(x) \mid p(\phi)^N v = 0 \textit{ for some positive integer } N\}.$$

Show that $V_{p(t)}$ is a subrepresentation, i.e., $V(a)V_{p(t)}(ta) \subseteq V_{p(t)}(ha)$ for all arrows $a \in Q_1$. Also show that V is the direct sum of all $V_{p(t)}$, where $p(t)$ runs over all irreducible polynomials in $\mathbb{F}[t]$.

Exercise 8.4.2. *Suppose that V is an indecomposable representation of V over $\mathbb{F}$.*

(1) *Show that for every $\phi \in \operatorname{End}_Q(V)$ there exists an irreducible polynomial $p(t)$ and a positive integer N such that $p(\phi)^N = 0$. (Use Exercise 8.4.1).*
(2) *Show that every $\phi \in \operatorname{End}_Q(V)$ is either a unit or nilpotent (Fitting Lemma).*
(3) *Suppose that $\phi_1, \phi_2 \in \operatorname{End}_Q(V)$ are nilpotent. Show that $\phi_1 + \phi_2$ is nilpotent. (Hint: Suppose not, then*
$$\phi_1\psi + \phi_2\psi = (\phi_1 + \phi_2)\psi = 1$$
for some $\psi \in \operatorname{End}_Q(V)$. Deduce that $\phi_1\psi, \phi_2\psi = 1 - \phi_1\psi$ are nilpotent and find a contradiction.)
(4) *Let $\mathfrak{m}$ the the ideal of all nilpotent elements in $\operatorname{End}_Q(V)$. Show that $\mathfrak{m}$ is an ideal.*
(5) *Show that $\operatorname{End}_Q(V)/\mathfrak{m}$ is a skew-field.*

Exercise 8.4.3. *Verify that the Krull-Remak-Schmidt theorem (Theorem 1.7.4) works over an arbitrary field. (Use Exercise 8.4.2.)*

8.5. Bibliographical Remarks

The original proof of Kac's theorem in [**53**] used characteristic $p > 0$ and the Weil conjectures. Here we follow the ideas of Crawley-Boevey and Van den Bergh [**20**] who found a characteristic zero proof that works for indivisible dimension vectors. The proof is based on deformed preprojective algebras which were introduced by Crawley-Boevey and Holland [**19**]. We then add some positive characteristic calculations to complete the proof of the general case.

It is still an open problem to find a proof of Kac's theorem that does not use the positive characteristic.

Kac later in his Lecture Notes [**55**] proposed a series of more precise conjectures on the structure of indecomposable modules for arbitrary quiver. He introduced a certain family of polynomials that would count the indecomposable representations of certain dimension vector over finite fields. One of the conjectures claimed that these polynomials have nonnegative coefficients. Some of Kac's conjectures were recently proved by Hausel, Letellier and Rodrigues-Villegas [**48**]. They linked the Kac polynomials to the intersection cohomology of Nakajima quiver varieties.

Chapter 9

Geometric Invariant Theory

The techniques based on Auslander-Reiten theory work very well for the algebras of finite and tame representation type. For wild algebras, the isomorphism classes of representations appear in families with many parameters. To deal with this, we will use techniques from Geometric Invariant Theory to construct moduli spaces for representations of quivers and algebras.

In this chapter we develop the basic notions of Geometric Invariant Theory. We start with actions of algebraic groups on affine varieties and their invariant rings. Linearly reductive groups are algebraic groups with a Reynolds operator. Invariant Theory for linearly reductive groups is particularly nice, because invariant rings are finitely generated. The quotient space of an action of a linearly reductive group on an affine variety is defined as the spectrum of the invariant ring. This quotient has nice geometric properties. In Geometric Invariant Theory, we consider linearly reductive group actions on projective varieties. Quotients can be constructed using the notion of stability. The Hilbert-Mumford criterion can often be used to find more explicit descriptions of stable points, without having to know the invariant ring explicitly.

Now Geometric Invariant Theory can be applied to the action of GL_α on $\mathrm{Rep}_\alpha(Q)$. To get an interesting quotient, we restrict ourselves to a subgroup of GL_α which is the kernel of a multiplicative character $\chi : \mathrm{GL}_\alpha \to \mathbb{C}^\star$. We prove the King's theorem describing the semi-stable points with respect to weight χ in terms of existence of some subrepresentations.

In Section 9.4 we prove the Sato-Kimura lemma that we will use in the next chapter.

9.1. Algebraic Group Actions

In this section we will give a brief introduction to algebraic groups and their actions. We will consider affine algebraic groups that are affine varieties with a group structure. These groups are the same as linear algebraic groups which are closed subgroups of the general linear groups.

Suppose that G is a group which also has the structure of an affine variety. Let $m : G \times G \to G$ be the multiplication map $(g, h) \mapsto gh$ and let $\iota : G \to G$ be the map that sents an element $g \in G$ to its inverse $g^{-1} \in G$.

Definition 9.1.1. The group G is called an **affine algebraic group** if m and ι are morphisms of affine varieties.

Example 9.1.2. Consider the additive group $\mathbb{G}_{\mathrm{a}} = (\mathbb{C}, +)$. The group multiplication $m : \mathbb{G}_{\mathrm{a}} \times \mathbb{G}_{\mathrm{a}} \to \mathbb{G}_{\mathrm{a}}$ is addition and is given by $m(s,t) = s+t$. The inverse map $\iota : \mathbb{G}_{\mathrm{a}} \to \mathbb{G}_{\mathrm{a}}$ is given by $\iota(s) = -s$. These maps are clearly morphisms of affine varieties, so $\mathbb{G}_{\mathrm{a}}$ is an affine algebraic group.

Example 9.1.3. Suppose that V is an n-dimensional $\mathbb{C}$-vector space. Consider the group $\mathrm{GL}(V) \subset \mathrm{End}(V)$ of invertible endomorphisms of V. If we choose a basis of V, then we can identify $\mathrm{End}(V)$ with the space $\mathrm{Mat}_{n,n}$ of $n \times n$ matrices. The ring of polynomial functions on $\mathrm{End}(V)$ is the polynomial ring

$$\mathbb{C}[\mathrm{End}(V)] \cong \mathbb{C}[\mathrm{Mat}_{n,n}] = \mathbb{C}[x_{1,1}, x_{1,2}, \dots, x_{n,n}],$$

where $x_{i,j}$ is the coordinate function for the (i,j)-entry of the matrix. Let $\det \in \mathbb{C}[V]$ be the determinant function. The set $\mathrm{GL}(V)$ is defined as the open subset of $\mathrm{End}(V)$ where det is nonzero. So the ring of regular functions on $\mathrm{GL}(V)$ is equal to the localization

$$\mathbb{C}[\mathrm{GL}(V)] \cong \mathbb{C}[\mathrm{Mat}_{n,n}]_{\det} = \mathbb{C}[x_{1,1}, x_{1,2}, \dots, x_{n,n}, \det^{-1}].$$

Clearly, the matrix product $m : (A, B) \mapsto AB$ defines a morphism $\mathrm{GL}(V) \times \mathrm{GL}(V) \to \mathrm{GL}(V)$ because the entries of AB are quadratic polynomials in the entries of A and B. For a matrix A, let $\widetilde{A}$ be the cofactor matrix of A. The inverse map $\iota : \mathrm{GL}(V) \to \mathrm{GL}(V)$ is a morphism of affine varieties because the entries of $A^{-1} = \det(A)^{-1}\widetilde{A}$ can be expressed as polynomials in the entries of A and $\det(A)^{-1}$. We conclude that $\mathrm{GL}(V)$ is an algebraic group.

Example 9.1.4. Let $\mathbb{C}^\star = \mathbb{C} \setminus \{0\}$ be the set of multiplicative units. The multiplicative group $\mathbb{G}_{\mathrm{m}} = (\mathbb{C}^\star, \cdot)$ can be identified with GL_1 and is an affine algebraic group.

Example 9.1.5. Suppose $G \subseteq \mathrm{GL}(V)$ is a subgroup that is Zariski closed. Then the multiplication and inverse map of $\mathrm{GL}(V)$ restrict to G. It is clear that these restrictions are again morphisms, so G is an affine algebraic group as well.

Example 9.1.6. Let $\mathrm{O}_n \subset \mathrm{GL}_n$ be the set of $n \times n$ orthogonal matrices. This subset is defined by $AA^t = I_n$ where A^t is the transpose of A and I_n is the $n \times n$ identity matrix. The entries of this matrix equation give n^2 quadratic equations for O_n. This shows that O_n is a Zariski closed subset of GL_n. So O_n is an affine algebraic group.

Definition 9.1.7. A group homomorphism $\varphi : G \to H$ between two affine algebraic groups is a **homomorphism of algebraic groups** if it is also a morphism of affine varieties. Such a homomorphism is an **isomorphism of algebraic groups** if φ is an isomorphism of affine varieties.

The following proposition will be proven in the Exercise 9.1.6.

Proposition 9.1.8. *If G is an affine algebraic group, then there exists a finite dimensional vector space W such that G is isomorphic, as an affine algebraic group, to a Zariski closed subgroup of* $\mathrm{GL}(W)$.

Because of the proposition, affine algebraic groups are also called **linear algebraic groups**. Smooth elliptic curves are examples of projective varieties with a group structure. These are considered algebraic groups, but they are not affine. We will only consider linear/affine algebraic groups here.

Definition 9.1.9. Suppose that G is an affine algebraic group, X is an affine variety, and $\mu : G \times X \to X$ is a left action. The action is called **regular** if μ is a morphism of affine varieties. An affine variety X with a regular action of G is called an **affine G-variety**.

Example 9.1.10. Left multiplication in the group G defines a left action of G on itself. This is clear, because the action map $\mu : G \times G \to G$ is equal to the group multiplication. Since G is an algebraic group, μ is a morphism of affine varieties.

Another action of G on itself is given by

$$\mu(g, h) = hg^{-1} = m(g, \iota(h)).$$

It is again clear that μ is a morphism of affine varieties, because m and ι are. So this action is regular as well.

Finally, the conjugation action of G on itself is also regular.

Definition 9.1.11. Suppose that G is an affine algebraic group. A **rational representation of** G is a homomorphism $\rho : G \to \mathrm{GL}(V)$ of affine algebraic groups, where V is some finite dimensional $\mathbb{C}$-vector space. Sometimes we will just say that V is a rational representation when it is clear which homomorphism ρ we are referring to.

If $\rho : G \to GL(V)$ is a rational representation of G, then we can define a regular action $\mu : G \times V \to V$ by $\mu(g, v) = \rho(g)v$ for all $g \in G$. Conversely, we have the following lemma.

Lemma 9.1.12. *If $\mu : G \times V \to V$ is a regular action of G and $\mu(g, -) : V \to V$ is linear for all $g \in G$, then there exists a rational representation $\rho : G \to \mathrm{GL}(V)$ such that $\mu(g, v) = \rho(g)v$.*

Proof. Suppose that $\mu : G \times V \to V$ is a regular action, and that $\rho(g) := \mu(g, -) : V \to V$ is linear for all $g \in G$. Then $\rho : G \to \mathrm{GL}(V)$ is a group homomorphism. Let us choose a basis $e_1, \dots, e_n$ of V and let $f_1, \dots, f_n \in V^\star$ be a dual basis. For every i, j we have that $f_i(\rho(g)e_j) = f_i(\mu(g, e_j))$ is a polynomial function in $g \in G$. This shows that ρ is a morphism of affine algebraic varieties. So ρ is a representation. $\square$

If G acts regularly on X, then G also acts on the affine coordinate ring $\mathbb{C}[X]$ of X as follows. If $f \in \mathbb{C}[X]$ and $g \in G$, then $g \cdot f \in \mathbb{C}[X]$ is defined by

$$(g \cdot f)(x) = f(g^{-1}x) = f(\mu(\iota(g), x))$$

for all $x \in X$. It is clear that $g \cdot f$ lies again in $\mathbb{C}[X]$. To see that this defines an action on $\mathbb{C}[X]$, we verify

$$\begin{aligned}(h \cdot (g \cdot f))(x) &= (g \cdot f)(h^{-1}x) = f(g^{-1}(h^{-1}x)) \\ &= f((g^{-1}h^{-1})x) = f((hg)^{-1}x) = ((hg) \cdot f)(x)\end{aligned}$$

for all $g, h \in G$, $x \in X$ and $f \in \mathbb{C}[X]$.

The following lemma is often quite useful and will be used in Proposition 9.1.15.

Lemma 9.1.13. *Suppose that G is an affine algebraic group acting regularly on an affine variety X. Suppose that $W \subseteq \mathbb{C}[X]$ is a finite dimensional subspace. Let GW be the span of all $g \cdot w$ with $g \in G$ and $w \in W$. Then GW is finite dimensional as well, and the action of G on GW is a representation of G.*

Proof. Let $\mu : G \times X \to X$ be the action of G on X. This corresponds to an algebra homomorphism

$$\mu^\star : \mathbb{C}[X] \to \mathbb{C}[G] \otimes \mathbb{C}[X].$$

If $f \in \mathbb{C}[X]$, $g \in G$ and $x \in X$, then we have

$$(g \cdot f)(x) = f(g^{-1}x) = f(\mu(g^{-1}, x)) = (\mu^\star(f))(g^{-1}, x).$$

If we write $\mu^\star(f) = \sum_{i=1}^s h_i \otimes f_i$, then we have

$$(g \cdot f)(x) = \sum_{i=1}^{s} h_i(g^{-1}) f_i(x)$$

and

$$g \cdot f = \sum_{i=1}^{s} h_i(g^{-1}) f_i. \tag{9.1}$$

We see that if $\mu^\star(f) \subseteq A \otimes B$, then $g \cdot f \in B$ for all $g \in G$. Since $\mu^\star(W) \subseteq A \otimes B$, we have $GW \subseteq B$. This shows that $V := GW$ is finite dimensional. The action of G on V is linear because of (9.1). So there exists a group homomorphism $\rho : G \to \mathrm{GL}(V)$ such that $g \cdot f = \rho(g) f$ for all $g \in G$ and $f \in V$. Choose $v \in V$ and $h \in V^\star$. We can extend h to an element in $\mathbb{C}[G]^\star$. We get that

$$h(\rho(g)v) = h(g \cdot v) = h(\mu(g, v))$$

is a polynomial function on $g \in G$. This shows that $\rho : G \to \mathrm{GL}(V)$ is a morphism of affine varieties and a homorphism of affine algebraic groups. $\square$

Definition 9.1.14. If G is a group acting on sets X and Y, then a map $\varphi : X \to Y$ is called G-**equivariant** if $\varphi(g \cdot x) = g \cdot \varphi(x)$ for all $g \in G$ and all $x \in X$.

The following proposition allows us to reduce some problems for algebraic group actions on affine varieties to the case where the affine G-variety is a representation.

Proposition 9.1.15. *If X is an affine G-variety, then there exists a representation V and a closed G-equivariant closed immersion $\varphi : X \hookrightarrow V$.*

Proof. Suppose that G is an affine algebraic group acting regularly on an affine variety X. We can choose a finite dimensional subspace $W \subseteq \mathbb{C}[X]$ such that W generates $\mathbb{C}[X]$ as a $\mathbb{C}$-algebra. Let $V = (GW)^\star$, the dual space of GW. If we choose a basis $z_1, \dots, z_n$ of GW, then there is a unique $\mathbb{C}$-algebra homomorphim

$$\psi : \mathbb{C}[z_1, \dots, z_n] = \mathbb{C}[V] \twoheadrightarrow \mathbb{C}[X]$$

such that the restriction of φ to GW is the inclusion $GW \subseteq \mathbb{C}[X]$. Since W generates $\mathbb{C}[X]$, it is clear that ψ is surjective. This ring homomorphism corresponds to a closed immersion $\varphi : X \hookrightarrow V$. The homomorphism ψ is G-equivariant, so φ is G-equivariant as well. $\square$

If a group G acts on a set V, then $V^G = \{v \in V \mid \forall g \in G\, g \cdot v = v\}$ denotes the subset of G-invariant elements. A subset $W \subseteq V$ is called G-**stable** if $g \cdot W \subseteq W$ for all $g \in G$.

Suppose that G acts regularly on an affine variety X. Recall that G also acts on $\mathbb{C}[X]$ as follows: If $f \in \mathbb{C}[X]$ and $g \in G$, then $g \cdot f \in \mathbb{C}[X]$ is defined by

$$(g \cdot f)(v) := f(g^{-1} \cdot v)$$

for all $v \in V$.

Definition 9.1.16. The **invariant ring** of the G-action on $\mathbb{C}[X]$ is $\mathbb{C}[X]^G$. It is easy to verify that $\mathbb{C}[X]^G$ is a $\mathbb{C}$-algebra.

Example 9.1.17. Suppose that $V = \mathbb{C}^n$ and the symmetric group S_n acts on V by permuting the coordinates. Then the invariant ring is

$$\mathbb{C}[x_1, x_2, \ldots, x_n]^{S_n} = \mathbb{C}[e_1, e_2, \ldots, e_n],$$

where

$$e_k = \sum_{i_1 < i_2 < \cdots < i_k} x_{i_1} x_{i_2} \cdots x_{i_k}.$$

Exercises.

Exercise 9.1.1. *Show that the symplectic group* Sp_{2n} *is an affine algebraic group.*

Exercise 9.1.2. *Show that every finite group has the structure of an affine algebraic group.*

Exercise 9.1.3. *Suppose that G is an affine algebraic group acting regularly on an affine variety X, and let $x \in X$. Consider the orbit map $\varphi : G \to X$ defined by $\varphi(g) = g \cdot x$.*

1. *Let $G_x = \varphi^{-1}(x)$ be the stabilizer group. Show that G_x is again an affine algebraic group.*
2. *Show that the orbit $Gx = \{g \cdot x \mid g \in G\} \subseteq X$ is locally closed, i.e., Gx is an open subset of its closure $\overline{Gx}$. (You can use the fact that $\varphi(G) = Gx$ is a constructible set, and that it contains some open subset U of $\overline{Gx}$.)*

Exercise 9.1.4. *Suppose that G is an affine algebraic group. Prove that G is nonsingular as follows: The set of nonsingular points of G is open and dense. Then use the action of G on itself to show that all points of G are nonsingular.*

Exercise 9.1.5. *Suppose that G is an affine algebraic group and H is a subgroup of G. Show that the Zariski closure $\overline{H}$ is an affine algebraic group.*

Exercise 9.1.6. *Suppose that G is an affine algebraic group. Then G acts on itself by left multiplication. There exists a rational representation $\rho : G \to \mathrm{GL}(V)$ and a closed G-equivariant immersion $\varphi : G \hookrightarrow V$ by*

Proposition 9.1.15. *Show that ρ is a closed immersion. In particular, G is isomorphic to a Zariski closed subgroup of* $\mathrm{GL}(V)$*, so G is a linear algebraic group.*

Exercise 9.1.7. *Suppose that $\rho : G \to \mathrm{GL}(V)$ is a rational representation of G. If $A \in \mathrm{GL}(V)$, let $A^\star \in \mathrm{GL}(V^\star)$ be the dual map. Define $\lambda : G \to \mathrm{GL}(V^\star)$ by $\lambda(g) = \rho(g^{-1})^\star \in \mathrm{GL}(V^\star)$. Show that λ is a rational representation. This is the* **dual representation**.

9.2. Linearly Reductive Groups

In the theory of Lie groups, compact Lie groups are particularly nice. Compact Lie groups have a **Haar measure**, a volume form that is invariant under the left and right action of the group on itself. In the theory of algebraic groups reductive groups play the role of compact Lie groups. Linearly reductive groups have a Reynolds operator, which can be thought of as averaging over the group. But the Reynolds operator can be defined in a purely algebraic way. One of the main results is Hilbert's Finiteness Theorem (Theorem 9.2.6 and Corollary 9.2.8) which tells us that invariant rings for linearly reductive groups are finitely generated. Also, the category of representations of a linearly reductive groups is semi-simple which means that every representation is a direct sum if irreducible representations (Proposition 9.2.11).

Suppose that G is an affine algebraic group. Consider the action of G on itself on the left. Then G also acts on $\mathbb{C}[G]$ as follows. If $h \in G$ and $f \in \mathbb{C}[G]$, then we have

$$(h \cdot f)(g) = f(h^{-1}g)$$

for all $g \in G$.

Definition 9.2.1. Suppose that G is an affine algebraic group. A linear map

$$\mathcal{R} : \mathbb{C}[G] \to \mathbb{C}$$

is called a **Reynolds operator** if it satisfies the following 2 axioms:

(1) $\mathcal{R}(1) = 1$;

(2) $\mathcal{R}(h \cdot f) = \mathcal{R}(f)$ for all $h \in G$ and all $f \in \mathbb{C}[G]$.

An algebraic group G is called **linearly reductive** if there exists a Reynolds operator $\mathcal{R} : \mathbb{C}[G] \to \mathbb{C}$.

The definition of a **reductive** algebraic group is different. But if one works over $\mathbb{C}$ as we do, then reductive algebraic groups are the same as linearly reductive algebraic groups.

Example 9.2.2. Suppose that G is a finite group. Define an operator $\mathcal{R} : \mathbb{C}[G] \to \mathbb{C}$ by

$$\mathcal{R}(f) = \frac{1}{|G|} \sum_{g \in G} f(g)$$

We verify that $\mathcal{R}$ is a Reynolds operator. The first axiom is satisfied:

$$\mathcal{R}(1) = \frac{1}{|G|} \sum_{g \in G} 1 = \frac{|G|}{|G|} = 1.$$

Suppose that $f \in \mathbb{C}[G]$ and $h \in G$. We have

$$\mathcal{R}(h \cdot f) = \frac{1}{|G|} \sum_{g \in G} f(h^{-1}g) = \frac{1}{|G|} \sum_{g \in G} f(g) = \mathcal{R}(f),$$

so the second axiom is satisfied as well. So finite groups are reductive.

Example 9.2.3. Consider the subgroup $\mathrm{U}_n \subseteq \mathrm{GL}_n$ of unitary matrices. The group U_n is a compact Lie group, which has an invariant Haar measure $d\mu$. Define $\mathcal{R} : \mathbb{C}[\mathrm{GL}_n] \to \mathbb{C}$ by

$$\mathcal{R}(f) = \int_{\mathrm{U}_n} f \, d\mu.$$

We can normalize the Haar measure $d\mu$ such that $\mathcal{R}(1) = 1$.

Define a function $F : G \to \mathbb{C}$ by $F(g) = \mathcal{R}(g \cdot f)$. Let $m^\star : \mathbb{C}[G] \to \mathbb{C}[G] \otimes \mathbb{C}[G]$ be the ring homomorphism corresponding to the multiplication $m : G \times G \to G$. If we write

$$m^\star(f) = \sum_{i=1}^{r} h_i \otimes f_i,$$

then we have

$$g \cdot f = \sum_{i=1}^{r} h_i(g^{-1}) f_i$$

for all $g \in G$, and

$$F(g) = \mathcal{R}(g \cdot f) = \sum_{i=1}^{r} h_i(g^{-1}) \mathcal{R}(f_i).$$

This shows that $F \in \mathbb{C}[G]$. We have $F(g) = f$ for all $g \in \mathrm{U}_n$. But because U_n lies Zariski dense in GL_n (see Exercise 9.2.1), we have $F(g) = f$ for all $g \in \mathrm{GL}_n$. We conclude that $\mathcal{R}(g \cdot f) = \mathcal{R}(f)$ for all $g \in \mathrm{GL}_n$. Both axioms for the Reynolds operator are satisfied, so GL_n is reductive. A similar argument, using Exercise 9.2.3 shows that SL_n is reductive.

Suppose that G is a reductive algebraic group and that $\mu : G \times X \to X$ is a regular action. This corresponds to a group homomorphism

$$\mu^\star : \mathbb{C}[X] \to \mathbb{C}[G] \otimes \mathbb{C}[X].$$

Using the Reynolds operator, we can also devine an operator $\mathcal{R} : \mathbb{C}[X] \to \mathbb{C}[X]^G$ which can be thought of as averaging a function $f \in \mathbb{C}[X]$ over the group. Define $\mathcal{R}_X : \mathbb{C}[X] \to \mathbb{C}[X]$ as the composition $(\mathcal{R} \otimes 1) \circ \mu^\star$:

$$\begin{array}{ccc} \mathbb{C}[X] & \xrightarrow{\mu^\star} & \mathbb{C}[G] \otimes \mathbb{C}[X] \\ & \searrow^{\mathcal{R}_X} & \downarrow{\scriptstyle \mathcal{R}\otimes 1} \\ & & C[X]. \end{array}$$

Lemma 9.2.4. *The operator $\mathcal{R}_X : \mathbb{C}[X] \to \mathbb{C}[X]$ has the following properties:*

(1) $\mathcal{R}_X(\mathbb{C}[X]) \subseteq \mathbb{C}[X]^G$*;*

(2) *if* $f \in \mathbb{C}[X]^G$ *then* $\mathcal{R}_X(f) = f$*;*

(3) $\mathcal{R}_X$ *is an* $\mathbb{C}[X]^G$*-module homomorphism;*

(4) *if* $W \subseteq \mathbb{C}[X]$ *is* G*-stable, then we have* $\mathcal{R}_X(W) = W^G$*.*

Proof. Suppose that $f \in \mathbb{C}[X]^G$. We have $\mu^\star(f) = 1 \otimes f$. If $h \in \mathbb{C}[X]$, and we write $\mu^\star(h) = \sum_{i=1}^r u_i \otimes h_i$, then we have

$$\mu^\star(fh) = \mu^\star(f)\mu^\star(h) = \sum_{i=1}^r u_i \otimes fh_i.$$

It follows that

$$\mathcal{R}_X(fh) = \sum_{i=1}^r \mathcal{R}(u_i) fh_i = f \sum_{i=1}^r \mathcal{R}(u_i) h_i = f\mathcal{R}_X(h).$$

This proves (3). If $f \in \mathbb{C}[X]^G$, then we have $\mathcal{R}_X(f) = (\mathcal{R} \otimes 1) \circ \mu^\star(f) = \mathcal{R}(1)f = f$. This proves (2). To show (1), consider the commutative diagram

$$\begin{array}{ccc} G \times G \times X & \xrightarrow{m \times 1} & G \times X \\ {\scriptstyle 1\times\mu}\downarrow & & \downarrow{\scriptstyle \mu} \\ G \times X & \xrightarrow[\mu]{} & X. \end{array}$$

This corresponds to a commutative diagram of algebra homomorphisms:

$$\begin{array}{ccc}
\mathbb{C}[X] & \xrightarrow{\mu^\star} & \mathbb{C}[G] \otimes \mathbb{C}[X] \\
{\scriptstyle \mu^\star}\big\downarrow & & \big\downarrow{\scriptstyle m^\star \otimes 1} \\
\mathbb{C}[G] \otimes \mathbb{C}[X] & \xrightarrow[1\otimes\mu^\star]{} & \mathbb{C}[G] \otimes \mathbb{C}[G] \otimes \mathbb{C}[X] \\
{\scriptstyle \mathcal{R}\otimes 1}\big\downarrow & & \big\downarrow{\scriptstyle \mathcal{R}\otimes 1 \otimes 1} \\
\mathbb{C}[X] & \xrightarrow[\mu^\star]{} & \mathbb{C}[G] \otimes \mathbb{C}[X]
\end{array} .$$

Suppose that $W \subseteq \mathbb{C}[X]$ is G-stable. If $h \in W$, then we can write

$$\mu(h) = \sum_{i=1}^{r} u_i \otimes h_i,$$

where $u_1, \ldots, u_r \in \mathbb{C}[G]$ are linearly independent and $h_1, \ldots, h_r \in \mathbb{C}[X]$ are linearly independent. If $h_i \notin W$, then we can choose $g \in G$ such that $u_i(g) \neq 0$. Then we have

$$g^{-1} \cdot h = \sum_{i=1}^{r} u_i(g) h_i \notin W$$

which contradicts the assumption that W is G-stable. So $h_1, \ldots, h_r \in W$. From the definition it is clear that $\mathcal{R}_X(h) \in W$. This proves (4). $\square$

Consider a graded commutative $\mathbb{C}$-algebra $R = \bigoplus_{d=0}^{\infty} R_d$, where $R_0 = \mathbb{C}$ and $R_d R_e \subseteq R_{d+e}$ for all d, e. An element $f \in R$ is called **homogeneous** of degree d if $f \in R_d$ for some d, and we write $\deg(f) = d$ if f is nonzero. Consider the homogeneous maximal ideal $\mathfrak{m} = \bigoplus_{d=1}^{\infty} R_d$. The following proposition will be used in the proof of Hilbert's Finiteness Theorem.

Proposition 9.2.5. *Suppose that $f_1, \ldots, f_r \in R$ are homogeneous generators of $\mathfrak{m}$, then R is, as an algebra, generated by $f_1, \ldots, f_r$.*

Proof. Let $S = \mathbb{C}[f_1, \ldots, f_r] \subseteq R$. We will prove that $S = R$.

Suppose that $h \in R_d$. By induction on d we will prove that $h \in S$. If $d = 0$, then $h \in \mathbb{C} \subseteq S$, so it is clear. Suppose that $d > 0$. Then $h \in \mathfrak{m}$ so we can write $h = \sum_{i=1}^{r} u_i f_i$. Without loss of generality, we may assume that u_i is homogeneous of degree $d - \deg(f_i)$. By the induction hypothesis, $u_1, \ldots, u_r \in S$. It follows that $h \in S$. $\square$

Theorem 9.2.6 (Hilbert's Finiteness Theorem)**.** *If V is an n-dimensional representation of a linearly reductive group G, then $\mathbb{C}[V]^G$ is finitely generated.*

Proof. First, note that $\mathbb{C}[V]_d$ is G-stable for all d. So $\mathcal{R}_V$ preserves degree: $\mathcal{R}_V(\mathbb{C}[V]_d) \subseteq \mathbb{C}[V]_d$. Let $\mathfrak{m} \subseteq \mathbb{C}[V]^G$ be the ideal generated by all homogenous invariants of positive degree, and let $I = \mathfrak{m}\mathbb{C}[V]$ be the ideal in $\mathbb{C}[V]$ generated by $\mathfrak{m}$. Because the ring $\mathbb{C}[V]$ is noetherian, we can choose finitely many homogeneous invariants $f_1, \dots, f_r \in \mathfrak{m}$ that generate the ideal I. We claim that $f_1, \dots, f_r$ generate the ideal $\mathfrak{m}$. To see this, suppose that $h \in \mathbb{C}[V]^G$. We can write

$$h = \sum_{i=1}^r a_i f_i,$$

where $a_1, \dots, a_r \in \mathbb{C}[V]$. Applying the Reynolds operator yields:

$$h = \mathcal{R}_V(h) = \mathcal{R}_V\Big(\sum_{i=1}^r a_i f_i\Big) = \sum_{i=1}^r \mathcal{R}_V(a_i) f_i.$$

So h lies in the ideal in $\mathbb{C}[V]^G$ generated by $f_1, \dots, f_r$.

From Proposition 9.2.5 it follows that $\mathbb{C}[V]^G$ is generated as an algebra by $f_1, \dots, f_r$. □

Proposition 9.2.7. *Suppose that $\varphi : X \to Y$ is a G-equivariant morphism of affine G-varieties, and G is linearly reductive. Then we have the following commutative diagram:*

$$\begin{array}{ccc} \mathbb{C}[Y] & \xrightarrow{\mathcal{R}_Y} & \mathbb{C}[Y] \\ \varphi^\star \downarrow & & \downarrow \varphi^\star \\ \mathbb{C}[X] & \xrightarrow[\mathcal{R}_X]{} & \mathbb{C}[X]. \end{array} \tag{9.2}$$

In particular, $\varphi^\star(\mathbb{C}[Y]^G) = \mathbb{C}[X]^G$.

Proof. Consider the diagram

$$\begin{array}{ccccc} \mathbb{C}[Y] & \xrightarrow{\mu} & \mathbb{C}[G] \otimes \mathbb{C}[Y] & \xrightarrow{\mathcal{R}\otimes 1} & \mathbb{C}[Y] \\ \varphi^\star \downarrow & & \downarrow 1\otimes\varphi_\star & & \downarrow \varphi^\star \\ \mathbb{C}[X] & \xrightarrow[\mu]{} & \mathbb{C}[G] \otimes \mathbb{C}[X] & \xrightarrow[\mathcal{R}\otimes 1]{} & \mathbb{C}[X]. \end{array}$$

The square on the left commutes because φ is G-equivariant. It is clear that the square on the right commutes, and the composition of the two squares is (9.2).

It follows that

$$\varphi^\star(\mathbb{C}[Y]^G) = \varphi^\star(\mathcal{R}_Y(\mathbb{C}[Y])) = \mathcal{R}_X(\varphi^\star(\mathbb{C}[Y])) = \mathcal{R}_X(\mathbb{C}[X]) = \mathbb{C}[X]^G.$$

□

Corollary 9.2.8. *If G is a linearly reductive algebraic group acting regularly on an affine variety X. Then $\mathbb{C}[X]^G$ is finitely generated.*

Proof. By Proposition 9.1.15 there exists a closed immersion $\varphi : X \hookrightarrow V$ where V is a representation of G. By Theorem 9.2.6, $\mathbb{C}[V]^G$ is finitely generated. By Proposition 9.2.7, $\varphi^\star(\mathbb{C}[V]^G) = \mathbb{C}[X]^G$. This shows that $\mathbb{C}[X]^G$ is finitely generated. □

In the remainder of the section we show that the representation theory of a linearly reductive group G is semi-simple. We first need some lemmas.

Lemma 9.2.9. *Suppose that V is a representation of a linearly reductive group G. If $v \in V^G$ is nonzero, then there exists a function $f \in (V^\star)^G$ with $f(v) \neq 0$. So the restriction of the pairing $V \times V^\star \to \mathbb{C}$ to a pairing $V^G \times (V^\star)^G \to \mathbb{C}$ is nondegenerate.*

Proof. Suppose that $v \in V^G$ is nonzero, and choose a linear function h on V such that $h(v) = 1$. Let $f = \mathcal{R}(h)$. The subspace $V^\star \subseteq \mathbb{C}[V]$ is G-stable and $h \in V^\star$, so $f = \mathcal{R}(h) \in V^\star$. The maximal ideal $\mathfrak{m}_v$ of all polynomials vanishing at v is G-stable and $h - 1 \in \mathfrak{m}_v$, so $f - 1 = \mathcal{R}(h) - \mathcal{R}(1) = \mathcal{R}(h-1) \in \mathfrak{m}_v$ and therefore $f(v) = 1$. □

If V is a representation of G, and $W \subseteq V$ is a G-stable subspace, then W is also a representation, and we call W a **subrepresentation** of V. A nonzero representation V is called an irreducible representation if $\{0\}$ and V are the only subrepresentations. The proof of Schur's lemma for algebraic groups below is the same as its counterpart for modules over an algebra.

Lemma 9.2.10 (Schur's Lemma)**.** *If V is an irreducible representation of an affine algebraic group, then* $\mathrm{Hom}(V,V)^G$ *is one-dimensional and spanned by the identity 1_V.*

Proof. Suppose that V is an irredicible representation of G. Assume that $\varphi \in \mathrm{Hom}(V,V)^G$ and let λ be an eigenvalue of φ. Then the kernel of $\varphi - \lambda 1_V$ is a nonzero subrepresentation of V. Since V is irredicible, the kernel of $\varphi - \lambda 1_V$ is equal to V itself. It follows that φ is equal to $\lambda 1_V$. This shows that $\mathrm{Hom}(V,V)^G$ is spanned by the identity map. □

Proposition 9.2.11. *Suppose that G is a linearly reductive group. Then every representation of G is a direct sum of irreducible representations.*

Proof. We prove the statement by induction on the dimension n of V. The cases $n = 0$ and $n = 1$ are obvious. Suppose that $n > 2$ and assume that every representation of dimension $< n$ is a direct sum of irreducible representations. Choose a nonzero subrepresentation W of V of minimal dimension. If $W = V$, then V is irreducible and we are done. Otherwise, W

is an irreducible proper subrepresentation of V. Let $\varphi \in \operatorname{Hom}(W,V)^G$ be the inclusion map. The spaces $\operatorname{Hom}(V,W)$ and $\operatorname{Hom}(V,W)$ are dual to each other, and the pairing is given by $\operatorname{Hom}(V,W) \times \operatorname{Hom}(W,V) \to \mathbb{C}$ is given by

$$(\varphi, \psi) \mapsto \operatorname{trace}(\varphi\psi).$$

By Lemma 9.2.9, the pairing $\operatorname{Hom}(V,W)^G \times \operatorname{Hom}(W,V)^G \to \mathbb{C}$ is nondegenerate. So there exists a map $\psi \in \operatorname{Hom}(V,W)^G$ such that $\operatorname{trace}(\psi\varphi) \neq 0$. The composition $\psi\varphi$ lies in $\operatorname{Hom}(W,W)^G$. Since $\operatorname{Hom}(W,W)^G$ is spanned by 1_W, and $\operatorname{trace}(\psi\varphi) \neq 0$ we have $\psi\varphi = \lambda 1_W$ for some nonzero $\lambda \in \mathbb{C}$. Let Z be the kernel of ψ. It is easy to verify that Z a G-stable and $V = W \oplus Z$. By the induction hypothesis, Z is a direct sum of irreducible representations, so V is a direct sum of irreducible representations as well. □

Exercises.

Exercise 9.2.1. *Show that the* $\mathrm{U}_n \subset \mathrm{GL}_n$ *is Zariski dense as follows.*

Let G *be the Zariski closure of* U_n. *This is an affine algebraic group by Exercise* 9.1.5.

(1) *Let* T *be the set of diagonal matrices, and let* $H \subseteq T$ *be the set of all matrices*

$$\begin{pmatrix} z_1 & & & \\ & z_2 & & \\ & & \ddots & \\ & & & z_n \end{pmatrix}$$

with $|z_1| = |z_2| = \cdots = |z_n| = 1$. *Show that* H *lies Zariski dense in* T.

(2) *Every element of* GL_n *has a singular value decomposition* $U_1 D U_2$ *where* $U_1, U_2 \in \mathrm{U}_n$ *and* $D \in T$. *Use this to show that* $H = G$.

Exercise 9.2.2. *What is a Zariski dense compact subgroup of* $\mathrm{O}_n(\mathbb{C})$*?*

Exercise 9.2.3. *Show that the subgroup* $\mathrm{SU}_n \subset \mathrm{SL}_n$ *is Zariski dense.*

Exercise 9.2.4. *Suppose that* V *is a rational representation of a linearly reductive group* G *and that* $W \subseteq V$ *is a subrepresentation. Show that there exists a subrepresentation* Z *of* V *such that* $V = W \oplus Z$.

Exercise 9.2.5. *Consider the representation* $\rho : \mathbb{G}_\mathrm{a} \to \mathrm{GL}_2(\mathbb{C})$ *defined by*

$$\rho(t) = \begin{pmatrix} 1 & t \\ 0 & 1 \end{pmatrix}.$$

Show that $\{(x,0) \mid x \in \mathbb{C}\}$ *is a subrepresentation that does not have a* $\mathbb{G}_\mathrm{a}$*-stable complement. Conclude that* $\mathbb{G}_\mathrm{a}$ *is not linearly reductive.*

Exercise 9.2.6. *Suppose that T is a torus group, i.e., T is isomorphic to $\mathbb{G}_{\mathrm{m}}^r$ for some r.*

(1) *Show that T contains an element g such that the group generated by g lies Zariski dense in T.*
(2) *Suppose that $\rho : T \to \mathrm{GL}(V)$ is a representation. Show that V has a one-dimenional subrepresentation. (Hint: Take a subspace of V spanned by an eigenvector of $\rho(g)$.)*
(3) *Show that T is linearly reductive.*
(4) *Show that every representation of T is a direct sum of 1-dimensional representations.*
(5) *Show that there exists a basis $e_1, e_2, \dots, e_n$ such that the matrix of $\rho(t)$ with respect to this basis is of the form*

$$\rho(t) = \begin{pmatrix} \chi_1(t) & & & \\ & \chi_2(t) & & \\ & & \ddots & \\ & & & \chi_n(t) \end{pmatrix},$$

where $\chi_1, \dots, \chi_n : T \to \mathbb{G}_{\mathrm{m}}$ are homomorphisms of algebraic groups (multiplicative characters).

Exercise 9.2.7. *Suppose that $\chi_1, \dots, \chi_n$ are distinct multiplicative characters of an algebraic group G, V is a representation of T and $v_1, \dots, v_n$ are nonzero and satisfy $g \cdot v_i = \chi_i(g)v_i$ for all $g \in G$. Show that $v_1, \dots, v_n$ are linearly independent.*

9.3. The Geometry of Quotients

In this section we assume that G is a linearly reductive group and X is an affine variety with a regular G-action. For a finitely generated commutative $\mathbb{C}$-algebra R we let $\mathrm{Spec}(R)$ be the **max spectrum**. As a set, $\mathrm{Spec}(R)$ consists of all maximal ideals of R. It is equipped with the Zariski topology. So $\mathrm{Spec}(R)$ is the affine variety that corresponds to the commutative ring R. The invariant ring $\mathbb{C}[X]^G$ is finitely generated, and corresponds to an affine variety $X/\!\!/G := \mathrm{Spec}(\mathbb{C}[X]^G)$. The inclusion $\mathbb{C}[X]^G \hookrightarrow \mathbb{C}[X]$ corresponds to a dominant morphism $\pi : X \to X/\!\!/G$. From the definitions it is clear that the map π is G-invariant:

$$\pi(g \cdot x) = \pi(x), \quad x \in X, g \in G.$$

This morphism is called the **categorical quotient** of X with respect to G. In this section, we will further investigate the geometric properties of the categorical quotient.

Lemma 9.3.1. *If $A_1, A_2 \subseteq X$ are closed and G-stable, then we have*

$$\overline{\pi(A_1)} \cap \overline{\pi(A_2)} = \overline{\pi(A_1 \cap A_2)}.$$

Proof. Let $J_k = I(A_k)$ for $k = 1, 2$. Then $\overline{\pi(A_k)}$ is the zero set of the ideal $J_k \cap \mathbb{C}[X]^G = J_k^G$. Because J_k is G-stable, we have $J_k^G = \mathcal{R}_X(J_k)$. It follows that $\overline{\pi(A_1)} \cap \overline{\pi(A_2)}$ is the zero set of

$$J_1^G + J_2^G = \mathcal{R}_X(J_1) + \mathcal{R}_X(J_2) = \mathcal{R}_X(J_1 + J_2) = (J_1 + J_2)^G = (J_1 + J_2) \cap \mathbb{C}[X]^G.$$

Since $J_1 + J_2$ is the vanishing ideal of $A_1 \cap A_2$, we have that $(J_1 + J_2) \cap \mathbb{C}[X]^G$ is the vanishing ideal of $\overline{\pi(A_1 \cap A_2)}$. □

Lemma 9.3.2. *If $A \subseteq X$ is closed and G-stable, then $\pi(A)$ is closed. In particular, $\pi : X \to X/\!\!/G$ is surjective.*

Proof. Suppose that $\overline{\pi(A)}$ is not closed. Choose $y \in \overline{\pi(A)} \setminus \pi(A)$. The set $B := \pi^{-1}(y)$ is G-stable and closed. It follows that

$$y \in \overline{\pi(A)} \cap \overline{\pi(B)} = \overline{\pi(A \cap B)} = \emptyset.$$

This is a contradiction, so $\pi(A)$ must be closed.

In particular $\pi(X)$ is closed and dense. We conclude that $\pi(X) = X/\!\!/G$. □

Corollary 9.3.3. *For $y \in X/\!\!/G$, $\pi^{-1}(y)$ contains a unique closed orbit.*

Proof. First we show that $\pi^{-1}(y)$ contains at least one closed orbit. Suppose that $\pi^{-1}(y)$ does not contain any closed orbit. Construct a sequence of points as follows. We choose $z_1 \in \pi^{-1}(y)$. For $i \geq 2$ we choose

$$z_i \in \overline{G \cdot z_{i-1}} \setminus G \cdot z_{i-1}.$$

Then we have an infinite sequence

$$\overline{G \cdot z_1} \supset \overline{G \cdot z_2} \supset \cdots$$

where none of the inclusions is an equality. This contradicts the noetherian property. We conclude that $\pi^{-1}(y)$ does contain a closed orbit.

If $G \cdot z_1$ and $G \cdot z_2$ are two distinct closed orbits in $\pi^{-1}(y)$, then we have

$$y \in \pi(G \cdot z_1) \cap \pi(G \cdot z_2) = \pi(G \cdot z_1 \cap G \cdot z_2) = \emptyset.$$

Contradiction. □

Suppose that V is a representation of a linearly reductive group G, and let $\pi : V \to V/\!\!/G$ be the quotient map. We denote $\pi(0) \in V/\!\!/G$ again by 0. **Hilbert's nullcone** is defined by

$$\mathcal{N} = \pi^{-1}(0).$$

As in the proof of Theorem 9.2.6, let $\mathfrak{m} \subseteq \mathbb{C}[V]^G$ be the ideal generated by all invariants of positive degree, and let $I = \mathfrak{m}\mathbb{C}[V]$ be the ideal of $\mathbb{C}[V]$ generated by all invariants of positive degree. Then the zero set of I is exactly $\mathcal{N}$. However, the ideal I does not need to be reduced so I may not be the vanishing ideal of $\mathcal{N}$.

The fiber $\mathcal{N}$ contains exactly one closed orbit, namely $\{0\}$. Every other orbit closure in $\mathcal{N}$ contains 0. So we have

$$\mathcal{N} = \{v \in V \mid 0 \in \overline{G \cdot v}\}.$$

Exercises.

Exercise 9.3.1. *What is the nullcone for the action of* $\mathrm{GL}_n(\mathbb{C}) \times \mathrm{GL}_m(\mathbb{C})$ *on* $\mathrm{Mat}_{m,n}(\mathbb{C})$ *by left-right multiplication? (Hint: Distinguish the cases* $m = n$ *and* $m \neq n$*.)*

Exercise 9.3.2. *What is the nullcone for the conjugation action of* $\mathrm{GL}_n(\mathbb{C})$ *on* $\mathrm{Mat}_{n,n}(\mathbb{C})$*?*

Exercise 9.3.3. *Suppose that* V *is a representation of a finite group* G*. Show that the nullcone is just* $\{0\}$*.*

Exercise 9.3.4. *Suppose that* G *is linearly reductive,* X *is an affine* G*-variety and* $\pi : X \to X/\!\!/G$ *is the quotient. Assuming that all the* G*-orbits in* X *have the same dimension, show that:*

(1) *all the* G*-orbits are closed;*
(2) *the fibers* $\pi^{-1}(y)$*,* $y \in X/\!\!/G$ *are exactly all the orbits.*

Exercise 9.3.5. *Suppose that* G *is linearly reductive acting regularly on an affine variety* X*. Define*

$$B = \{(v, w) \in V \times V \mid f(v) = f(w) \text{ for all } f \in \mathbb{C}[X]^G\}.$$

Show that $(v, w) \in B$ *if an only if* $\overline{G \cdot v} \cap \overline{G \cdot w} \neq \emptyset$*.*

9.4. Semi-Invariants and the Sato-Kimura Lemma

In this section we introduce the ring of semi-invariants. Semi-invariants for quivers will be studied in the next Chapter. They also play a role in the construction of GIT quotients for quiver representations with respect to weights in Section 9.8.

Suppose that V is a representation of a linear algebraic group G. A **multiplicative character** of G is a homomorphism of algebraic groups $\chi : G \to \mathbb{G}_{\mathrm{m}}$. We say that $f \in \mathbb{C}[V]$ is a **semi-invariant** of weight χ if $g \cdot f = \chi(g)f$ for all $g \in G$. The set of all semi-invariants of weight χ is a subspace of $\mathbb{C}[V]$ and will be denoted by $\mathbb{C}[V]_\chi$. If f_1 and f_2 are two semi-invariants of weight χ_1 and χ_2, respectively, then $f_1 f_2$ is a semi-invariant of weight $\chi_1\chi_2$. The **ring of semi-invariants** is $\mathrm{SI}(G,V) := \bigoplus_\chi \mathbb{C}[V]_\chi$.

Lemma 9.4.1. *Suppose that G is reductive. Let $[G,G]$ be the smallest algebraic subgroup of G containing all commutators $ghg^{-1}h^{-1}$ with $g,h \in G$. Then the ring of semi-invariants is equal to $\mathbb{C}[V]^{[G,G]}$.*

Proof. Since $\chi(ghg^{-1}h^{-1}) = \chi(g)\chi(h)\chi(g)^{-1}\chi(h)^{-1} = 1$ for all $g, h \in G$, we see that every semi-invariant is invariant under $[G,G]$. The space $\mathbb{C}[V]^{[G,G]}$ is a sum of finite dimensional irreducible representations of G. Let $W \subseteq \mathbb{C}[V]^{[G,G]}$ be a finite dimensional irreducible subrepresentation. If $g \in G$, then there exists a nonzero $f \in W$ and a scalar λ such that $g \cdot f = \lambda f$ for some scalar λ. For $h \in G$ we have $g\cdot(h\cdot f) = gh\cdot f = hg \cdot f = h\cdot(\lambda f) = \lambda(h\cdot f)$. Since W is spanned by all $h \cdot f$, $h \in G$ because it is irreducible, we have $g \cdot f = \lambda f$ for all $f \in W$. The value of the scalar λ depends on the choice of g. So we have a function $\chi : G \to \mathbb{G}_{\mathrm{m}}$ such that $g \cdot f = \chi(g) f$ for all $g \in G$ and all $f \in W$. It is easy to verify that χ is a homomorphism of algebraic groups. So we have $W \subseteq \mathbb{C}[V]_\chi$. We conclude that $\mathbb{C}[V]^{[G,G]}$ is contained in $\bigoplus_\chi \mathbb{C}[V]_\chi$. $\square$

Lemma 9.4.2 (Sato-Kimura). *Let V be a representation of a connected linear algebraic group G and that the action has a dense open orbit. Then the ring of semi-invariant is a polynomial ring.*

Proof. Let $O \subset V$ be the dense open orbit. Let $Z = V \setminus Q$. We can write Z as a union of irreducible components

$$Z = Z_1 \cup \ldots \cup Z_m.$$

Assume that we reorder the components so $\operatorname{codim} Z_i = 1$ for $1 \le i \le t$ and $\operatorname{codim} Z_i > 1$ for $i > t$. Let f_i be the irreducible polynomial vanishing on Z_i for $1 \le i \le t$. We claim that $f_1, f_2, \ldots, f_t$ generate $\mathrm{SI}(G,V)$ and are algebraically independent. Since each $g \in G$ fixes Z_i, we have $g \cdot f_i = \chi_i(g) f_i$ for some constants $\chi_i(g)$. It follows immediately that $\chi_i : G \to \mathbb{G}_{\mathrm{m}}$ is a multiplicative character for $i = 1, 2, \ldots, t$. This shows that $f_i \in \mathbb{C}[V]_{\chi_i}$ for all i.

Suppose that $\chi_1, \chi_2, \ldots, \chi_t$ are linearly dependent, say $\chi_1^{a_1}\chi_2^{a_2}\cdots\chi_t^{a_t} = 1$ for $a_1, \ldots, a_t \in \mathbb{Z}$, not all 0. Then $f = f_1^{a_1} f_2^{a_2} \cdots f_t^{a_t}$ is a rational G-invariant function. It is constant on the dense orbit O, so f is constant. This contradicts the unique factorization in the polynomial ring $\mathbb{C}[V]$. So $\chi_1, \chi_2, \ldots, \chi_t$ are independent. Now Exercise 9.2.7 implies that all the monomials

$$f_1^{a_1} f_2^{a_2} \cdots f_t^{a_t}, \quad a_1, a_2, \ldots, a_t \ge 0$$

are linearly independent, because they all have different characters. We conclude that $f_1, f_2, \ldots, f_t$ are algebraically independent.

It remains to show that $f_1, \ldots, f_t$ generate the ring of semi-invariants. Let h be a semi-invariant of some fixed weight χ. We will show that h lies in the ring generated by $f_1, \ldots, f_t$. Since G is connected, each irreducible factor of h is a semi-invariant, so we can assume that h is irreducible. Consider the set of zeros of h. It is a G-fixed set of codimension 1. Thus it has to contain one of the components Z_i for $1 \leq i \leq t$. This means h is divisible by f_i for some i, and therefore it is equal to f_i, up to a constant. $\square$

Example 9.4.3. Let $V = \mathrm{Mat}_{n,n}(\mathbb{C})$ be the space of $n \times n$ matrices, let B_+ be the group of upper triangular matrices, and B_- be the group of lower triangular matrices. The group $G = B_+ \times B_-$ acts on V by left-right multiplication.

$$(A_+, A_-) \cdot B = A_+ C A_-^{-1}, \quad (A_+, A_-) \in G, \ B \in V.$$

The orbit of the identity matrix is equal to B_+B_- and lies dense in V. By the Sato-Kimura lemma the ring of semi-invariants is a polynomial ring. Let $p_i(B)$ be the determinant of the submatrix consisting of the first i rows and first i columns of the matrix B. Then $p_1, p_2, \ldots, p_n$ generate the ring of semi-invariants.

Example 9.4.4. For a quiver Q of finite representation type, and a dimension vector β, the group GL_β acts on $\mathrm{Rep}_\beta(Q)$. Since there are finitely many indecomposable representations, there are only finitely many isomorphism classes in $\mathrm{Rep}_\beta(Q)$, so there are only finitely many orbits. One of the orbits has to be dense. The ring of semi-invariants, denoted by $\mathrm{SI}(Q, \beta)$, is a polynomial ring by the Sato-Kimura lemma. The group $[\mathrm{GL}_\beta, \mathrm{GL}_\beta]$ is equal to $\mathrm{SL}_\beta = \prod_{x \in Q_0} \mathrm{SL}_{\beta(x)}$, so we have $\mathrm{SI}(Q, \beta) = \mathbb{C}[\mathrm{Rep}_\beta(Q)]^{\mathrm{SL}_\beta}$ by Lemma 9.4.1. Rings of semi-invariants for quivers will be studied in more detail in the next chapter.

Exercises.

Exercise 9.4.1. *Let Q be the $\mathbf{A}_n$ type quiver*

$$1 \longrightarrow 2 \longrightarrow \cdots \longrightarrow n$$

and β is the dimension vector $\beta = (k, k, \ldots, k)$. Describe the ring of semi-invariants $\mathrm{SI}(Q, \beta)$.

Exercise 9.4.2. *Adapt the proof of the Sato-Kimura lemma so that it is also valid for linear algebraic groups G that are not connected.*

9.5. Geometric Invariant Theory

In this section we discuss how to construct quotients for group actions on projective variety.

Often one has to omit some points of the variety in order to construct a more interesting quotient. To illustrate this, suppose that X is an irreducible affine variety with a regular action of the multiplicative group $\mathbb{G}_{\mathrm{m}}$ with the following properties:

(1) the action has exactly one fixed point with we will denote by $0 \in X$;

(2) every other orbit has 0 in its closure.

In this case, the quotient $X/\!\!/\mathbb{G}_{\mathrm{m}}$ is just a point, which is not very interesting. A more interesting quotient can be constructed by omitting 0.

Proposition 9.5.1. *There exists a projective variety $\mathbb{P}(X)$ and a morphism of varieties $\gamma_X : X \setminus \{0\} \to \mathbb{P}(X)$ with the following properties:*

(1) *for every $y \in \mathbb{P}(X)$, $\gamma_X^{-1}(y)$ is exactly one orbit;*

(2) *a subset $A \subseteq \mathbb{P}(X)$ is open (resp. closed) if and only if $\gamma_X^{-1}(A)$ is an open (resp. closed) subset of $X \setminus \{0\}$;*

(3) *if $A \subseteq \mathbb{P}(X)$ is an open affine subset, then $\gamma_X^{-1}(A)$ is open and affine and the restriction of γ_X to $\gamma_X^{-1}(A) \to A$ is exactly the affine quotient of $\gamma_X^{-1}(A)$ with respect to $\mathbb{G}_{\mathrm{m}}$.*

Proof. The $\mathbb{G}_{\mathrm{m}}$-action on $\mathbb{C}[X]$ gives $\mathbb{C}[X]$ a grading $\mathbb{C}[X] = \bigoplus_{d \in \mathbb{Z}} \mathbb{C}[X]_d$. There cannot be both positive and negative weights because then there would be nontrivial invariants. So without loss of generality we may assume that $\mathbb{C}[X]$ is positively graded $\mathbb{C}[X] = \bigoplus_{d=0}^{\infty} \mathbb{C}[X]_d$. The projective space $\mathbb{P}(X) = \mathrm{Proj}(\mathbb{C}[X])$ is exactly the projective variety with the homogeneous coordinate ring $\mathbb{C}[X]$. The details are left to the reader. $\square$

As we will see in Section 9.8, we will need to take out certain points to construct nice quotients of $\mathrm{Rep}_\alpha(Q)$ with respect to GL_α. For this construction, we need to study quotients of reductive groups acting on *projective* varieties.

Suppose that V is an n-dimensional representation of a linearly reductive group G. Then $\mathbb{P}(V)$ is the usual $(n-1)$-dimensional projective space. The group G also acts on $\mathbb{P}(V)$ and we would like to construct a quotient with nice geometric properties (similar to the quotients of affine varieties). Let $\pi : V \to V/\!\!/G$ be the affine quotient, and let $\mathcal{N} = \pi^{-1}(0)$ be the nullcone. Define $V^{\mathrm{ss}} = V \setminus \mathcal{N}$. Elements of V^{ss} are called **semi-stable** points. So the following are equivalent:

(1) $v \in V$ is semi-stable;

(2) the orbit closure $\overline{G \cdot v}$ does not contain 0;

(3) there exists a homogeneous invariant $f \in \mathbb{C}[V]^G$ with $f(v) \neq 0$.

Define $\mathbb{P}(V)^{\mathrm{ss}} = \gamma_V(V^{\mathrm{ss}})$. Since V^{ss} is a union of orbits, we also have $V^{\mathrm{ss}} = \gamma_V^{-1}(\mathbb{P}(V)^{\mathrm{ss}})$. The nullcone is closed, so V^{ss} is open. Therefore, $\mathbb{P}(V)^{\mathrm{ss}}$ is open. The map $\gamma_V : V \setminus \{0\} \to \mathbb{P}(V)$ restricts to a map $V^{\mathrm{ss}} \to \mathbb{P}(V)^{\mathrm{ss}}$. Since the actions of $\mathbb{G}_{\mathrm{m}}$ and G on V commute, we also have a regular action of $\mathbb{G}_{\mathrm{m}}$ on $V/\!\!/G$. Now $\{\pi(0)\}$ is a closed orbit and every other $\mathbb{G}_{\mathrm{m}}$-orbit contains this orbit in its closure. We denote $\pi(0)$ again by 0 and we have a $\mathbb{G}_{\mathrm{m}}$-quotient $\gamma_{V/\!\!/G} : V/\!\!/G \setminus \{0\} \to \mathbb{P}(V/\!\!/G)$. We now can define a map $\widetilde{\pi} : \mathbb{P}(V)^{\mathrm{ss}} \to \mathbb{P}(V/\!\!/G)$ as follows. If $y \in \mathbb{P}(V)^{\mathrm{ss}}$, then $\gamma_V^{-1}(y)$ is a $\mathbb{G}_{\mathrm{m}}$-orbit. Choose $x \in \gamma_V^{-1}(y)$ and define $\widetilde{\pi}(y) = \gamma_{V/\!\!/G}(\pi(x))$. The map $\widetilde{\pi}$ is well defined because the composition $\gamma_{V/\!\!/G} \circ \pi$ is constant on $\mathbb{G}_{\mathrm{m}}$-orbits (it is even constant on $\mathbb{G}_{\mathrm{m}} \times G$-orbits).

We have the following diagram:

$$\begin{array}{ccc} V^{\mathrm{ss}} & \xrightarrow{\pi} & V/\!\!/G \setminus \{0\} \\ {\scriptstyle\gamma_V}\downarrow & & \downarrow{\scriptstyle\gamma_{V/\!\!/G}} \\ \mathbb{P}(V)^{\mathrm{ss}} & \xrightarrow[\widetilde{\pi}]{} & \mathbb{P}(V/\!\!/G). \end{array}$$

Lemma 9.5.2. *For a subset $A \subseteq \mathbb{P}(V/\!\!/G)$ we have*

$$A \text{ is closed} \Leftrightarrow \widetilde{\pi}(A) \text{ is closed.}$$

Proof.

$$\begin{aligned} A \text{ is closed} &\Leftrightarrow \gamma_{V/\!\!/G}^{-1}(A) \text{ is closed} \\ &\Leftrightarrow \gamma_V^{-1}(\widetilde{\pi}^{-1}(A)) = \pi^{-1}(\gamma_{V/\!\!/G}^{-1}(A)) \text{ is closed} \Leftrightarrow \widetilde{\pi}^{-1}(A) \text{ is closed.} \end{aligned}$$

□

Lemma 9.5.3. *Suppose that $x_1, x_2 \subseteq \mathbb{P}(V)^{\mathrm{ss}}$ such that the G-orbits $G \cdot x_1$ and $G \cdot x_2$ are distinct and closed. Then we have $\widetilde{\pi}(x_1) \neq \widetilde{\pi}(x_2)$.*

Proof. Let $A_1 = \gamma_V^{-1}(G \cdot x_1)$ and $A_2 = \gamma_V^{-1}(G \cdot x_2)$. Since

$$(G \cdot x_1) \cap (G \cdot x_2) = \emptyset,$$

we have $A_1 \cap A_2 = \emptyset$. Let $\overline{A_1}, \overline{A_2}$ be the closures of $A_1, A_2 \subseteq V^{\mathrm{ss}}$ in V respectively. Clearly, $\overline{A_i} \setminus A_i$ is contained in $\pi^{-1}(0)$ for $i = 1, 2$, so we have $\overline{A_1} \cap \overline{A_2} \subseteq \pi^{-1}(0)$. It follows that $\pi(\overline{A_1}) \cap \pi(\overline{A_2}) = \pi(\overline{A_1} \cap \overline{A_2}) \subseteq \{0\}$ and $\pi(A_1) \cap \pi(A_2) = \emptyset$. Since $\pi(A_1)$ and $\pi(A_2)$ are G-stable and closed, we get

$$\{\widetilde{\pi}(x_1)\} \cap \{\widetilde{\pi}(x_2)\} = \gamma_{V/\!\!/G}(\pi(A_1)) \cap \gamma_{V/\!\!/G}(\pi(A_2)) = \emptyset,$$

so $\widetilde{\pi}(x_1) \neq \widetilde{\pi}(x_2)$. □

Definition 9.5.4. A nonzero vector $v \in V$ is **stable** if the orbit $G \cdot v$ is closed and of dimension $\dim G$.

Let V^{s} be the set of stable vectors. It is easy to see that $V^{\mathrm{s}} \subseteq V^{\mathrm{ss}}$. We will show that V^{s} is an open set.

Lemma 9.5.5. *The set V^{s} is open.*

Proof. Define the closed set $Z \subset G \times V$ by

$$Z = \{(g, v) \mid g \cdot v = v\}.$$

Let $p : Z \to G \times V$ be the restriction of the projection $G \times V \to V$. For a closed subset $A \subseteq Z$, the number $\dim_z(A)$ is the largest dimension of all irreducible components of A that contain z. The function θ defined by $\theta(z) = \dim_z p^{-1}(p(z))$ is semi-continuous. Let $\gamma : V \to G \times V$ be the section defined by $\gamma(v) = (e, v)$. Then the function

$$v \in V \mapsto \theta(\gamma(v)) = \dim_{(e,v)} p^{-1} p(\gamma(v)) = \dim_{(e,v)} p^{-1}(v) = \dim G_v$$

is semi-continuous. In particular, the set $A = \{v \in V \mid \dim G_v > 0\}$ is closed.

Suppose that $w \in V^{\mathrm{s}}$ is G-stable. We will show that V^{s} contains an open neighborhood of $G \cdot w$. Since $G \cdot w$ and A are disjoint, closed and stable under the G-action, there exists an invariant $f \in \mathbb{C}[V]^G$ such that f is constant 0 on A and constant 1 on $G \cdot w$ by Lemma 9.3.1. The set $U = \{v \in V \mid f(v) \neq 0\}$ is clearly open. If $v \in U$, then we have $v \notin A$ and G_v is finite. Suppose the orbit of v is not closed. Choose $x \in \overline{G \cdot v} \setminus G \cdot v$. We have $\dim G \cdot x < \dim G \cdot v$ and $\dim G_x > 0$. So $x \in A$ and therefore $f(v) = f(x) = 0$. This contradicts $v \in U$. Therefore, the orbit of x must be closed and x is G-stable. It follows that $U \subseteq V^{\mathrm{s}}$ and U is an open neighborhood of w in V^{s}. We conclude that V^{s} is open. □

Let $\mathbb{P}(V)^{\mathrm{s}} = \gamma_V(V^{\mathrm{s}})$. It is clear that V^{s} is stable under the $\mathbb{G}_{\mathrm{m}}$ action, so $V^{\mathrm{s}} = \gamma_V^{-1}(\mathbb{P}(V)^{\mathrm{s}})$. This shows that $\mathbb{P}(V)^{\mathrm{s}}$ is open. If we restrict the quotient $\widetilde{\pi} : \mathbb{P}(V)^{\mathrm{ss}} \to \mathbb{P}(V /\!\!/ G)$ to $\mathbb{P}(V)^{\mathrm{s}}$, then $\widetilde{\pi}$ will separate all the orbits in $\mathbb{P}(V)^{\mathrm{s}}$.

Exercises.

Exercise 9.5.1. *Show that if $A_1, A_2 \subseteq \mathbb{P}(V)^{\mathrm{ss}}$ are G-stable and closed, then $\widetilde{\pi}(A_1) \cap \widetilde{\pi}(A_2) = \widetilde{\pi}(A_1 \cap A_2)$.*

9.6. The Hilbert-Mumford Criterion

A very useful tool for determining stable and semi-stable points is the Hilbert-Mumford criterion. We will use it to show in Section 9.7 that nonzero semi-stable points for the action of GL_α on $\mathrm{Rep}_\alpha(Q)$ are exactly the nonzero semi-stable representations. It is used to prove King's stability criterion for quiver representations in Section 9.8. In this section, we give some classical examples of applications of the Hilbert-Mumford criterion.

Definition 9.6.1. A **one-parameter subgroup** of an algebraic group G is a homomorphism $\lambda : \mathbb{G}_{\mathrm{m}} \to G$ of algebraic groups.

Proposition 9.6.2 (Hilbert-Mumford criterion). *Suppose that X is an affine G-variety, $x \in X$ and $G \cdot y$ is a closed orbit contained in the orbit closure $\overline{G \cdot x}$. Then there exists a one-parameter subgroup $\lambda : \mathbb{G}_{\mathrm{m}} \to G$ such that $\lim_{t \to 0} \lambda(t) \cdot x = y$.*

The following example is a classical application of the Hilbert-Mumford criterion:

Example 9.6.3 (binary forms). Let

$$V_d = \{a_0 y^d + a_1 x y^{d-1} + \cdots + a_d x^d \mid a_0, a_1, \ldots, a_d \in \mathbb{C}\}$$

be the $(n+1)$-dimensional space of homogeneous polynomials of degree d in the variables x and y. The group SL_2 acts on V_d by substitution:

$$\begin{pmatrix} a & b \\ c & d \end{pmatrix} \cdot f(x, y) = f(ax + cy, bx + dy).$$

An element of V_d is called a **binary form** of degree d. We will determine the nullcone for the representation V_d of SL_2. Suppose that $f(x, y) \in V_d$ is nonzero and lies in the nullcone $\mathcal{N}$. By the Hilbert-Mumford criterion, there exists a one-parameter subgroup $\lambda : \mathbb{G}_{\mathrm{m}} \to \mathrm{SL}_2$ is a one-parameter subgroup. The homomorphism λ corresponds to a nontrivial two-dimensional representation of the multiplicative group $\mathbb{G}_{\mathrm{m}}$. By Exercise 9.2.6, we have

$$\lambda(t) = C^{-1} \mu(t) C,$$

where

$$\mu(t) = \begin{pmatrix} t^k & 0 \\ 0 & t^l \end{pmatrix}$$

for some invertible matrix $C \in \mathrm{GL}_2$ and some integers k and l. We can choose C to lie in SL_2 by multiplying it with a scalar matrix. We must have $l = -k \neq 0$ because $\lambda(t)$ is contained in SL_2 for all t. We also can choose C such that $k > 0$ (if $k < 0$, multiply C on the left with the matrix that switches the coordinates). Let $\widetilde{f}(x, y) = C \cdot f(x, y)$. Then we have

$$0 = \lim_{t \to 0} \lambda(t) \cdot f(x, y) = \lim_{t \to 0} (C^{-1} \mu(t) C) \cdot f(x, y) = C^{-1} \lim_{t \to 0} \mu(t) \cdot \widetilde{f}(x, y).$$

It follows that $\lim_{t \to 0} \mu(t) \cdot \widetilde{f}(x, y) = 0$. If we write

$$\widetilde{f}(x, y) = a_0 y^d + a_1 x y^{d-1} + \cdots + a_d x^d \in V_d,$$

then we get

$$0 = \mu(t) \cdot \widetilde{f}(x, y) = \lim_{t \to 0} a_0 y^d t^{-kd} + a_1 x y^{d-1} t^{k(2-d)} + \cdots a_d x^d t^{kd}.$$

So we must have $a_0 = a_1 = \cdots = a_{\lceil d/2 \rceil} = 0$. This shows that the polynomials $\widetilde{f}(x, y)$ has a root of multiplicity at least $\lceil d/2 \rceil + 1$ and, consequently the same is true for $f(x, y)$. We leave it to the reader to verify that all the steps can be reversed and that every polynomial with a root of multiplicity at least $\lceil d/2 \rceil$ does lie in the nullcone. So the nullcone $\mathcal{N}$ for the representation V_d consists of all homogeneous polynomials (including the zero polynomial) which have a root of multiplicity at least $\lceil d/2 \rceil + 1$. The set of semi-stable points is $V_d^{\mathrm{ss}} = V_d \setminus \mathcal{N}$.

We claim that the set V_d^{s} of stable forms consists of all forms $f(x, y) \in V_d$ for which all the roots have multiplicity less than $d/2$. Suppose that all the roots of $f(x, y) \in V_d$ have multiplicity less than $d/2$. From the Hilbert-Mumford criterion follows that the orbit is closed. Moreover, $f(x, y)$ must have at least 3 distinct roots and it is easy to see that $f(x, y)$ has a finite stabilizer. So $f(x, y)$ is stable. Conversely, suppose that $f(x, y)$ is stable. Then all the roots of $f(x, y)$ have multiplicity $\leq d/2$. If $f(x, y)$ has a root of multiplicity exactly $d/2$, then, using a one-parameter subgroup we see that $x^{d/2} y^{d/2}$ lies in the orbit closure of $f(x, y)$. But the stabilizer of $x^{d/2} y^{d/2}$ has dimension 1. So the orbit of $f(x, y)$ is not closed, or the stabilizer of $f(x, y)$ is not finite. This is a contradiction because $f(x, y)$ was assumed to be stable. We conclude that all the roots of $f(x, y)$ have multiplicity less than $d/2$.

The following example will be closer to the applications to invariants and semi-invariants of quiver representations.

Example 9.6.4. Suppose that V is an n-dimensional vector space and that $G = \mathrm{GL}(V)$ acting on $\mathrm{End}(V)$ by conjugation, then

$$A \cdot B = ABA^{-1}$$

for $A \in \mathrm{GL}(V)$ and $B \in \mathrm{End}(V)$. Let us determine the nullcone $\mathcal{N}$ for this representation. Suppose that $B \in \mathcal{N}$. Then there exists a one-parameter subgroup $\lambda : \mathbb{G}_{\mathrm{m}} \to \mathrm{GL}(V)$ such that $\lim_{t \to 0} \lambda(t) \cdot B = 0$. By Exercise 9.2.6, we can choose a basis of V such that the matrix of $\lambda(t)$ is given by

$$\lambda(t) = \begin{pmatrix} t^{k_1} & & & \\ & t^{k_2} & & \\ & & \ddots & \\ & & & t^{k_n} \end{pmatrix}$$

with $k_1 \leq k_2 \leq \cdots \leq k_n$. If B is given by the matrix $(b_{i,j})$, then we have

$$\lambda(t) \cdot B = \begin{pmatrix} b_{1,1} & t^{k_1-k_2} b_{1,2} & \cdots & t^{k_1-k_n} b_{1,n} \\ t^{k_2-k_1} b_{2,1} & b_{2,2} & \cdots & t^{k_2-k_n} b_{2,n} \\ \vdots & \vdots & \ddots & \vdots \\ t^{k_n-k_1} b_{n,1} & t^{k_n-k_2} b_{n,2} & \cdots & b_{n,n} \end{pmatrix}.$$

Since $\lim_{t\to 0} \lambda(t) \cdot B = 0$, we have that $b_{i,j} = 0$ for $i > j$. In other words, B is an upper triangular matrix. So B is nilpotent. Conversely, every nilpotent matrix is upper triangular with respect to some basis, and can be shown to lie in the nullcone. The nullcone for the the representation $\operatorname{End}(V)$ of $\operatorname{GL}(V)$ is the set of all nilpotent matrices.

Exercises.

Exercise 9.6.1. *Consider the action of* $\operatorname{GL}(V)$ *on* $\operatorname{End}(V)^m$. *Show that the nullcone consists of all* m*-tuples* $(B_1, B_2, \ldots, B_m)$ *such that there exists an invertible matrix* A *such that* $AB_1A^{-1}, \ldots, AB_mA^{-1}$ *are all upper triangular.*

9.7. GIT for Quiver Representations

In this section we will study Geometric Invariant Theory (GIT) for quiver representations. Let Q be quiver (possibly with oriented cycles). Recall that a representation V of Q is called simple or irreducible if it has exactly two subrepresentations, namely 0 and V itself. To every vertex $x \in Q_0$ we can associate a simple representation, but if Q has oriented cycles, then there are more simple representations. Suppose that α is a dimension vector. We will study the geometry of the group $\operatorname{GL}_\alpha = \prod_{x\in Q_0} \operatorname{GL}_{\alpha(x)}$ acting on $\operatorname{Rep}_\alpha(Q) = \prod_{x\in Q_1} \operatorname{Mat}_{\alpha(ha),\alpha(ta)}$.

Lemma 9.7.1. *Suppose that* $V \in \operatorname{Rep}_\alpha(Q)$ *and* $V_1 \subseteq V$ *is a subrepresentation. Then* $V_1 \oplus (V/V_1)$ *lies in the orbit closure of* V.

Proof. For every vertex $x \in Q_0$, choose a complement $V_2(x)$ for the subspace $V_1(x)$ of $V(x) = \mathbb{C}^{\alpha(x)}$. We define $\lambda(t,x) : V(x) \to V(x)$ as a linear map that is multiplication by t on $V_1(x)$ and the identity on $V_2(x)$. Let $\lambda(t) = (\lambda(t,x), x \in \mathbb{Q}_0)$ be the one-parameter subgroup. We can write

$$V(a) = \begin{pmatrix} V(a)_{1,1} & V(a)_{1,2} \\ V(a)_{2,1} & V(a)_{2,2} \end{pmatrix},$$

where $V(a)_{i,j} \in \operatorname{Hom}(V_j(ta), V_i(ha))$. Because V_1 is a subrepresentation, we have $V(a)_{2,1} = 0$ for all a. The one-parameter subgroup $\lambda(t)$ acts on

$\mathrm{Hom}(V_i(ta), V_j(ta))$ as multiplication by t^{j-i}. We have

$$(\lambda(t)\cdot V)(a) = \begin{pmatrix} V(a)_{1,1} & tV(a)_{1,2} \\ 0 & V(a)_{2,2} \end{pmatrix}.$$

Let $W = \lim_{t\to 0}\lambda(t)\cdot V$. We have

$$W(a) = \begin{pmatrix} V(a)_{1,1} & 0 \\ 0 & V(a)_{2,2} \end{pmatrix}.$$

We see that V_2 is a subrepresentation of W isomorphic to V/V_1. So W is isomorphic to the representation $V_1 \oplus V_2 \cong V_1 \oplus (V/V_1)$. □

Corollary 9.7.2. *Suppose that*

$$0 = V_0 \subset V_1 \subset V_2 \subset \cdots \subset V_m = V$$

is a filtration of V. Then $(V_1/V_0) \oplus (V_2/V_1) \oplus \cdots \oplus (V_m/V_{m-1})$ lies in the orbit closure of V.

Proof. See Exercise 9.7.1. □

Lemma 9.7.3. *Suppose that $V, W \in \mathrm{Rep}_\alpha(Q)$ such that there exists a one-parameter subgroup $\lambda : \mathbb{G}_\mathrm{m} \to \mathrm{GL}_\alpha$ such that $\lim_{t\to 0}\lambda(t)\cdot V = W$. Then there exists a filtration*

$$0 = V_0 \subset V_1 \subset V_2 \subset \cdots \subset V_m = V$$

such that W is isomorphic to $(V_1/V_0) \oplus (V_2/V_1) \oplus \cdots (V_m/V_{m-1})$.

Proof. We can write

$$\lambda(t) = (\lambda(t,x) \mid x \in Q_0),$$

where $\lambda(t,x) \in \mathrm{GL}_{\alpha(x)}$ for all x. We can write $\mathbb{C}^{\alpha(x)} = V(x) = \bigoplus_{i\in\mathbb{Z}} W_i(x)$ such that $\lambda(t,x)$ acts on $W_i(x)$ by multiplying with t^{-i}. We can write $V(a) = \sum_{i,j} V(a)_{i,j}$ with $V(a)_{i,j} \in \mathrm{Hom}(W_j(ta), W_i(ha))$ for all i and all $a \in Q_1$. The one-parameter subgroup $\lambda(t)$ acts on $\mathrm{Hom}(W_j(ta), W_i(ha))$ by multiplication by t^{j-i}. Since $\lim_{t\to 0}\lambda(t)\cdot V = W$, we have that $V_{i,j} = 0$ if $i > j$. If we define $W_i(a) = V(a)_{i,i}$, then W_i is a representation of Q and from the block diagonal form of the $W(a)$'s follows that $W = W_1 \oplus W_2 \oplus \cdots \oplus W_m$. From the upper triangular block form of V follows that V_{i+1}/V_i is isomorphic to W_i for all i. □

Proposition 9.7.4. *A representation $V \in \mathrm{Rep}_\alpha(Q)$ lies in the nullcone if and only if there exists a filtration*

$$0 = V_0 \subset V_1 \subset V_2 \subset \cdots \subset V_m = V$$

of representations such that each quotient V_i/V_{i-1} is isomorphic to one of the simple representations S_x, $x \in Q_0$.

Proof. Assume that $V \in \operatorname{Rep}_\alpha(Q)$ has such a filtration. For $i = 1, 2, \ldots, m$. By Corollary 9.7.2, $W = (V_1/V_0) \oplus (V_2/V_1) \oplus \cdots \oplus (V_m/V_{m-1})$ lies in the orbit closure of V but W is just equal to $0 \in \operatorname{Rep}_\alpha(Q)$. This shows that V lies in the nullcone.

Conversely, suppose that $V \in \operatorname{Rep}_\alpha(Q)$ lies in the nullcone. By the Hilbert-Mumford criterion, there exists a one-parameter subgroup $\lambda : \mathbb{G}_m \to \mathrm{GL}_\alpha$ such that $\lim_{t\to 0} \lambda(t) \cdot V = 0$. By Lemma 9.7.3, there exists a filtration

$$0 = W_0 \subset W_1 \subset \cdots \subset W_r = V$$

such that $(W_1/W_0) \oplus (W_2/W_1) \oplus \cdots \oplus (W_r/W_{r-1})$ is isomorphic to the zero representation $\bigoplus_{x\in Q_0} S_x^{\alpha(x)}$. For each i, the representation W_i/W_{i-1} is a direct sum of the S_x's. From this it is clear that we can choose a filtration

$$V_0 = 0 \subset V_1 \subset \cdots \subset V_m = V$$

that is a refinement of the filtration $W_0 \subset W_1 \subset \cdots \subset W_r$ such that all the quotients are of the form S_x for some $x \in Q_0$. □

Corollary 9.7.5. *If the quiver Q has no oriented cycles, then the nullcone is equal to the entire space $\operatorname{Rep}_\alpha(Q)$ and the ring of invariants $\mathbb{C}[\operatorname{Rep}_\alpha(Q)]^{\mathrm{GL}_\alpha}$ is equal to $\mathbb{C}$.*

Proof. This is Exercise 9.7.2. □

Lemma 9.7.6. *For a representation V there exists a filtration*

$$V_0 = 0 \subset V_1 \subset V_2 \subset \cdots \subset V_m = V$$

such that every quotient V_i/V_{i-1} is simple. Morover, the list of representation $V_1/V_0, V_2/V_1, \ldots, V_m/V_{m-1}$ is uniquely determined, up to ordering, by V (and independent of the choice of filtration).

Proof. Choose a filtration,

$$V_0 = 0 \subset V_1 \subset V_2 \subset \cdots \subset V_m = V,$$

that is maximal, i.e., if W is a subrepresentation of V with $V_i \subseteq W \subseteq V_{i+1}$ for some i, then $W = V_i$ or $W = V_{i+1}$. Suppose that V_{i+1}/V_i is not simple for some i. We will prove a contradiction. There exists a nontrivial subrepresentation Z of V_{i+1}/V_i and this pulls back to a proper subrepresentation W of V_{i+1} that contains V_i properly. Contradiction. This shows that $V_1/V_0, \ldots, V_m/V_{m-1}$ are all simple.

Now we prove the uniqueness. Suppose that

$$0 = W_0 \subset W_1 \subset \cdots \subset W_r = V$$

is another filtration with simple quotients. Choose an i such that V_{i+1} contains W_1, but V_i does not. The map $W_1 \to V_{i+1}/V_i$ between simple representations given by $w \mapsto w + V_i$ is nonzero, so it must be an isomorphism.

From the injectivity and surjectivity follows that $W_1 \oplus V_i = V_{i+1}$. We have two filtrations of W/W_1: the filtration

$$0 \subset W_2/W_1 \subset W_3/W_1 \subset \cdots \subset W_r/W_1 = V/W_1$$

has simple quotients $W_2/W_1, W_3/W_2, \ldots, W_r/W_1$ and

$$\begin{aligned} 0 \subset (W_1 + V_1)/W_1 \subset \cdots \subset (W_1 + V_i)/W_1 &= V_{i+1}/W_1 \\ &\subset V_{i+2}/W_1 \subset \cdots \subset V_m/W_1 = V/W_1 \end{aligned}$$

with simple quotients $V_1, V_2/V_1, \ldots, V_i/V_{i-1}, V_{i+2}/V_{i+1}, \ldots, V_m/V_{m-1}$. Now the lemma follows by induction. $\square$

Proposition 9.7.7. *The orbit of $V \in \operatorname{Rep}_\alpha(Q)$ is closed if and only if V is a direct sum of simple representations. In particular, a nonzero representation $V \in \operatorname{Rep}_\alpha(Q)$ is semi-stable with respect to GL_α if and only if it is semi-simple.*

Proof. Suppose that the orbit of $V \in \operatorname{Rep}_\alpha(Q)$ is closed. There exists a filtration

$$0 = V_0 \subset V_1 \subset V_2 \subset \cdots \subset V_m = V$$

such that all quotients are simple. The representation $W = (V_1/V_0) \oplus (V_2/V_1) \oplus \cdots \oplus (V_m/V_{m-1})$ lies in the closure of the orbit of V. Since the orbit is closed, V and W lie in the same orbit and are therefore isomorphic. This shows that V is a direct sum of simple representations.

Conversely, suppose that V is a direct sum of simple representations, say $V = V_1 \oplus V_2 \oplus \cdots \oplus V_m$. Let $\mathrm{GL}_\alpha \cdot W$ be the unique closed orbit in the orbit closure of V. By the Hilbert-Mumford criterion, there exists a one-parameter subgroup $\lambda : \mathbb{G}_{\mathrm{m}} \to \mathrm{GL}_\alpha$ such that $\lim_{t\to 0} \lambda(t) \cdot V = W$. By Lemma 9.7.3 there exists a filtration

$$0 = U_0 \subset U_1 \subset \cdots \subset U_r = V$$

such that $W = (U_1/U_0) \oplus (U_2/U_1) \oplus \cdots (U_r/U_{r-1})$. This shows that W has a filtration such that the quotients are $W_1, W_2, \ldots, W_m$. Since the orbit of W is closed, W is a sum of simple representations (this is the direction of the proposition that we already proved). This proves that W is isomorphic to $W_1 \oplus W_2 \oplus \cdots \oplus W_m = V$. This proves that the orbit of V is closed. $\square$

The one-dimensional subgroup

$$T_\alpha = \{(t1_\alpha(x), x \in Q_0) \mid t \in \mathbb{C}^\star\}$$

of GL_α acts trivially on $\operatorname{Rep}_\alpha(Q)$. This means that no element of $\operatorname{Rep}_\alpha(Q)$ has a finite stabilizer and there are no stable representations for the action of the group GL_α. Define $\mathrm{PGL}_\alpha := \mathrm{GL}_\alpha / T_\alpha$. The group PGL_α also acts on $\operatorname{Rep}_\alpha(Q)$. For the notion of semi-stability, it does not matter whether we work with PGL_α or GL_α. But we will work with the group PGL_α to have a more interesting notion of stability.

Proposition 9.7.8. *A nonzero representation $V \in \mathrm{Rep}_\alpha(Q)$ is stable if and only if it is simple.*

Proof. Suppose that V is simple. By Schur's lemma, the stabilizer of V in PGL_α is trivial, and the orbit is closed by Proposition 9.7.7. So V is stable. Conversely, suppose that V is stable. Then the orbit of V is closed, so V is a sum of simple representations by Proposition 9.7.7, say $V = V_1 \oplus V_2 \oplus \cdots \oplus V_r$. Then $\dim \mathrm{Hom}_Q(V, V)$ is at least r. The stabilizer of V in GL_α is $\mathrm{Hom}_Q(V, V)^\times$ which has dimension at least r. So the stabilizer of V in PGL_α has dimension at least $r - 1$, but at the same time this stabilizer is finite because V is stable. It follows that $r = 1$ and $V = V_1$ are simple. □

Exercises.

Exercise 9.7.1. *Proof Corollary* 9.7.2, *for example by generalizing the argument, or by using Lemma* 9.7.1 *and induction.*

Exercise 9.7.2. *Prove Corollary* 9.7.5.

Exercise 9.7.3. *Let Q be a quiver and $\mathfrak{m}$ be the ideal in $\mathbb{C}Q$ generated by all arrows $a \in Q_1$. A representation V is nilpoent if and only if $\mathfrak{m}^N V = 0$ for some integer N. Show that nullcone in $\mathrm{Rep}_\alpha(Q)$ is the set of nilpotent representations of dimension α.*

9.8. GIT Quotients with Respect to Weights

Recall that a **multiplicative character** of an algebraic group G is a one-dimensional representation $\chi : G \to \mathrm{GL}_1(\mathbb{C}) = \mathbb{G}_\mathrm{m}$. A multiplicative character of $\chi : \mathrm{GL}_n \to \mathbb{G}_\mathrm{m}$ is of the form $\chi(A) = \det(A)^k$ for some $k \in \mathbb{Z}$ (see Exercise 9.8.1). Suppose that Q is quiver. A function $\sigma : Q_0 \to \mathbb{Z}$ is called a **weight** for the quiver Q. For a weight $\sigma \in \mathbb{Z}^{Q_0}$, we define a character χ_σ by

$$\chi_\sigma\big(A(x),\ x \in Q_0\big) = \prod_{x \in Q_0} \det(A(x))^{\sigma(x)}.$$

Every multiplicative character of GL_α is of the form χ_σ for some weight σ (see Exercise 9.8.2).

In this section we will assume that Q has no oriented cycles. By Corollary 9.7.5, there are no semi-stable elements in $\mathrm{Rep}_\alpha(Q)$ for the action of GL_α and our GIT approach fails. However, using a small modification we can still apply GIT theory and obtain nontrivial quotients. We will choose some character χ_σ which we will view as a one-dimensional representation, and consider the variety $\mathbb{P}(\mathrm{Rep}_\alpha(Q)\oplus\chi_\sigma)$ that contains $\mathrm{Rep}_\alpha(Q)$ as an open subset. Now we can apply Geometric Invariant Theory.

Definition 9.8.1. Suppose that σ is a nonzero weight. A representation $V \in \mathrm{Rep}_\alpha(Q)$ is called **σ-semi-stable** if $(V,1) \in \mathrm{Rep}_\alpha(Q)\oplus\chi_\sigma$ is GL_α-semi-stable. We say that $V \in \mathrm{Rep}_\alpha(Q)$ is **σ-stable** if $(V,1)$ is PGL_α-stable.

For a weight $\sigma \in \mathbb{Z}^{Q_0}$ and a dimension vector $\alpha \in \mathbb{N}^{Q_0}$ is a dimension vector, then we define $\sigma(\alpha) = \sum_{x\in Q_0}\sigma(x)\alpha(x)$.

Lemma 9.8.2. *If $V \in \mathrm{Rep}_\alpha(Q)$ is σ-semistable, then $\sigma(\alpha)=0$.*

Proof. Consider the one-parameter subgroup $\lambda : \mathbb{G}_\mathrm{m} \to \mathrm{GL}_\alpha$ defined by

$$\lambda(t) = (t\cdot 1_{\alpha(x)}, x\in Q_0)$$

of GL_α. We have

$$\chi_\alpha(\lambda(t)) = \prod_{x\in Q_0}\det(t\cdot 1_{\alpha(x)})^{\sigma(x)} = \prod_{x\in Q_0} t^{\sigma(x)\alpha(x)} = t^{\sigma(\alpha)}$$

and

$$\lambda(t)\cdot(V,1) = (V,\chi_\sigma(\lambda(t))) = (V,t^{\sigma(\alpha)}).$$

Suppose that $\sigma(\alpha)\neq 0$. We have $\lim_{t\to 0}\lambda(t)\cdot(V,1) = (V,0)$ or $\lim_{t\to\infty}\lambda(t)\cdot(V,1) = (V,0)$. The pair $(V,0)$ lies in the nullcone because we assumed that Q has no oriented cycles. It follows that $(V,1)$ lies in the nullcone as well. Therefore, $(V,1)$ is not GL_α-stable, and by definition, V is not σ-semi-stable. □

A subrepresentation W of V is called proper if $W\neq V$ and W is not zero-dimensional. The following criterion for (semi)-stability was given by Alistair King.

Proposition 9.8.3 (King's criterion).

(1) *A representation $V\in\mathrm{Rep}_\alpha(Q)$ is σ-semi-stable if and only if $\sigma(\alpha)=0$ and $\sigma(\underline{\dim}\, W)\leq 0$ for all subrepresentations W of V.*

(2) *A representation $V\in\mathrm{Rep}_\alpha(Q)$ is σ-stable if and only if $\sigma(\alpha)=0$ and $\sigma(\underline{\dim}\, W)< 0$ for all proper subrepresentations W of V.*

Proof. Suppose that $\sigma(\dim W)\geq 0$ for some subrepresentation W of V. Let $\beta = \underline{\dim}\, W$. For every vertex $x\in Q_0$, choose a complement $Z(x)$ such that $V(x) = W(x)\oplus Z(x)$. We define a one-parameter subgroup $\lambda(t)$ by

$$\lambda(t) = \left(\begin{pmatrix} t1_{\beta(x)} & 0 \\ 0 & 1_{\alpha(x)-\beta(x)}\end{pmatrix},\ x\in Q_0\right). \tag{9.3}$$

For $a \in Q_1$ we can write

$$V(a) = \begin{pmatrix} W(a) & U(a) \\ 0 & Z(a) \end{pmatrix},$$

where $W(a) \in \operatorname{Hom}(W(ta), W(ha))$, $U(a) \in \operatorname{Hom}(Z(ta), W(ha))$ and $Z(a) \in \operatorname{Hom}(Z(ta), Z(ha))$. Note that Z is a representation isomorphic to V/W. We have

$$\lambda(t) \cdot V(a) = \begin{pmatrix} W(a) & tU(a) \\ 0 & Z(a) \end{pmatrix}.$$

Taking the limit gives $\lim_{t\to 0} \lambda(t) \cdot V = W \oplus Z$. We have $\chi_\alpha(\lambda(t)) = t^{\sigma(\beta)}$. Suppose that $\sigma(\beta) > 0$. Then we get $\lim_{t\to 0} \chi_\alpha(\lambda(t)) = 0$ and

$$\lim_{t\to 0} \lambda(t) \cdot (V, 1) = (W \oplus Z, 0),$$

which lies in the nullcone. Therefore, $(V, 1)$ lies in the nullcone, and V is not σ-semi-stable. This shows that if V is σ-semistable, then $\sigma(\underline{\dim}\, W) \leq 0$ for all subrepresentations W of V.

Suppose that $\sigma(\beta) = 0$ and W is a proper subrepresentation. Then we have

$$\lim_{t\to 0} \lambda(t) \cdot (V, 1) = (W \oplus Z, 1).$$

If V is not isomorphic to $W \oplus Z$, then the GL_α-orbit of $(V, 1)$ is not closed and V is not σ-stable. Suppose that V is isomorphic to $W \oplus Z$. The group

$$\begin{pmatrix} t1_{\beta(x)} & 0 \\ 0 & s1_{\alpha(x)-\beta(x)} \end{pmatrix}, \quad t, s \in \mathbb{C}^\star$$

is a two-dimensional subgroup of GL_α that stabilizes $(W \oplus Z, 1) \cong (V, 1)$. So the stabilizer of $(V, 1)$ in PGL_α is not finite, so V is not σ-stable. We have shown that if V is σ-stable, then we have $\sigma(\underline{\dim}\, W) < 0$ for every proper subrepresentation W of V.

Conversely, suppose that $(V, 1)$ is not σ-semi-stable. By the Hilbert-Mumford criterion, there exists a one-parameter subgroup $\lambda(t) : \mathbb{G}_{\mathrm{m}} \to \mathrm{GL}_\alpha$ such that $\lim_{t\to 0} \lambda(t) \cdot (V, 1) = (0, 0)$. For every $x \in Q_0$ we have a decomposition $V(x) = \bigoplus_{i \in \mathbb{Z}} W_i(x)$ where $\lambda(t)$ acts on $W_i(x)$ by multiplication by t^{-i}. For $a \in Q_1$ we can write $V(a) = \sum_{i,j} W(a)_{i,j}$ where $W(a)_{i,j} \in \operatorname{Hom}(W(x)_j, W(x)_i)$. The element $\lambda(t)$ acts on $\operatorname{Hom}(W(x)_j, W(x)_i)$ by multiplication by t^{j-i}. Since $\lim_{t\to 0} \lambda(t) \cdot V(a) = 0$, we get $W(a)_{i,j} = 0$ for $i \geq j$. Define $V_j(x) = \bigoplus_{i \leq j} W(x)_i$. Then V_j is a subrepresentation of V.

Let $\beta_j = \underline{\dim} V_j$, so that $\beta_j - \beta_{j-1} = \underline{\dim} W_j$ for all j. We have

$$\chi_\alpha(\lambda(t)) = \prod_{x \in Q_0} \Big(\prod_{i\in\mathbb{Z}} (t^{-i})^{\beta_i(x)-\beta_{i-1}(x)}\Big)^{\sigma(x)}$$
$$= \prod_{i\in\mathbb{Z}} t^{-i(\sigma(\beta_i)-\sigma(\beta_{i-1}))} = t^{\sum_{i\in\mathbb{Z}}\sigma(\beta_i)}.$$

Because $\lim_{t\to 0}\chi_\alpha(\lambda(t)) = 0$, we have $\sum_{i\in\mathbb{Z}}\sigma(\beta_i) > 0$. It follows that $\sigma(\beta_i) > 0$ for some i. So if we take $W = V_i$, then W is a subrepresentation of V such that $\sigma(\underline{\dim} W) > 0$. This shows that, if V is σ-semi-stable, then we have $\sigma(\underline{\dim} W) \leq 0$ for every subrepresentation W of V.

Suppose that $(V, 1)$ is σ-semi-stable, but not σ-stable. We will show that there exist a proper subrepresentation W of V with $\sigma(\underline{\dim} W) = 0$. Suppose that $\operatorname{Hom}_Q(V, V)$ is not one-dimensional. Then there exists a nonzero non-invertible element $\phi \in \operatorname{Hom}_Q(V, V)$. Let K and W be the kernel and image of ϕ, respectively, and set $\beta = \underline{\dim} W$. Because W is a subrepresentation of V, we have $\sigma(\beta) \leq 0$. Because K is a subrepresentation of V of dimension $\alpha - \beta$, we have $\sigma(\alpha - \beta) \leq 0$. Since $\sigma(\beta) + \sigma(\alpha - \beta) = \sigma(\alpha) = 0$, we get $\sigma(\beta) = 0$. Now suppose that $\operatorname{Hom}_Q(V, V)$ is one-dimensional. Then the stabilizer of $(V, 1)$ in PGL_α is trivial. Since $(V, 1)$ is not PGL_α-stable, the GL_α-orbit of $(V, 1)$ is not closed.

Then there exists a one-parameter subgroup $\lambda : \mathbb{G}_{\mathrm{m}} \to \mathrm{GL}_\alpha$ such that

$$\lim_{t\to 0} \lambda(t)\cdot(V, 1) = (U, s)$$

for some $U \in \operatorname{Rep}_\alpha(Q)$ and $s \in \mathbb{C}$ such that (U, s) does not lie in the orbit $(V, 1)$. Since $(V, 1)$ is semi-stable, it does not lie in the nullcone, so (U, s) does not lie in the nullcone. This shows that $s \neq 0$. So we get $\chi_\alpha(\lambda(t)) = 1$ and $s = 1$. We have $\lim_{t\to 0}\lambda(t)\cdot V = U$ such that U and V are not isomorphic. As in part (1), we can define subpresentations V_i, $i \in \mathbb{Z}$ such that $\chi_\alpha(\lambda(t)) = t^{\sum_i \sigma(\beta_i)}$. Because $\chi_\alpha(\lambda(t)) = 1$, we have $\sum_i \sigma(\beta_i) = 0$. Because $\sigma(\beta_i) \leq 0$ for all i, we get $\sigma(\beta_i) = 0$ for all i. For some i, V_i must be a proper subrepresentation because otherwise $U = \lim_{t\to 0}\lambda(t)\cdot V$ would be isomorphic to V. If we take $W = V_i$, then W is a proper subrepresentation with $\sigma(\underline{\dim} W) = 0$. We have shown that if V is a σ-stable representation, then $\sigma(\underline{\dim} W) < 0$ for every proper subrepresentation W of V. □

Consider the quotient map

$$\widetilde{\pi} : \mathbb{P}\Big((\operatorname{Rep}_\alpha(Q) \oplus \chi_\sigma)^{\mathrm{ss}}\Big) \twoheadrightarrow \mathbb{P}\Big((\operatorname{Rep}_\alpha(Q) \oplus \chi_\sigma) /\!\!/ \mathrm{GL}_\alpha\Big).$$

Let us denote the set of σ-semistable representations by $\operatorname{Rep}_\alpha(Q)^{\mathrm{ss}}_\sigma$, and the set of σ-stable representations by $\operatorname{Rep}_\alpha(Q)^{\mathrm{s}}_\sigma$. The set of semistable points with respect to GL_α is

$$(\mathrm{Rep}_\alpha(Q) \oplus \chi_\sigma)^{\mathrm{ss}} = \mathrm{Rep}_\alpha(Q)^{\mathrm{ss}}_\sigma \times \mathbb{C}^\star.$$

So we can make the identification

$$\widetilde{\pi} : \mathbb{P}\Big((\mathrm{Rep}_\alpha(Q) \oplus \chi_\sigma)^{\mathrm{ss}}\Big) = \mathrm{Rep}_\alpha(Q)^{\mathrm{ss}}_\sigma.$$

If we define

$$\mathrm{Rep}_\alpha(Q) /\!\!/_\sigma \mathrm{GL}_\alpha = \mathbb{P}\Big((\mathrm{Rep}_\alpha(Q) \oplus \chi_\sigma) /\!\!/ \mathrm{GL}_\alpha\Big)$$

and set $\pi_\sigma = \widetilde{\pi}$, then we have constructed a quotient map

$$\pi_\sigma : \mathrm{Rep}_\alpha(Q)^{\mathrm{ss}}_\sigma \longrightarrow\!\!\!\!\!\rightarrow \mathrm{Rep}_\alpha(Q) /\!\!/_\sigma \mathrm{GL}_\alpha .$$

Lemma 9.8.4. *Suppose that Q is a quiver with no oriented cycles and α is a dimension vector. Than there exists a weight τ such that the one-parameter subgroup*

$$\mu(t) = (t^{\tau(x)} 1_{\alpha(x)},\ x \in Q_0)$$

has the property that $\lim_{t\to 0} \mu(t) \cdot V = 0$ for all $V \in \mathrm{Rep}_\alpha(Q)$.

Proof. The proof is Exercise 9.8.5. □

Suppose that α is a dimension vector and σ is a weight with $\sigma(\alpha) = 0$. We define $\mathrm{GL}_{\alpha,\sigma}$ as the kernel of $\chi_\sigma : \mathrm{GL}_\alpha \to \mathbb{G}_{\mathrm{m}}$. Note that $\mathrm{GL}_{\alpha,\sigma}$ contains T_α. We define $\mathrm{PGL}_{\alpha,\sigma} = \mathrm{GL}_{\alpha,\sigma} / T_\alpha$.

Lemma 9.8.5. *Suppose that $\mu : \mathbb{G}_{\mathrm{m}} \to \mathrm{GL}_\alpha$ is a one-parameter subgroup as in Lemma 9.8.4 and $\sigma \in \mathbb{Z}^{Q_0}$ is a weight with $\sigma(\alpha) = 0$ and $\chi_\sigma(\mu(t)) = t^{-l}$ with $l < 0$. Then a representation V is σ-(semi-)stable if and only if it is (semi-)stable with respect to $\mathrm{PGL}_{\alpha,\sigma}$.*

Proof. Without loss of generality we may assume that $V(a) \neq 0$ for all $a \in Q_1$ because otherwise we could restrict to a quiver with fewer arrows. Suppose that V is not σ-semi-stable. Then there exists a one-parameter subgroup $\lambda : \mathbb{G}_{\mathrm{m}} \to \mathrm{GL}_\alpha$ such that $\lim_{t\to 0} \lambda(t) \cdot (V, 1) = (0, 0)$. So $\chi_\sigma(\lambda(t)) = t^k$ where $k > 0$. We have

$$\mu(t) \cdot V = (t^{\tau(ha)-\tau(ta)} V(a), a \in Q_1).$$

From $\lim_{t\to 0} \mu(t) \cdot V = 0$ it follows that $\tau(ha) > \tau(ta)$ for all $a \in Q_1$. Since $\lim_{t\to 0} \lambda(t) \cdot V = 0$, only positive exponents of t appear in $\lambda(t) \cdot V(a)$. So only positive exponents of t appear in

$$(\lambda(t^l)\mu(t^k)) \cdot V(a) = t^{\tau(ha)-\tau(ta)} \lambda(t) \cdot V(a).$$

It follows that $\lim_{t\to 0} \lambda(t^l)\mu(t^k) \cdot V = 0$. We have $\chi_\sigma(\lambda(t^l)\mu(t^k)) = t^{kl} \cdot t^{-kl} = 1$, so $\lambda(t^l)\mu(t^k)$ is a one-parameter subgroup of $\mathrm{GL}_{\alpha,\sigma}$. Conversely, suppose that V is not $\mathrm{PGL}_{\alpha,\sigma}$-semi-stable. Then there exists a one-parameter subgroup $\lambda : \mathbb{G}_{\mathrm{m}} \to \mathrm{PGL}_{\alpha,\sigma}$ such that $\lim_{t\to 0} \lambda(t) \cdot V = 0$. It follows that $\lim_{t\to\infty} \lambda(t) \cdot (V, 1) = (0, 1)$, so $(0, 1)$ lies in the PGL_α-orbit closure of $(V, 1)$ and $(0, 0)$ lies in the orbit close of $(0, 1)$. So $(0, 0)$ lies in the orbit closure of $(V, 1)$ and $(V, 1)$ is not σ-semi-stable.

We leave it to the reader to show that V is σ-stable if and only if V is $\mathrm{PGL}_{\alpha,\sigma}$-stable (see Exercise 9.8.6). □

Example 9.8.6. Consider the generalized Kronecker quiver with vertices $Q_0 = \{1, 2\}$ and arrows $\{a_1, a_2, \dots, a_m\}$ going from 1 to 2. Let $\alpha = (n, n)$ and $\sigma = (1, -1)$. Then we have $\mathrm{Rep}_\alpha(Q) = \mathrm{Mat}^m_{n,n}$. A representation $V \in \mathrm{Rep}_\alpha(Q)$ is σ-semi-stable if and only if for every subspace W of $V(1) = \mathbb{C}^n$ we have

$$\dim \textstyle\sum_{i=1}^m V(a_i)(W) \geq \dim W.$$

Indeed, if $V \in \mathrm{Rep}_\alpha(Q)$ is σ-semi-stable and $W \subseteq V(1)$, then we define a subrepresentation of Z by $Z(1) = W$ and $Z(2) = \sum_{i=1}^m V(a_i)(W)$. By King's criterion, we have

$$\sigma(\underline{\dim}\, W) = \dim Z(1) - \dim Z(2) = \dim W - \dim \textstyle\sum_{ii=1}^m V(a_i)(W) \leq 0.$$

On the other hand, if V is not σ-semi-stable, then it has a subrepresentation Z with $\sigma(\underline{\dim}\, Z) > 0$. If $W = Z(1)$ then $\sum_{ii=1}^m V(a_i)(W) \subseteq Z(2)$ and

$$\dim W - \dim \textstyle\sum_{ii=1}^m V(a_i)(W) \geq \dim Z(1) - \dim Z(2) = \sigma(\underline{\dim}\, Z) > 0.$$

Exercises.

Exercise 9.8.1. *If $\chi : \mathrm{GL}_n \to \mathbb{G}_{\mathrm{m}}$ is a character, then we may view χ as a unit in the ring $\mathbb{C}[\mathrm{GL}_n] \cong \mathbb{C}[x_{1,1}, x_{1,2}, \dots, x_{n,n}]_{\det}$, where* det *is the determinant function and $\mathbb{C}[x_{1,1}, x_{1,2}, \dots, x_{n,n}]_{\det}$ is the localization of the polynomial ring $\mathbb{C}[x_{1,1}, x_{1,2}, \dots, x_{n,n}]$ with respect to the element* det. *What are the units in $\mathbb{C}[\mathrm{GL}_n]$? Show that a character $\chi : \mathrm{GL}_n \to \mathbb{G}_{\mathrm{m}}$ is of the form $\chi = \det^k$ for some $k \in \mathbb{Z}$.*

Exercise 9.8.2. *Show that every multiplicative character of GL_α is of the form χ_σ for some weight $\sigma \in \mathbb{Z}^{Q_0}$.*

Exercise 9.8.3.

(1) *Suppose that V and W are σ-semi-stable representations, and $\phi : V \to W$ is a morphism. Show that the kernel and image of ϕ are σ-semistable representations.*

(2) *Suppose that $0 \to W \to V \to Z \to 0$ is an exact sequence of representations, and W and Z are σ-semi-stable. Show that V is σ-semi-stable.*

This exercise shows if we restrict to σ-semi-stable representations, then we get a full subcategory that is closed under extensions, taking kernels and taking cokernels.

Exercise 9.8.4. *Show that if $\phi : V \to W$ is a nonzero morphism between σ-stable representations, then ϕ is an isomorphism.*

Exercise 9.8.5. *Prove Lemma* 9.8.4.

Exercise 9.8.6. *Finish the proof of Lemma* 9.8.5.

Exercise 9.8.7. *Show that if V is a σ-stable representation for some weight σ, then V is indecomposable.*

Exercise 9.8.8. *Suppose that $V \in \mathrm{Rep}_\alpha(Q)$, σ and τ are weights with $\sigma(\alpha) = \tau(\alpha) = 0$ and V is σ-stable. Show that for a positive integer $k \gg 0$, V is $(k\sigma + \tau)$-stable.*

Exercise 9.8.9. *Suppose that $0 < k < m$. Let $Q = \theta(m)$ be the generalized Kronecker quiver with m arrows $a_1, \ldots, a_m$ from vertex 1 to vertex 2. Take $\alpha = (k, 1)$ and $\sigma = (1, -k)$.*

(1) *Show that a representation V is σ-semi-stable if and only if*

$$\begin{pmatrix} V(a_1) \\ V(a_2) \\ \vdots \\ V(a_m) \end{pmatrix} : \mathbb{C}^k \longrightarrow \mathbb{C}^m$$

is injective.

(2) *Show that every σ-semi-stable representation is σ-stable.*

(3) *Show that the projective variety $\mathrm{Rep}_\alpha(Q) /\!\!/_\sigma \mathrm{GL}_\alpha$ parametrizes the k-dimensional subspaces of $\mathbb{C}^n$. This is the grassmannian variety* $\mathrm{Grass}(k, m)$.

9.9. Bibliographical Remarks

The basic facts about linear algebraic groups we discussed in Sections 9.1, 9.2 are proved in the book of Humphreys [**50**]. The application of the GIT to quiver representations, describing semistable point with respect to a given weight in terms of quiver representations (Proposition 9.8.3) is due to King ([**57**]). It is really a direct application of Hilbert-Mumford criterion to the case of the action of GL_α on $\mathrm{Rep}_\alpha(Q)$. The basic material regarding Geometric Invariant Theory is quite standard and similar to the book by Mumford [**65**].

Chapter 10

Semi-invariants of Quiver Representations

Geometric Invariant Theory was introduced because it allows us to describe the unstable points without calculating the ring of invariants. In general, computing generators of an invariant ring can be computationally intractible.

In the case of quiver representations the situation is a little different. In Section 10.2 will discuss the Le Bruyn-Procesi theorem that gives a qualitative description of GL_α-invariants for the action on $\mathrm{Rep}_\alpha(Q)$ in terms of the traces of oriented cycles.

Next we consider the ring of semi-invariants $\mathrm{SI}(Q,\gamma) := \mathbb{C}[\mathrm{Rep}_\gamma(Q)]^{\mathrm{SL}_\gamma}$ for a quiver Q without oriented cycles. Schofield constructed for a representation V with $\langle \underline{\dim}\, V, \gamma\rangle = 0$ a semi-invariant $c^V \in \mathrm{SI}(Q,\gamma)$. The main result for semi-invariants is Theorem 10.7.1 which states that the Schofield semi-invariants span $\mathrm{SI}(Q,\gamma)$. The proof of this result is in Section 10.8.

This approach allows us to prove saturation for the ring of semi-invariants $\mathrm{SI}(Q,\gamma)$: If there is a nonzero semi-invariant of weight $n\sigma$, then there is a nonzero semi-invariant of weight σ. The Main Theorem allows us to describe the rings of semi-invariants explicitly in the finite representation type and tame cases. We exhibit many examples of such calculations. We also give an application to the representation theory of general linear group by proving saturation for the Littlewood-Richardson coefficients.

10.1. Background from Classical Invariant Theory

A crucial tool for studying invariants of quiver representations is the First Fundamental Theorem of Invariant Theory for $\mathrm{GL}(V)$. This fundamental theorem is a classical result about tensor invariants. Using the technique of polarization, one obtains generators of the invariant ring

$$\mathbb{C}[(V^\star)^r \oplus V^s]^{\mathrm{GL}(V)} \cong \mathbb{C}[\mathrm{Hom}(W,V) \oplus \mathrm{Hom}(V,Z)]^{\mathrm{GL}(V)},$$

where $W = \mathbb{C}^s$ and $Z = \mathbb{C}^r$, and $\mathrm{Hom}(W,V) \oplus \mathrm{Hom}(V,Z)$ can be identified with a representation space of the equioriented $\mathbf{A}_3$ quiver

$$\circ \longrightarrow \circ \longrightarrow \circ\,.$$

Let $V = \mathbb{C}^n$ be an n-dimensional vector space. We will first study the invariant theory of $\mathrm{GL}(V)$. The space of multi-linear functions $(V^\star)^d \times V^e \to \mathbb{C}$ can be identified with

$$V^{\otimes(d,e)} = V^{\otimes d} \otimes (V^\star)^{\otimes e} = \underbrace{V \otimes \cdots \otimes V}_{d} \otimes \underbrace{V^\star \otimes \cdots \otimes V^\star}_{e}.$$

An element $t \cdot 1_V \in \mathrm{GL}(V)$ acts on $V^{\otimes(d,e)}$ by the scalar t^{d-e}. So if $d \neq e$, then there are no $\mathrm{GL}(V)$-invariant multilinear functions. If $d = e$, then there are invariant multilinear functions. Let Perm_d be the set of all permutations of $\{1, 2, \ldots, d\}$. For a permutation $\sigma \in \mathrm{Perm}_d$ we define a multilinear map $M_\sigma : (V^\star)^d \times V^d \to \mathbb{C}$ by

$$M_\sigma(f_1, f_2, \ldots, f_d, v_1, v_2, \ldots, v_d) = \prod_{i=1}^{d} f_i(v_{\sigma(i)}).$$

The function M_σ is $\mathrm{GL}(V)$-invariant.

Theorem 10.1.1 (Fundamental Theorem of Invariant Theory). *The space of* $\mathrm{GL}(V)$*-invariant multilinear functions* $(V^\star)^d \times V^d \to \mathbb{C}$ *is spanned by all* M_σ, $\sigma \in \mathrm{Perm}_d$.

For the proof we refer the reader to [**45**].

Suppose that $f \in \mathbb{C}[W]$ is a homogeneous polynomial of degree d on the vector space W. If $w_1, w_2, \ldots, w_d \in W$, then we can write

$$f(t_1 w_1 + t_2 w_2 + \cdots + t_d w_d) = \sum_{0 \leq k_1, k_2, \ldots, k_d \leq d} f_{k_1,\ldots,k_d}(w_1, \ldots, w_d) t_1^{k_1} t_2^{k_2} \cdots t_d^{k_d},$$

where $f_{k_1,\ldots,k_d} \in \mathbb{C}[W^d]$ is a function of multi-degree $(k_1, k_2, \ldots, k_d)$.

In particular, $\mathrm{Pol}(f) := f_{1,1,\dots,1}$ is multi-linear. If we set $w_1 = w_2 = \cdots = w_d = w$, then we get

$$\sum_{0\leq k_1,\dots,k_d\leq d} \binom{d}{k_1,k_2,\dots,k_d} f(w)t_1^{k_1}\cdots t_d^{k_d} = (t_1+\cdots+t_d)^d$$

$$= f((t_1+\cdots+t_d)w) = \sum_{0\leq k_1,k_2,\dots,k_d\leq d} f_{k_1,\dots,k_d}(w,\dots,w)t_1^{k_1}t_2^{k_2}\cdots t_d^{k_d}.$$

Comparing coefficients shows that

$$\mathrm{Pol}(f)(w,w,\dots,w) = \binom{d}{1,1,\dots,1} f(w) = d!f(w).$$

The function $\mathrm{Pol}(f)$ is called the **polarization** of f.

Suppose now that W is a representation of a linear algebraic group G. It is easy to see that if $f \in \mathbb{C}[W]^G$ is invariant, then so is $\mathrm{Pol}(f)$. From the discussion above the following corollary follows.

Corollary 10.1.2. *For a representation W of a linear algebraic group G and a homogeneous invariant $f \in \mathbb{C}[W]$ of degree d there exists a multi-linear G-invariant function $F : W^d \to \mathbb{C}$ such that $f(w) = F(w,w,\dots,w)$ for $w \in W$.*

We leave it to the reader to prove the following generalization (Exercise 10.1.5).

Corollary 10.1.3. *For representations $W_1, W_2, \dots, W_r$ of a linear algebraic group G and a multi-homogeneous invariant*

$$f \in \mathbb{C}[W_1 \oplus W_2 \oplus \cdots \oplus W_r]$$

of multi-degree $(d_1, d_2, \dots, d_r)$ there exists a multi-linear G-invariant function

$$F : W_1^{d_1} \times W_2^{d_2} \times \cdots \times W_r^{d_r} \to \mathbb{C}$$

such that

$$f(w_1,\dots,w_r) = F(w_1,\dots,w_1,w_2,\dots,w_2,\dots,w_r,\dots,w_r)$$

for $w_i \in W_i$ for all i.

Let us consider the action of $\mathrm{GL}(V)$ on $(V^\star)^r \times V^s$. If $1 \leq i \leq r$ and $1 \leq j \leq s$, then we define an invariant function $H_{i,j} \in \mathbb{C}[(V^\star)^r \times V^s]^{\mathrm{GL}(V)}$ by

$$H_{i,j}(u_1,u_2,\dots,u_r,v_1,v_2,\dots,v_s) = u_i(v_j).$$

Theorem 10.1.4. *The invariant ring $\mathbb{C}[(V^\star)^r \oplus V^s]^{\mathrm{GL}(V)}$ is generated by all $H_{i,j}$.*

Proof. The ring $\mathbb{C}[(V^\star)^r \oplus V^s]$ is multi-graded. Suppose that $f \in \mathbb{C}[(V^\star)^r \oplus V^s]^{\mathrm{GL}(V)}$ is homogeneous of degree $(d_1, d_2, \dots, d_r, e_1, e_2, \dots, e_s)$ and let $d = \sum_{i=1}^r d_i$, $e = \sum_{i=1}^s e_i$. There exists an invariant multi-linear function $F : (V^\star)^d \times V^e \to \mathbb{C}$ such that

$$F(\underbrace{u_1, \dots, u_1}_{d_1}, \dots, \underbrace{u_r, \dots, u_r}_{d_r}, \underbrace{v_1, \dots, v_1}_{e_1}, \dots, \underbrace{v_s, \dots, v_s}_{e_s}) \\ = f(u_1, \dots, u_r, v_1, \dots, v_s).$$

If F is nonzero, then we must have $d = e$. By the Fundamental Theorem of Invariant Theory, F is a linear combination of the functions M_σ, $\sigma \in \mathrm{Perm}_d$. So $F(u_1 \dots, u_d, v_1, \dots, v_d)$ is a polynomial in the $H_{i,j}$ with $1 \le i, j \le d$. By substitution, $f(u_1, \dots, u_r, v_1, \dots, v_r)$ is a polynomial in the $H_{i,j}$ with $1 \le i \le r$ and $1 \le j \le s$. □

If we take $Z = \mathbb{C}^r$ and $W = \mathbb{C}^s$, then we can identify $(V^\star)^r$ with $\mathrm{Hom}(V, Z)$ and V^s with $\mathrm{Hom}(W, Z)$. The invariant bilinear map

$$\pi : \mathrm{Hom}(W, V) \times \mathrm{Hom}(V, Z) \to \mathrm{Hom}(W, Z)$$

is given by the matrix $(H_{i,j})$. The ring homomorphism

$$\pi^\star : \mathbb{C}[\mathrm{Hom}(W, Z)] \to \mathbb{C}[\mathrm{Hom}(W, V) \times \mathrm{Hom}(V, Z)]$$

correponding to ϕ maps $\mathbb{C}[\mathrm{Hom}(W, Z)]$ onto the invariant ring

$$\mathbb{C}[\mathrm{Hom}(W, V) \times \mathrm{Hom}(V, Z)]^{\mathrm{GL}(V)}$$

because the $H_{i,j}$'s generate $\mathbb{C}[\mathrm{Hom}(W, V) \times \mathrm{Hom}(V, Z)]$. Let $Y \subseteq \mathrm{Hom}(W, Z)$ be the subset of all linear maps of rank at most $\min\{r, s, n\}$. Then the morphism $\pi : \mathrm{Hom}(W, V) \times \mathrm{Hom}(V, Z) \to Y$ is the categorical quotient with respect to $\mathrm{GL}(V)$.

Exercises.

Exercise 10.1.1. *Show that if* $\max\{r, s\} \le n = \dim V$, *then the ring* $\mathbb{C}[(V^\star)^r \oplus V^s]^{\mathrm{GL}(V)}$ *is a polynomial ring.*

Exercise 10.1.2. *Suppose that* $r, s > n = \dim V$, $1 \le i_1 < i_2 < \cdots < i_{n+1} \le r$ *and* $1 \le j_1 < j_2 < \cdots < j_{n+1} \le s$. *Let* $H_{i,j} \in \mathbb{C}[(V^\star)^r \times V^s]^{\mathrm{GL}(V)}$ *as in Theorem* 10.1.4. *Show that*

$$\det \begin{pmatrix} H_{i_1,j_1} & H_{i_1,j_2} & \cdots & H_{i_1,j_{n+1}} \\ H_{i_2,j_1} & H_{i_2,j_2} & \cdots & H_{i_2,j_{n+1}} \\ \vdots & \vdots & \ddots & \vdots \\ H_{i_{n+1},j_1} & H_{i_{n+1},j_2} & \cdots & H_{i_{n+1},j_{n+1}} \end{pmatrix} = 0.$$

(The Second Fundamental Theorem of Invariant Theory states that these relations generate the ideal of relations among $H_{1,1}, H_{1,2}, \dots, H_{r,s}$.*)*

Exercise 10.1.3. *Suppose that* $d \leq n = \dim V$. *The group* $\mathrm{GL}(V)$ *acts on* $\mathrm{End}(V^{\otimes d}) = \mathrm{Hom}(V^{\otimes d}, V^{\otimes d})$. *Show that the space* $\mathrm{End}(V^{\otimes d})^{\mathrm{GL}(V)}$ *of invariant endomorphism is a subalgebra of* $\mathrm{End}(V^{\otimes d})$ *and that it is ismorphic to the group algebra of* Perm_d.

Exercise 10.1.4. *Suppose that* V *is an* n*-dimensional vector space equipped with a nondegenerate symmetric bilinear form* $\langle \cdot, \cdot \rangle$. *Let* $\mathrm{O}(V)$ *be the orthogonal group. If* $P = \{\{i_1, j_1\}, \{i_2, j_2\}, \dots, \{i_d, j_d\}\}$ *is a partition of the set* $\{1, 2, \dots, 2d\}$, *then we can define an* $\mathrm{O}(V)$*-invariant multi-linear form* $M_P : V^{2d} \to \mathbb{C}$ *by*

$$M_P(v_1, v_2, \dots, v_{2d}) = \prod_{k=1}^{d} \langle v_{i_k}, v_{j_k} \rangle.$$

The First Fundamental Theorem of Invariant Theory for the orthogonal group states that the space of $\mathrm{O}(V)$*-invariant multi-linear functions* $V^e \to \mathbb{C}$ *is* 0 *iff* e *is odd, and spanned by all* M_P *where* P *is a partition of* $\{1, 2, \dots, 2d\}$ *into pairs if* $e = 2d$. *Define* $H_{i,j} : V^d \to \mathbb{C}$ *by* $H_{i,j}(v_1, \dots, v_n) = \langle v_i, v_j \rangle$. *Show that the invariant ring* $\mathbb{C}[V^r]^{\mathrm{O}(V)}$ *is generated by all* $H_{i,j}$ *with* $1 \leq i, j \leq r$.

Exercise 10.1.5. *Proof Corollary* 10.1.3.

10.2. The Le Bruyn-Procesi Theorem

Using the First Fundamental Theorem of Invariant Theory from the previous section, we will now give a description of the invariant ring $\mathbb{C}[\mathrm{Rep}_\alpha(Q)]^{\mathrm{GL}_\alpha}$ due to Le Bruyn and Procesi for a quiver Q and a dimension vector α. The proof uses repeated contractions that come from the First Fundamental Theorem and the trace map which is a contraction from $V^\star \otimes V \cong \mathrm{End}(V)$ to $\mathbb{C}$.

Suppose that V, W, Z are finite dimensional vector spaces. We can identify $\mathrm{Hom}(W, V)$ with $W^\star \otimes V$. Let $\partial : V^\star \otimes V \to \mathbb{C}$ be the linear map given by $\partial(u \otimes v) = u(v)$. We have a linear map

$$\varphi = 1_Z \otimes \partial \otimes 1_{W^\star} : Z \otimes V^\star \otimes V \otimes W^\star \to Z \otimes W^\star.$$

After identifications, we can view φ as a linear map

$$\mathrm{Hom}(V, Z) \otimes \mathrm{Hom}(W, V) \to \mathrm{Hom}(W, Z)$$

and it is given by composition of linear maps: $\varphi(A \otimes B) = AB$.

Let us study invariant linear functions $\mathrm{Hom}(V, V)^{\otimes d} \to \mathbb{C}$. If we identify $\mathrm{Hom}(V, V)$ with $V^\star \otimes V$, then $\partial : V^\star \otimes V \to \mathbb{C}$ corresponds to the trace map trace : $\operatorname{Hom}(V,V) \to \mathbb{C}$. For a permutation $\sigma \in \operatorname{Perm}_d$, let $\partial_\sigma : V^{\otimes(d,d)} \to \mathbb{C}$ be the linear map corresponding to

$$M_\sigma : (V^\star)^d \times V^d \to \mathbb{C}.$$

We define a linear equivariant isomorphism

$$\Phi : V^{\otimes(d,d)} \to \operatorname{Hom}(V,V)^{\otimes d}$$

by

$$\Phi(u_1 \otimes \cdots \otimes u_d \otimes v_1 \otimes \cdots \otimes v_d) = (v_1 u_1) \otimes \cdots \otimes (v_d u_d).$$

The space of invariant linear functions $\operatorname{Hom}(V,V)^{\otimes d} \to \mathbb{C}$ is spanned by all $M_\sigma \circ \Phi^{-1}$, with $\sigma \in \operatorname{Perm}_d$. The multi-linear invariant functions $\operatorname{Hom}(V,V)^{\otimes d} \to \mathbb{C}$ correspond to products of traces of words in $A_1, \dots, A_d$ such that each matrix appears exactly once. For example, if $d = 6$ and $\sigma = (1\ 3\ 6)(2\ 5) = (1\ 3\ 6)(2\ 5)(4)$, then we have

$$\partial_\sigma(\Phi^{-1}(A_1 \otimes A_2 \otimes \cdots \otimes A_6)) = \operatorname{trace}(A_1 A_3 A_6)\operatorname{trace}(A_2 A_5)\operatorname{trace}(A_4).$$

More generally, we can study the invariant linear functions

$$\operatorname{Hom}(V,V)^{\otimes d} \otimes V^{\otimes(e,e)} \to \mathbb{C}.$$

We define a linear equivariant isomorphism

$$\Psi : V^{\otimes(d+e,d+e)} \to \operatorname{Hom}(V,V)^{\otimes d} \otimes V^{\otimes(e,e)}$$

by

$$\begin{aligned}
&\Psi(u_1 \otimes \cdots \otimes u_{d+e} \otimes v_1 \otimes \cdots \otimes v_{d+e}) \\
&\quad = (v_1 u_1) \otimes \cdots \otimes (v_d u_d) \otimes u_{d+1} \otimes \cdots \otimes u_{d+e} \otimes v_{d+1} \otimes \cdots \otimes v_{d+e}.
\end{aligned}$$

The space of invariant linear functions $\operatorname{Hom}(V,V)^{\otimes d} \otimes V^{\otimes(e,e)} \to \mathbb{C}$ is spanned by $M_\sigma \circ \Psi^{-1}$, where $\sigma \in \operatorname{Perm}_{d+e}$. For example, if $d = 3$ and $e = 3$ and $\sigma = (1\ 3)(2\ 5\ 4\ 6)$, then

$$\begin{aligned}
&M_\sigma(\Phi^{-1}(A_1 \otimes A_2 \otimes A_3 \otimes u_4 \otimes u_5 \otimes u_6 \otimes v_4 \otimes v_5 \otimes v_6)) \\
&\qquad = \operatorname{trace}(A_1 A_3)\operatorname{trace}(A_2 (v_5 u_5)(v_4 u_4)(v_6 u_6)) \\
&\qquad\qquad = \operatorname{trace}(A_1 A_3) \cdot (u_6 A_2 v_5) \cdot (u_5 v_4) \cdot (u_4 v_6).
\end{aligned}$$

Theorem 10.2.1 (Procesi). *Let $(A_1, \dots, A_p, u_1, \dots, u_r, v_1, \dots, v_s)$ be a general element of the representation $\operatorname{Hom}(V,V)^p \oplus (V^\star)^r \oplus V^s$. The invariant ring $\mathbb{C}[\operatorname{Hom}(V,V)^p \oplus (V^\star)^r \otimes V^s]^{\operatorname{GL}(V)}$ is generated by all invariants of the form:*

(1) $\operatorname{trace}(A_{i_1} A_{i_2} \cdots A_{i_k})$ *with* $1 \le i_1, \dots, i_k \le p$;

(2) $u_l A_{i_1} A_{i_2} \cdots A_{i_k} v_j$ *where* $1 \le i_1, \dots, i_k \le p$, $1 \le l \le r$ *and* $1 \le j \le s$.

Proof. The ring $\mathbb{C}[\mathrm{Hom}(V,V)^p \oplus (V^\star)^r \oplus V^s]$ is multi-graded. Suppose that $f \in \mathbb{C}[\mathrm{Hom}(V,V)^p \oplus (V^\star)^r \oplus V^s]^{\mathrm{GL}(V)}$ is homogeneous of degree $(c_1,\dots,c_p,d_1,d_2,\dots,d_r,e_1,e_2,\dots,e_s)$ and let $d=\sum_{i=1}^r d_i$, $e=\sum_{i=1}^s e_i$, $c=\sum_{i=1}^p c_i$. If f is nonzero, then $d=e$. There exists an invariant multi-linear function $F : \mathrm{Hom}(V,V)^c \times (V^\star)^d \times V^d \to \mathbb{C}$ such that

$$F(\underbrace{A_1,\dots,A_1}_{c_1},\dots,\underbrace{A_p,\dots,A_p}_{c_p},\underbrace{u_1,\dots,u_1}_{d_1},\dots,\underbrace{u_r,\dots,u_r}_{d_r},\underbrace{v_1,\dots,v_1}_{e_1},\dots,$$
$$\underbrace{v_s,\dots,v_s}_{e_s}) = f(A_1,\dots,A_p,u_1,\dots,u_r,v_1,\dots,v_s).$$

If F is nonzero, then we must have $d=e$. By the Fundamental Theorem of Invariant Theory, F is a linear combination of the functions $M_\sigma \circ \Psi^{-1}$, $\sigma \in \mathrm{Perm}_{c+d}$. But

$$M_\sigma \circ \Psi^{-1}(\underbrace{A_1,\dots,A_1}_{c_1},\dots,\underbrace{A_p,\dots,A_p}_{c_p},\underbrace{u_1,\dots,u_1}_{d_1},\dots,$$
$$\underbrace{u_r,\dots,u_r}_{d_r},\underbrace{v_1,\dots,v_1}_{e_1},\dots,\underbrace{v_s,\dots,v_s}_{e_s})$$

is a product of invariants of the form (1) and (2) in the theorem. □

Let us now consider an arbitary quiver Q, and let U be a generic α-dimensional representation.

Theorem 10.2.2. *For a dimension vector $\alpha \in \mathbb{N}^{Q_0}$, the ring of invariants $\mathbb{C}[\mathrm{Rep}_\alpha(Q)]^{\mathrm{GL}_\alpha}$ is generated by all* $\mathrm{trace}(U(p))$ *where p is a cyclic path.*

Proof. We will prove the theorem by induction on n, the number of vertices. The case $n=1$ follows from Procesi's theorem. Suppose that $n>1$. Let us choose a vertex $x \in Q_0$. For a representation $U \in \mathrm{Rep}_\alpha(Q)$, define

$$X = \bigoplus_{\substack{a\in Q_1\\ ha\neq x\neq ta}} \mathrm{Hom}(U(ta),U(ha)),$$

$$W = \bigoplus_{\substack{a\in Q_1\\ ha=x}} U(ta) \text{ and } Z = \bigoplus_{\substack{a\in Q_1\\ ta=x}} U(ha).$$

Let $V = U(x) \cong \mathbb{C}^{\alpha(x)}$. As a representation of $\mathrm{GL}(V)=\mathrm{GL}_{\alpha(x)}$, we have an isomorphism

$$\mathrm{Rep}_\alpha(Q) \cong \mathrm{Hom}(V,V)^r \oplus \mathrm{Hom}(V,Z) \oplus \mathrm{Hom}(W,V) \oplus X,$$

where r is the number of loops at vertex x (the number of arrows $a \in Q_1$ with $ha=ta=x$), and $\mathrm{GL}(V)$ acts trivially on X, W and Z. We can write $\mathrm{GL}_\alpha = \mathrm{GL}_{\alpha'}$ where α' is the restriction to $Q_0' := Q_0 \setminus \{x\}$.

From Theorem 10.2.1 follows that the invariant ring $K[\mathrm{Rep}_\alpha(Q)]^{\mathrm{GL}(V)}$ is generated by invariants of the form

(1) $\mathrm{trace}(U(p))$ where p is a cyclic path in the quiver Q and
(2) the entries of $U(q)$ where q is a path in Q that does not start or end at x.

Since the invariant ring is finitely generated, we can choose cyclic paths $p_1, p_2, \ldots, p_r$, and paths $q_1, q_2, \ldots, q_s$ not starting or ending at x, such that the invariant ring is generated by $\mathrm{trace}(U(p_i))$, $i = 1, 2, \ldots, r$ and the entries of $U(q_j)$, $j = 1, 2, \ldots, s$. Let Q' be the quiver (Q'_0, Q'_1) where $Q'_0 = Q_0 \setminus \{x\}$ as before, and $Q'_1 = \{b_1, b_2, \ldots, b_s\}$ where b_i is an arrow with the same head and tail as the path q_i for all i. We can view $(U(q_1), \ldots, U(q_s))$ as a representation of Q' of dimension α'. So we have a morphism

$$\pi : \mathrm{Rep}_\alpha(Q) \to \mathrm{Rep}_{\alpha'}(Q) \oplus \mathbb{C}^r$$

defined by

$$\pi(U) = (U(q_1), \ldots, U(q_s)), (\mathrm{trace}\, U(p_1), \ldots, \mathrm{trace}\, U(p_r)).$$

The invariant ring $\mathbb{C}[\mathrm{Rep}_\alpha(Q)]^{\mathrm{GL}(V)}$ is generated by all polynomial functions on $\mathrm{Rep}_{\alpha'}(Q') \oplus \mathbb{C}^r$ pulled back via φ. In other words, if we restrict the codomain of π to its image, then π is the categorical quotient map. Note that π is $\mathrm{GL}_{\alpha'}$-equivariant, where $\mathrm{GL}_{\alpha'}$ acts trivially on the $\mathbb{C}^r$. Generators of the invariant ring

$$\mathbb{C}[\mathrm{Rep}_\alpha(Q)]^{\mathrm{GL}_\alpha} = (\mathbb{C}[\mathrm{Rep}_\alpha(Q)]^{\mathrm{GL}(V)})^{\mathrm{GL}'_\alpha}$$

are obtained by pulling back invariant functions in

$$\mathbb{C}[\mathrm{Rep}_{\alpha'}(Q') \oplus \mathbb{C}^r]^{\mathrm{GL}_{\alpha'}} = \mathbb{C}[\mathrm{Rep}_{\alpha'}(Q')]^{\mathrm{GL}_{\alpha'}}[z_1, z_2, \ldots, z_r]$$

via φ. By induction, we know that $\mathbb{C}[\mathrm{Rep}_{\alpha'}(Q')]^{\mathrm{GL}_{\alpha'}}$ is generated by traces of cyclic paths. If we pull back traces of cyclic paths in Q' via φ, then we get traces if cyclic paths in $U(q_1), \ldots, U(q_s)$. If $b_{i_1} b_{i_2} \cdots b_{i_t}$ is a cyclic path in Q', then $q = q_{i_1} q_{i_2} \cdots q_{i_t}$ is a cyclic path in Q and

$$\mathrm{trace}(U(q_{i_1})U(q_{i_2}) \cdots U(q_{i_t})) = \mathrm{trace}(U(q))$$

is a trace of a cyclic path in Q. These invariants, together with $\mathrm{trace}(U(p_i))$ for $i = 1, 2, \ldots, r$, generate $\mathbb{C}[\mathrm{Rep}_\alpha(Q)]^{\mathrm{GL}_\alpha}$. So the invariant ring

$$\mathbb{C}[\mathrm{Rep}_\alpha(Q)]^{\mathrm{GL}_\alpha}$$

is generated by traces of cyclic paths in Q. □

Exercises.

Exercise 10.2.1. *Suppose that V is an n-dimensional vector space. Show that $\mathbb{C}[\mathrm{Hom}(V,V)]^{\mathrm{GL}(V)}$ is generated by the functions* $\mathrm{trace}(A)$, $\mathrm{trace}(A^2)$, $\dots$, $\mathrm{trace}(A^n)$. *(Hint: Take the trace of the Cayley-Hamilton identity.)*

Exercise 10.2.2. *We can view* $\mathrm{Rep}_\alpha(Q)$ *as a representation of* GL_α*: Show that for a representation $V \in \mathrm{Rep}_\alpha(Q)$ the following statements are equivalent:*

(1) *V is nilpotent (see Exercise 9.7.3);*

(2) *V has a filtration $V_0 \subseteq V_1 \subseteq V_2 \subseteq \cdots \subseteq V_r = V$ such that every quotient V_i/V_{i-1} is one-dimensional;*

(3) *for every cyclic path p, $V(p)$ is nilpotent.*

10.3. Background from the Representation Theory of GL_n

In this section we will study the representation theory of GL_n and the related Schur functors. The properties of Schur functors, such as the tensor product formulas using Littlewood-Richardson coefficients and the Cauchy formulas can be used to computations in rings of semi-invariants, which will be explored in the next section.

Suppose that $V = \mathbb{C}^n$ is an n-dimensional vector space, and let $\mathrm{GL}_n = \mathrm{GL}(V)$ be the general linear group. A rational representation of $\mathrm{GL}(V)$ is a morphism $\rho : \mathrm{GL}(V) \to \mathrm{GL}(W)$ of algebraic groups, where W is a finite dimensional $\mathbb{C}$-vector space. If we choose a basis in W, and the dimension of W is m, then we can write

$$\rho = \begin{pmatrix} \rho_{1,1} & \rho_{1,2} & \cdots & \rho_{1,m} \\ \rho_{2,1} & \rho_{2,2} & \cdots & \rho_{2m} \\ \vdots & \vdots & \ddots & \vdots \\ \rho_{m,1} & \rho_{m,2} & \cdots & \rho_{m,m} \end{pmatrix},$$

where $\rho_{i,j} \in \mathbb{C}[x_{1,1}, x_{1,2}, \dots, x_{n,n}, \det^{-1}] = \mathbb{C}[\mathrm{GL}(W)]$ for all i, j.

Definition 10.3.1. The representation ρ is **polynomial** if $\rho_{i,j} \in \mathbb{C}[x_{1,1}, x_{1,2}, \dots, x_{n,n}]$ for all i, j. A polynomial representation ρ is **homogeneous** of degree d if the entry functions $\rho_{i,j}$ are homogeneous polynomial functions of degree d for all i, j.

Let $B \subseteq \mathrm{GL}_n$ be the set of upper triangular matrix and let $U \subseteq B$ be the subgroup of upper triangular matrices with 1's on the diagonal. The subgroup of $T \subseteq B$ consists of all diagonal matrices. The group B is a semi-direct product $B = T \ltimes U$. The subsets T, U, B are closed subgroups.

As before, let V be a rational representation of GL_n. Then V, treated as a representation of T decomposes to one-dimensional representations (characters) of T. This leads to **the weight space decomposition**

$$V = \bigoplus_{\chi \in \mathbb{Z}^n} V_\chi,$$

where

$$V_\chi = \{v \in V \mid \forall t \in T \; t \cdot v = t_1^{\chi_1} \dots t_n^{\chi_n} v\},$$

and

$$t = \begin{pmatrix} t_1 & & & \\ & t_2 & & \\ & & \ddots & \\ & & & t_n \end{pmatrix}$$

is a general diagonal matrix. Elements of V_χ are called weight vectors with weight χ.

A nonzero vector $v \in V$ is a **highest weight vector** of weight $\mu = (\mu_1, \dots, \mu_n)$ if v is U-invariant and is a weight vector of weight μ. A weight $\mu = (\mu_1, \dots, \mu_n) \in \mathbb{Z}^n$ is a **dominant weight** if $\mu_1 \geq \dots \geq \mu_n$. Weights of highest weight vectors are always dominant. We call V a **highest weight module** of weight μ if V is generated as a GL_n-module by a highest weight vector of weight μ. Every rational representation V of GL_n has a highest weight vector. The next two results can be found in [**45**].

Theorem 10.3.2. *The isomorphism classes of irreducible representations of GL_n are in bijection with the set Λ^+ of dominant weights. To each weight μ this bijection associates the (unique up to isomorphism) irreducible highest weight module of weight μ.*

Proposition 10.3.3.

(1) *For every rational representation V there is a natural number m such that $V \otimes \det^m$ is polynomial;*

(2) *every polynomial representation V of GL_n decomposes to a direct sum of homogeneous polynomial representations.*

Next we describe the irreducible representations of GL_n that are polynomial of degree d. Let $V = \mathbb{C}^n$ be the natural representation, i.e., $\mathrm{GL}_n = \mathrm{GL}(V)$.

The d-fold tensor product

$$V^{\otimes d} = \underbrace{V \otimes V \otimes \dots \otimes V}_{d},$$

the d-th symmetric power S^dV and the d-th exterior power $\bigwedge^d V$of V are irreducible representations of $\mathrm{GL}(V)$. Using these we can build a more complex **Schur representation** $\mathbb{S}_\lambda V$ as follows.

A **partition** with r parts is a sequence $\lambda = (\lambda_1, \lambda_2, \ldots, \lambda_r)$ of positive integers with $\lambda_1 \geq \lambda_2 \geq \cdots \geq \lambda_r$. If $d = |\lambda| = \lambda_1 + \cdots + \lambda_r$, then we say that λ is a partition of d. We also identify $\lambda = (\lambda_1, \lambda_2, \ldots, \lambda_r)$ with $(\lambda_1, \lambda_2, \ldots, \lambda_r, 0, 0, \ldots, 0)$. So a partition of at most m parts is a sequence $(\lambda_1, \ldots, \lambda_m)$ of nonnegative integers with $\lambda_1 \geq \lambda_2 \geq \cdots \geq \lambda_m \geq 0$. We sometimes will use shorthand notation: a^b means repeating b times the number a. For example, we have $(5^2, 3, 2^3, 1^4) = (5, 5, 3, 2, 2, 2, 1, 1, 1, 1)$.

Let λ be a partition. The **Young diagram** $D(\lambda)$ of λ is the arrangement of boxes with λ_i boxes in the i-th row, aligned in such way that the first boxes in each row from a column.

Example 10.3.4. Let $\lambda = (4, 2, 1)$. We have

$D(\lambda) =$ 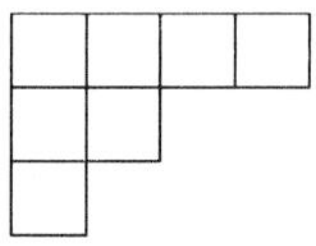

and $|\lambda| = 4 + 2 + 1 = 7$.

The **conjugate partition** λ' to the partition λ is the partition we get when we read the column lengths of $D(\lambda)$.

Example 10.3.5. For the partition $\lambda = (5, 4, 1, 1)$ of 11, we have $\lambda' = (3, 2, 1, 1)$. The diagram $D(\lambda')$ is obtained from $D(\lambda)$ by flipping it about the diagonal:

$D(\lambda) =$ 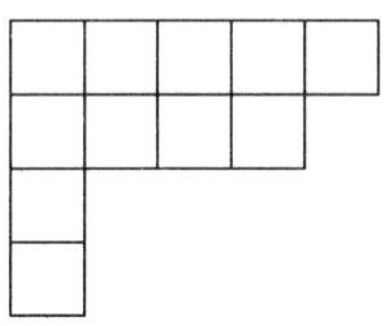 and $D(\lambda') =$ 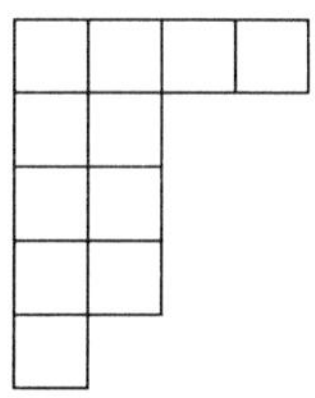

Let $\lambda = (\lambda_1, \ldots, \lambda_m)$ be a partition, and let $\lambda' = (\lambda'_1, \lambda'_2, \ldots, \lambda'_r)$ be its conjugate. We define a $\mathrm{GL}(V)$-equivariant map

$$\Phi_\lambda : \bigwedge^{\lambda'_1} V \otimes \ldots \otimes \bigwedge^{\lambda'_s} V \to S^{\lambda_1} V \otimes \ldots \otimes S^{\lambda_n} V$$

as the composition

$$\bigwedge^{\lambda'_1} V \otimes \cdots \otimes \bigwedge^{\lambda'_s} V \xrightarrow{\ \Delta\ } V^{\otimes d} \xrightarrow{\ m\ } S^{\lambda_1} V \otimes \cdots \otimes S^{\lambda_n} V \ .$$

Here Δ is a tensor product of the diagonal embeddings

$$\Delta_i : \bigwedge^{\lambda'_i} V \to V^{\otimes \lambda'_i}$$

and m is the product of the multiplication maps

$$m_j : V^{\otimes \lambda_j} \to S^{\lambda_j} V.$$

We define $\mathbb{S}_\lambda V$ is the image of the map Φ_λ.

We think of the factors in $V^{\otimes d}$ as corresponding to boxes in the Young diagram of λ. The diagonal map Δ_i maps $\bigwedge^{\lambda'_i} V$ into the factors corresponding to the i-th column of λ. Similarly, m_j multiplies the factors corresponding to the j-th row of λ.

If $\lambda'_1 = m$ (the number of parts in λ) is larger than $n = \dim V$, then $\bigwedge^{\lambda'_1} V = 0$ and $\mathbb{S}_\lambda V = 0$. If W is another vector space, then we have a commutative diagram,

$$\begin{array}{ccccc}
\bigwedge^{\lambda'_1} V \otimes \cdots \otimes \bigwedge^{\lambda'_s} V & \xrightarrow{\Delta} & V^{\otimes d} & \xrightarrow{m} & S^{\lambda_1} V \otimes \cdots \otimes S^{\lambda_n} V \\
\downarrow & & \downarrow & & \downarrow \\
\bigwedge^{\lambda'_1} W \otimes \cdots \otimes \bigwedge^{\lambda'_s} W & \xrightarrow{\Delta} & W^{\otimes d} & \xrightarrow{m} & S^{\lambda_1} W \otimes \cdots \otimes S^{\lambda_n} W
\end{array}$$

which induces a linear map from $\mathbb{S}_\lambda V = \operatorname{im} \Phi_\lambda^V$ to $\mathbb{S}_\lambda W = \operatorname{im} \Phi_\lambda^W$. One can verify that $\mathbb{S}_\lambda$ is a functor from the category of vector spaces to itself, and it is called the **Schur functor**.

Let λ and μ be two partitions. Assume that $\mu \subset \lambda$, i.e., $D(\mu) \subset D(\lambda)$. We define the skew diagram λ/μ to be $D(\lambda/\mu) := D(\lambda) \setminus D(\mu)$.

Example 10.3.6. For example, the skew diagram $D((4,4,2,1)/(2,1,1))$ is

<table>
<tr><td>•</td><td>•</td><td> </td><td> </td></tr>
<tr><td>•</td><td> </td><td> </td><td> </td></tr>
<tr><td>•</td><td> </td></tr>
<tr><td> </td></tr>
</table>

A **tableau** T of shape λ/μ is a function from boxes of $D(\lambda/\mu)$ to $\mathbb{Z}_{>0}$ sending the box (i,j) in the i-th row and j-th column of $D(\lambda/\mu)$ to $t(i,j)$. We think of this function as filling the diagram $D(\lambda/\mu)$ with numbers, putting $t(i,j)$ into the box (i,j).

A tableau T is **column-standard** if we have $t(i,j) < t(i+1,j)$ and $t(i,j) \leq t(i,j+1)$ provided both boxes are in $D(\lambda/\mu)$. The **content** of a tableau T is the sequence $(1^{a_1}, 2^{a_2}, \ldots, n^{a_n})$ where a_i is the number of boxes filled with the number i.

Example 10.3.7. The tableau

<table>
<tr><td>•</td><td>1</td><td>1</td><td>2</td></tr>
<tr><td>1</td><td>2</td></tr>
<tr><td>3</td></tr>
</table>

is a column standard tableau of shape $(4,2,1)/(1)$ of content $(1^3,2^2,3)$.

The main result on representation theory of GL_n is as follows.

Theorem 10.3.8. *The Schur representations twisted by powers of determinant give all irreducible representations of the group* GL_n. *More precisely:*

(1) *For a partition $\lambda = (\lambda_1, \lambda_2, \ldots, \lambda_r)$ with $r \leq n$ parts, the Schur representation $\mathbb{S}_\lambda V$ is the irreducible representation of GL_n with highest weight λ. The U-invariant of weight λ in $\mathbb{S}_\lambda V$ is the image by Φ_λ of the product of U-invariants $e_1 \wedge e_2 \wedge \ldots \wedge e_{\lambda'_i} \in \bigwedge^{\lambda'_i} V$.*

(2) *For the dominant integral weight $\mu = (\mu_1, \ldots, \mu_n)$ the corresponding irreducible representation of GL_n is $\mathbb{S}_\lambda V \otimes \det^{\mu_n}$ where $\lambda = (\lambda_1, \ldots, \lambda_n) = (\mu_1 - \mu_n, \ldots, \mu_{n-1} - \mu_n, 0)$.*

One of the key results is the **Littlewood-Richardson rule** giving the decomposition of tensor product into irreducible representations. Before we state the rule we need one more definition.

Let us write $[1,n]$ for the set $\{1,2,\ldots,n\}$. A word $w = (w_1, \ldots, w_m)$ with $w_1, w_2, \ldots, w_m \in [1,n]$ for $1 = 1, \ldots, m$ is a **lattice permutation** if for each k, $1 \leq k \leq m$ and for each $j = 1, \ldots, n-1$ we have

$$|\{i \in [1,k] \mid w_i = j\}| \geq |\{i \in [1,k] \mid w_i = j+1\}|.$$

Example 10.3.9. The word $(1,1,2,2,1,2)$ is a lattice permutation and $(1,2,2,1,1,2)$ is not (because there are more 2's than 1's among the first 3 entries).

Theorem 10.3.10. *Let λ, μ be two partitions into at most n parts. We have*

$$\mathbb{S}_\lambda V \otimes \mathbb{S}_\mu V = \bigoplus_{|\nu|=|\lambda|+|\mu|} (\mathbb{S}_\nu V)^{c^\nu_{\lambda,\mu}},$$

where $c^\nu_{\lambda,\mu}$ is the number of skew column-standard diagrams D of shape ν/λ, of content $1^{\mu_1}, \ldots, n^{\mu_n}$ such that the associated word $w(D)$ of D (obtained by reading D by rows, from right to left) is a lattice permutation.

The numbers $c^\nu_{\lambda,\mu}$ are called **Littlewood-Richardson coefficients**.

Example 10.3.11. Here are the diagrams giving the representation occurring in $\mathbb{S}_{(3,1)}V \otimes \mathbb{S}_{(2,1)}V$.

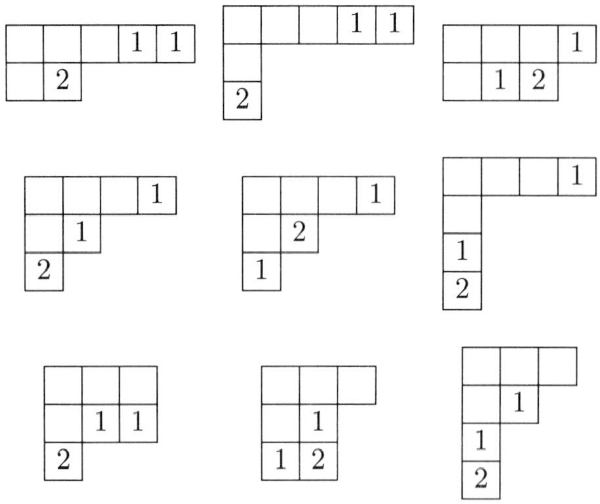

Note that the representation $\mathbb{S}_{(4,2,1)}V$ occurs in the tensor product with multiplicity 2.

The Littlewood-Richardson rule has an easy consequence which is very useful when calculating the semi-invariants.

Lemma 10.3.12. *Let $\lambda = (\lambda_1, \dots, \lambda_n)$ and $\mu = (\mu_1, \dots, \mu_n)$ be two highest weights for $\mathrm{GL}_n = \mathrm{GL}(V)$. The tensor product $\mathbb{S}_\lambda V \otimes \mathbb{S}_\mu V$ contains a semi-invariant for $\mathrm{GL}(V)$ if and only if*

$$\lambda_1 + \mu_n = \lambda_2 + \mu_{n-1} = \cdots = \lambda_n + \mu_1.$$

In such case the tensor product contains exactly one $\mathrm{GL}(V)$-semi-invariant of weight $\lambda_1 + \mu_n$.

Proof. See Exercise 10.3.1. □

Theorem 10.3.13 (Cauchy formulas). *Let $\mathcal{P}(d)$ denote the set of partitions of d. We have decompositions*

$$S^d(V \otimes W) = \bigoplus_{\lambda \in \mathcal{P}(d)} \mathbb{S}_\lambda V \otimes \mathbb{S}_\lambda W,$$

$$\bigwedge^d(V \otimes W) = \bigoplus_{\lambda \in \mathcal{P}(d)} \mathbb{S}_\lambda V \otimes \mathbb{S}_{\lambda'} W.$$

Suppose that $\lambda = (\lambda_1, \lambda_2, \dots, \lambda_n)$ is a dominant weight. If V is an n-dimensional representation, then we denote the irreducible representation of $\mathrm{GL}(V)$ with heighest weight by $\mathbb{S}_\lambda V$. If $\lambda_n \geq 0$, then we $\mathbb{S}_\lambda$ is a functor form the category of vector spaces to itself. If $\lambda_n < 0$, then we cannot view $\mathbb{S}_\lambda$ as a functor because $\mathbb{S}_\lambda V$ is not even defined if V is not n-dimensional.

Exercises.

Exercise 10.3.1. *Prove Lemma* 10.3.12 *as follows using the Littlewood-Richardson rule. Show that $\mathbb{S}_\lambda V \otimes \mathbb{S}_\mu V$ can only have a semi-invariant if $|\lambda| + |\mu|$ is divisible by n. Write $|\lambda| + |\mu| = nm$. Show that the dimension of $(\mathbb{S}_\lambda V \otimes \mathbb{S}_\mu V)^{\mathrm{SL}(V)}$ is equal to $c^{m^n}_{\lambda,\mu}$. Prove that there is only one skew column-standard diagram of shape m^n/λ for which the associated word is a lattice permutation.*

10.4. Semi-invariants and Representation Theory

For a quiver Q and a dimension vector α we have an action of GL_α on $\mathrm{Rep}_\alpha(Q)$ and an invariant ring $\mathrm{I}(Q,\alpha) = \mathbb{C}[\mathrm{Rep}_\alpha(Q)]^{\mathrm{GL}_\alpha}$. If Q has no oriented cycles, then this invariant ring is trivial. To have more invariants, we consider the action of a normal subgroup, namely

$$\mathrm{SL}_\alpha = \prod_{x \in Q_0} \mathrm{SL}_{\alpha(x)} \subset \mathrm{GL}_\alpha\,.$$

We will study the invariant ring

$$\mathrm{SI}(Q,\alpha) = \mathbb{C}[\mathrm{Rep}_\alpha(Q)]^{\mathrm{SL}_\alpha}.$$

Recall that for every weight $\sigma \in \mathbb{Z}^{Q_0}$ we have defined a multiplicative character $\chi_\sigma : \mathrm{GL}_\alpha \to \mathbb{G}_\mathrm{m}$ by

$$\chi_\alpha(A(x) \mid x \in Q_0) = \prod_{x \in Q_0} \det(A(x))^{\sigma(x)}.$$

The space of semi-invariants of weight σ is

$$\mathrm{SI}(Q,\alpha)_\sigma = \{f \in \mathbb{C}[\mathrm{Rep}_\alpha(Q)] \mid \forall A \in \mathrm{GL}_\alpha \;\; A \cdot f = \chi_\sigma(A)\}.$$

Note that $\mathrm{SI}(Q,\alpha)_\sigma$ is contained in $\mathrm{SI}(Q,\alpha)$.

Lemma 10.4.1. *We have*

$$\mathrm{SI}(Q,\alpha) = \bigoplus_\sigma \mathrm{SI}(Q,\alpha)_\sigma.$$

Proof. The group $\mathrm{GL}_\alpha / \mathrm{SL}_\alpha$ is a torus acting on $\mathrm{SI}(Q,\alpha)$. This shows that every $f \in \mathrm{SI}(Q,\alpha)$ is a sum of semi-invariants of various weights. The sum on the right-hand side is direct, because semi-invariants of different weights are linearly independent (see Exercise 9.2.7). □

The dimensions of the weight spaces of the rings of semi-invariants can be calculated using standard representation theory. Recall that if E is an n-dimensional vector space, then the irreducible one-dimensional polynomial representations of $\mathrm{GL}(E)$ are $\det^k = \mathbb{S}_{(k^n)}(E)$ with $k \in \mathbb{Z}$. If $\lambda = (\lambda_1, \ldots, \lambda_n)$ is dominant, then

$$\lambda^\star := (-\lambda_n, -\lambda_{n-1}, \ldots, -\lambda_1)$$

is also dominant and $\mathbb{S}_{\lambda^\star} E$ is isomorphic to the dual space $(\mathbb{S}_\lambda E)^\star \cong \mathbb{S}_\lambda E^\star$ as a representation of $\mathrm{GL}(E)$, which are the dominant integral weights

$$\lambda = (\lambda_1, \ldots, \lambda_n), \lambda_i \in \mathbb{Z}, \lambda_1 \geq \cdots \geq \lambda_n.$$

The corresponding representation, $\mathbb{S}_\lambda E$, is a Schur functor tensored with a power of determinantal character. The only one-dimensional representations are the powers of the determinant, i.e., the representations with highest weights $\lambda = (m^n)$, $m \in \mathbb{Z}$.

For a vector space W, the symmetric algebra on W is

$$S(W) \cong \bigoplus_{d \geq 0} S^d(W),$$

where $S^d(W)$ is the d-th symmetric power. The ring of polynomial functions on W is $\mathbb{C}[W] = S(W^\star)$. We have a natural isomorphism

$$S(W \oplus Z) = S(W) \otimes S(Z).$$

If $\mathcal{P} = \bigcup_d \mathcal{P}(d)$ is the set of all partitions, then it follows from Cauchy's formula (10.3.13) that

$$S(W \otimes Z) = \bigoplus_{d=0}^{\infty} S^d(W \otimes Z) = \bigoplus_{d=0}^{\infty} \bigoplus_{\lambda \in \mathcal{P}(d)} \mathbb{S}_\lambda W \otimes \mathbb{S}_\lambda Z = \bigoplus_{\lambda \in \mathcal{P}} \mathbb{S}_\lambda W \otimes \mathbb{S}_\lambda Z.$$

Let Q be a quiver and let β be a dimension vector. Denote by $V(x)$ a vector space of dimension $\beta(x)$ ($x \in Q_0$). The coordinate ring of the representation space $\mathrm{Rep}_\beta(Q)$ has the following decomposition to irreducible representations of GL_β:

$$\begin{aligned}(10.1)\quad \mathbb{C}[\mathrm{Rep}_\beta(Q)] &= \mathbb{C}[\textstyle\bigoplus_{a \in Q_1} \mathrm{Hom}(V(ta), V(ha))] \\ &= S(\textstyle\bigoplus_{a \in Q_1} \mathrm{Hom}(V(ta), V(ha))^\star) = \textstyle\bigotimes_{a \in Q_1} S(\mathrm{Hom}(V(ta), V(ha))^*) \\ &= \textstyle\bigotimes_{a \in Q_1} S(V(ta) \otimes V(ha)^*).\end{aligned}$$

Applying Cauchy's formula (10.3.13) we get the decomposition

$$\mathbb{C}[\mathrm{Rep}_\beta(Q)] = \bigoplus_{\lambda : Q_1 \to \mathcal{P}} \bigotimes_{a \in Q_1} \mathbb{S}_{\lambda(a)} V(ta) \otimes \mathbb{S}_{\lambda(a)} V(ha)^*, \tag{10.2}$$

where we sum into the set of functions λ from Q_1 to the set $\mathcal{P}$ of partitions.

A semi-invariant of weight σ is the occurrence of the representation

$$\bigotimes_{x\in Q_0} \mathbb{S}_{\sigma(x)^{\beta(x)}} V(x)$$

in $\mathbb{C}[\mathrm{Rep}_\beta(Q)]$. This means the dimension of the weight space can be calculated using Littlewood-Richardson rule (10.3.10).

$$\text{(10.3)} \quad \dim\ \mathrm{SI}(Q,\beta)_\sigma = \mathrm{mult}\left(\bigotimes_{x\in Q_0} \mathbb{S}_{\sigma(x)^{\beta(x)}} V(x); \mathbb{C}[\mathrm{Rep}_\beta(Q)]\right)$$

$$= \mathrm{mult}\left(\bigotimes_{x\in Q_0} \mathbb{S}_{\sigma(x)^{\beta(x)}} V(x); \bigoplus_{\lambda:Q_1\to\mathcal{P}} \bigotimes_{a\in Q_1} \mathbb{S}_{\lambda(a)} V(ta) \otimes \mathbb{S}_{\lambda(a)} V(ha)^*\right).$$

10.5. Examples for Dynkin Quivers

Let us look at some examples of semi-invariants for Dynkin quivers.

Example 10.5.1. Let Q is be the $\mathbf{A}_1$ type quiver

$$1 \longrightarrow 2\ .$$

Suppose that $\beta = (n, m)$ with $n \neq m$ and σ is a weight. If $\sigma(\beta) \neq 0$, then there are no semi-invariants of weight σ. Suppose that $\sigma(\beta) = 0$. Then we can write $\sigma = (km, -kn)$ for some rational number k for which km and kn are integers. Now $\dim \mathrm{SI}(Q, \beta)_\sigma$ is the multiplicity of

$$\text{(10.4)} \qquad \mathbb{S}_{((km)^n)} V(1) \otimes \mathbb{S}_{((-kn)^m)} V(2) = \mathbb{S}_{((km)^n)} V(1) \otimes \mathbb{S}_{((kn)^m)} V(2)^\star$$

in

$$\text{(10.5)} \qquad \mathbb{C}[\mathrm{Rep}_\beta(Q)] = \bigoplus_{\lambda\in\mathcal{P}} \mathbb{S}_\lambda V(1) \otimes \mathbb{S}_\lambda V(2)^\star.$$

For the representation (10.4) to appear in (10.5), we need $\lambda = ((km)^n) = ((kn)^m)$. Since $n \neq m$ we have $\mathrm{SI}(Q, \beta)_\sigma = 0$.

Now assume that $\beta = (n, n)$. To have a nontrivial weight, we need $\sigma = (k, -k)$ for some $k \in \mathbb{Z}$. Now $\dim \mathrm{SI}(Q, \beta)_\sigma$ is the multiplicity of

$$\text{(10.6)} \qquad \mathbb{S}_{(k^n)} V(1) \otimes \mathbb{S}_{((-k)^n)} V(2) = \mathbb{S}_{(k^n)} V(1) \otimes \mathbb{S}_{(k^n)} V(2)^\star$$

in (10.5). Clearly, if $k \geq 0$, then the representation (10.6) appears in (10.5) exactly once, namely for $\lambda = (k^n)$. This proves that $\dim \mathrm{SI}(Q, \beta)_\sigma = 1$. For $k < 0$ the representation (10.6) does not appear in (10.5) and $\dim \mathrm{SI}(Q, \beta)_\sigma = 0$. The polynomial function $\det V(a)$ is a semi-invariant sending a representation V to the determinant of the map $V(a)$. The weight of $\det V(a)$ is $(1, -1)$ because it corresponds to the representation $\bigwedge^n V(1) \otimes \bigwedge^n V(2)^*$ inside of $\mathbb{C}[\mathrm{Rep}_\beta(Q)]$. For $k \geq 0$, $(\det V(a))^k$ is a semi-invariant of weight $(k, -k)$. So $\det(V(a))^k$, $k \geq 0$ span the space of all invariants and we have $\mathrm{SI}(Q, \beta) = \mathbb{C}[\det V(a)]$.

In terms of Young diagrams, Lemma 10.3.12 states that $\mathbb{S}_{(k^n)}V$ appears in $\mathbb{S}_\lambda V \otimes \mathbb{S}_\mu V$ if μ rotated over 180° and λ can be put toghether to form an $n \times k$ box. For example, if $\lambda = (5,3,1,0)$ and $\mu = (6,5,3,1)$, then $\mathbb{S}_{(6^4)}V$ appears in $\mathbb{S}_\lambda V \otimes \mathbb{S}_\mu V$:

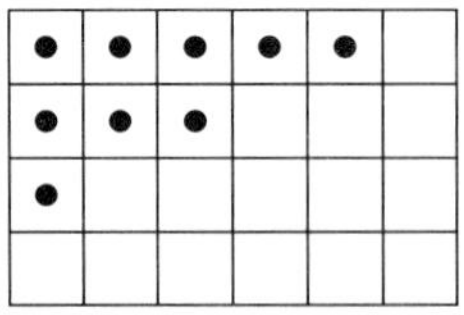

Example 10.5.2. Let Q is be the $\widehat{\mathbf{A}}_1$ type quiver

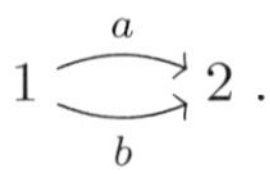

Suppose that $\beta = (n,n)$ and $\sigma = (k,-k)$. Now $\dim \mathrm{SI}(Q,\beta)_\sigma$ is the multiplicity of

$$\mathbb{S}_{(k^n)}V(1) \otimes \mathbb{S}_{(k^n)}V(2)^\star \tag{10.7}$$

in

$$\begin{aligned} \mathbb{C}[\mathrm{Rep}_\beta(Q)] &= \bigoplus_{\lambda,\mu \in \mathcal{P}} \mathbb{S}_\lambda V(1) \otimes \mathbb{S}_\lambda V(2)^\star \otimes \mathbb{S}_\mu V(1) \otimes \mathbb{S}_\mu V(2)^\star \\ &= \bigoplus_{\lambda,\mu \in \mathcal{P}} \Big(\mathbb{S}_\lambda V(1) \otimes \mathbb{S}_\mu V(1)\Big) \otimes \Big(\mathbb{S}_\lambda V(2)^\star \otimes \mathbb{S}_\mu V(2)^\star\Big). \end{aligned} \tag{10.8}$$

The representation $\mathbb{S}_{(k^n)}V(1)$ appears in $\mathbb{S}_\lambda V(1) \otimes \mathbb{S}_\mu V(1)$ if and only if λ and μ fit together in the box k^n, i.e., we have $\lambda_i = k - \mu_{n-i}$ for all i. So for every λ contained in the box k^n there exists exactly one μ such that $\mathbb{S}_{(k^n)}V(1)$ appears in $\mathbb{S}_\lambda V(1) \otimes \mathbb{S}_\mu V(1)$. For such choice of λ and μ the representation (10.7) appears in

$$\Big(\mathbb{S}_\lambda V(1) \otimes \mathbb{S}_\mu V(1)\Big) \otimes \Big(\mathbb{S}_\lambda V(2)^\star \otimes \mathbb{S}_\mu V(2)^\star\Big).$$

It follows that the multiplicity of (10.7) in (10.8) is exactly the number of partitions that fit into (k^n). We can think of a partition inside a $k \times n$ box as a path from the lower left corner to the upper right corner by only going right and up. There are $\binom{n+k}{k}$ such paths. We conclude that

$$\dim \mathrm{SI}(Q,\beta)_\sigma = \binom{n+k}{k}.$$

We will describe $\mathrm{SI}(Q,\beta)$. We can write

$$\det(V(a) + tV(b)) = p_0 + p_1 t + \cdots + p_n t^n,$$

where $p_0, p_1, \ldots, p_n$ are semi-invariants of weight $(1,-1)$. It turns out that $p_0, p_1, \ldots, p_n$ are algebraically independent (see Exercise 10.5.1), and $\mathrm{SI}(Q,\beta)$

contains the polynomial ring $\mathbb{C}[p_0, p_1, \ldots, p_n]$. There are $\binom{k+n}{k}$ linearly independent monomials of degree k in $p_0, p_1, \ldots, p_n$ and these are all semi-invariants of weight $(k, -k)$. It follows that $\mathrm{SI}(Q, \beta) = \mathbb{C}[p_0, p_1, \ldots, p_n]$.

Example 10.5.3. Let us consider the quiver

$$1 \longrightarrow 2 \longrightarrow \cdots \longrightarrow n \xrightarrow{(n)} n+1$$

with n arrows from n to $n+1$ and choose the dimension vector $\beta = (1, 2, 3, \ldots, n, 1)$. Suppose that $\lambda = (\lambda_1, \lambda_2, \ldots, \lambda_n)$ is a partition and define a weight by

$$\sigma = (\lambda_1 - \lambda_2, \lambda_2 - \lambda_3, \ldots, \lambda_{n-1} - \lambda_n, \lambda_n, -|\lambda|).$$

Choose $V(i) \cong \mathbb{C}^i$ for $i = 1, 2, \ldots, n$ and we take $V(n+1) = \mathbb{C}$.

$$\begin{aligned} \mathrm{Rep}(Q, \beta) &= \mathrm{Hom}(V(1), V(2)) \oplus \cdots \oplus \mathrm{Hom}(V(n-1), V(n)) \\ &\quad \oplus \mathrm{Hom}(V(n), V(n+1))^n \\ &\cong \mathrm{Hom}(V(1), V(2)) \oplus \cdots \oplus \mathrm{Hom}(V(n-1), V(n)) \oplus \mathrm{Hom}(V(n), W^\star), \end{aligned}$$

where $W \cong \mathbb{C}^n$. To find $\dim \mathrm{SI}(Q, \beta)_\sigma$, we find the multiplicity of

$$\mathbb{S}_{(\lambda_1 - \lambda_2)} V(1) \otimes \mathbb{S}_{((\lambda_2 - \lambda_3)^2)} V(2) \otimes \cdots \mathbb{S}_{(\lambda_n^n)} V(n) \tag{10.9}$$

in

$$\begin{aligned} \mathbb{C}[\mathrm{Rep}_\beta(Q)] = \bigoplus_{\mu^{(1)}, \ldots, \mu^{(n)}} & \mathbb{S}_{\mu^{(1)}} V(1) \otimes \mathbb{S}_{\mu^{(1)}} V(2)^\star \otimes \mathbb{S}_{\mu^{(2)}} V(2) \\ & \otimes \mathbb{S}_{\mu^{(2)}} V(3)^\star \otimes \cdots \otimes \mathbb{S}_{\mu^{(n)}} V(n) \otimes \mathbb{S}_{\mu^{(n)}} W. \end{aligned} \tag{10.10}$$

Suppose that (10.9) appears in

$$\mathbb{S}_{\mu^{(1)}} V(1) \otimes \mathbb{S}_{\mu^{(1)}} V(2)^\star \otimes \mathbb{S}_{\mu^{(2)}} V(2) \otimes \mathbb{S}_{\mu^{(2)}} V(3)^\star \otimes \cdots \otimes \mathbb{S}_{\mu^{(n)}} V(n) \otimes \mathbb{S}_{\mu^{(n)}} W.$$

Then we have $\mu^{(1)} = (\lambda_1 - \lambda_2)$. Because $\mathbb{S}_{((\lambda_2 - \lambda_3)^2)} V(2)$ appears in $\mathbb{S}_{\mu^{(1)}} V(2)^\star \otimes \mathbb{S}_{\mu^{(2)}} V(2)$ we get $\mu^{(2)} = (\lambda_1 - \lambda_3, \lambda_2 - \lambda_3)$. Continuing like this we get

$$\mu^{(i)} = (\lambda_1 - \lambda_{i+1}, \lambda_2 - \lambda_{i+1}, \ldots, \lambda_i - \lambda_{i+1})$$

for $i = 1, 2, \ldots, n-1$ and $\mu^{(n)} = \lambda$. It follows that the multiplicity of (10.9) in (10.10) is $\dim \mathbb{S}_\lambda W$. In fact, we can identify $\mathrm{SI}(Q, \beta)_\sigma$ with $\mathbb{S}_\lambda W$.

Example 10.5.4. For integers $p, q, r \geq 2$ we define a quiver $T_{p,q,r}$ by

$$\begin{array}{ccccccccccccc}
x_1 & \longrightarrow & x_2 & \longrightarrow \cdots \longrightarrow & x_{p-1} & \longrightarrow & w & \longleftarrow & y_{q-1} & \longleftarrow \cdots \longleftarrow & y_2 & \longleftarrow & y_1 \\
 & & & & & & \uparrow & & & & & & \\
 & & & & & & z_{r-1} & & & & & & \\
 & & & & & & \uparrow & & & & & & \\
 & & & & & & \vdots & & & & & & \\
 & & & & & & \uparrow & & & & & & \\
 & & & & & & z_2 & & & & & & \\
 & & & & & & \uparrow & & & & & & \\
 & & & & & & z_1 & & & & & & .
\end{array}$$

and we use the convention $x_p = y_q = z_r = w$. Let us take $p = q = r = n$ and choose the dimension vector

$$\beta = \begin{array}{ccccccccc}
1 & 2 & \cdots & n-1 & n & n-1 & \cdots & 2 & 1 \\
 & & & & n-1 & & & & \\
 & & & & \vdots & & & & \\
 & & & & 2 & & & & \\
 & & & & 1 & & & & .
\end{array}$$

Suppose that $\lambda = (\lambda_1, \ldots, \lambda_n)$, $\mu = (\mu_1, \ldots, \mu_n)$ and $\nu = (\nu_1, \ldots, \nu_n)$ are partitions with $|\lambda|+|\mu|+|\nu| = mn$ for some integer m. We define a weight σ by $\sigma(x_i) = \lambda_i - \lambda_{i+1}$, $\sigma(y_i) = \mu_i - \mu_{i+1}$ and $\sigma(z_i) = \nu_i - \nu_{i+1}$ for $i = 1, 2, \ldots, n$ and $\sigma(w) = \lambda_n + \mu_n + \nu_n - m$.

We have $\operatorname{Rep}_\beta(Q) = X \oplus Y \oplus Z$, where

$$\begin{aligned}
X &= \bigoplus_{i=1}^{n-1} \operatorname{Hom}(V(x_i), V(x_{i+1})), \\
Y &= \bigoplus_{i=1}^{n-1} \operatorname{Hom}(V(y_i), V(y_{i+1})), \\
Z &= \bigoplus_{i=1}^{n-1} \operatorname{Hom}(V(z_i), V(z_{i+1})),
\end{aligned}$$

and $\mathrm{GL}_\beta = \mathrm{GL}_X \times \mathrm{GL}_Y \times \mathrm{GL}_Z \times \mathrm{GL}(W)$, where

$$\begin{aligned} \mathrm{GL}_X &= \prod_{i=1}^{n-1} \mathrm{GL}(V(x_i)), \\ \mathrm{GL}_Y &= \prod_{i=1}^{n-1} \mathrm{GL}(V(y_i)), \\ \mathrm{GL}_Z &= \prod_{i=1}^{n-1} \mathrm{GL}(V(z_i)), \end{aligned}$$

and $W = V(w)$. We have

$$\mathbb{C}[\mathrm{Rep}_\alpha(Q)] = C[X] \otimes \mathbb{C}[Y] \otimes \mathbb{C}[Z].$$

If we restrict σ to $\{x_1, \dots, x_{n-1}\}$, then the space of semi-invariants in $\mathbb{C}[X]$ with respect to GL_X of weight σ can be identified with the space $\mathbb{S}_{\lambda'} W$. Therefore, the space of semi-invariants in $\mathbb{C}[\mathrm{Rep}_\alpha(Q)] = C[X] \otimes \mathbb{C}[Y] \otimes \mathbb{C}[Z]$ with respect to the group $\mathrm{GL}_X \times \mathrm{GL}_Y \times \mathrm{GL}_Z$ of weight σ is isomorphic to

$$\mathbb{S}_\lambda W \otimes \mathbb{S}_\mu W \otimes \mathbb{S}_\nu W.$$

It follows that

$$\mathrm{SI}(Q, \beta)_\sigma \cong (\mathbb{S}_\lambda W \otimes \mathbb{S}_\mu W \otimes \mathbb{S}_\nu W)^{\mathrm{SL}(W)}.$$

Exercises.

Exercise 10.5.1. *Show that* $p_0, p_1, \dots, p_n$ *in Exercise* 10.5.2 *are algebraically independent.*

Exercise 10.5.2. *Consider the quiver* $T_{p,q,r}$ *and let* β *be a dimension vector with* $0 < \beta(x_1) < \beta(x_2) < \cdots < \beta(x_p)$, $0 < \beta(y_1) < \beta(y_2) < \cdots < \beta(y_q)$ *and* $0 < \beta(z_1) < \beta(z_2) < \cdots < \beta(z_r)$ *and let* σ *be a weight with* $\sigma(\beta) = 0$ *and let* $\sigma(x_i) \geq 0$ *for* $i < p$, $\sigma(y_j) \geq 0$ *for* $j < q$ *and* $\sigma(z_k) \geq 0$ *for* $k < r$. *Show that*

$$\mathrm{SI}(Q, \beta)_\sigma \cong (\mathbb{S}_\lambda W \otimes \mathbb{S}_\lambda W \otimes \mathbb{S}_\mu W)^{\mathrm{SL}(W)},$$

where W *is a vector space of dimension* $\beta(w)$ *for certain partitions* $\lambda = \lambda(\beta, \sigma)$, $\mu = \mu(\beta, \sigma)$ *and* $\nu = \nu(\beta, \sigma)$.

Exercise 10.5.3. *For a partition* λ, *let* $j(\lambda)$ *be the number of jumps. For example, a rectangle* (λ^k) *with* $\lambda, k > 0$ *has* 1 *jump, a "fat hook"* $(\lambda_1^{k_1}, \lambda_2^{k_2})$ *with* $\lambda_1 > \lambda_2 > 0$ *and* $k_1, k_2 > 0$ *has two jumps, etc. Show that if*

$$\frac{1}{j(\lambda)+1} + \frac{1}{j(\mu)+1} + \frac{1}{j(\nu)} > 1,$$

then we have $c^\nu_{\lambda,\mu} \leq 1$. *(Hint: Use Exercise* 10.5.2 *and note that* $T_{p,q,r}$ *is of finite type whenever* $1/p + 1/q + 1/r > 1$.*)*

10.6. Schofield Semi-invariants

We will now discuss a method of Schofield that constructs semi-invariants from representations. The main result in this chapter states that these semi-invariants span the ring of semi-invariants.

Suppose that Q is a quiver without oriented cycles, and α, β are dimension vectors. For representations $V \in \operatorname{Rep}_\alpha(Q)$ and $W \in \operatorname{Rep}_\beta(Q)$ we defined in Chapter 1, (1.1) a linear map
(10.11)

$$\bigoplus_{x \in Q_0} \operatorname{Hom}_{\mathbb{C}}(V(x), W(x)) \xrightarrow{d_W^V} \bigoplus_{a \in Q_1} \operatorname{Hom}_{\mathbb{C}}(V(ta)\,, W(ha))$$

by the formula

$$d_W^V(\phi(x), x \in Q_0) = (\phi(ha)V(a) - W(a)\phi(ta), a \in Q_1)$$

whose kernel and cokernel were $\operatorname{Hom}_Q(V, W)$ and $\operatorname{Ext}_Q(V, W)$, respectively. The left-hand side of (10.11) has dimension $\sum_{x \in Q_0} \alpha(x)\beta(x)$ and the right-hand side has dimension $\sum_{a \in Q_1} \alpha(ta)\beta(ha)$. The difference is

$$\begin{aligned} &\dim \operatorname{Hom}_Q(V, W) - \dim \operatorname{Ext}_Q(V, W) \\ &\qquad = \sum_{x \in Q_0} \alpha(x)\beta(x) - \sum_{a \in Q_1} \alpha(ta)\beta(ha) = \langle \alpha, \beta \rangle. \end{aligned}$$

The matrix d_W^V is square if and only if $\langle \alpha, \beta \rangle = 0$. Assume that $\langle \alpha, \beta \rangle = 0$. We define a polynomial c on $\operatorname{Rep}_\alpha(Q) \times \operatorname{Rep}_\beta(Q)$ by $c(V, W) = \det(d_W^V)$. For fixed $V \in \operatorname{Rep}_\alpha(Q)$, we view $c^V = c(V, \cdot)$ as a polynomial function on $\operatorname{Rep}_\beta(Q)$, and for fixed W we view $c_W = c(\cdot, W)$ as a polynomial function on $\operatorname{Rep}_\alpha(Q)$. An element of the group SL_α and SL_β act on the matrix d_W^V by left and right multiplication with matrices of determinant 1. From this it is clear that the polynomial c is $\operatorname{SL}_\alpha \times \operatorname{SL}_\beta$-invariant. For $V \in \operatorname{Rep}_\alpha(Q)$, c^V is SL_α invariant, so $c^V \in \operatorname{SI}(Q, \alpha)$. For $W \in \operatorname{Rep}_\beta(Q)$ we have that $c_W \in \operatorname{SI}(Q, \beta)$. These semi-invariants are called **determinantal semi-invariants** or **Schofield semi-invariants** and they will play a key role in our approach.

If V and V' are isomorphic representations, then they lie in the same GL_α-orbit. Since GL_α acts by scalars, c^V and $c^{V'}$ are the same up to a scalar. For a representation V without a preferred basis in each of the spaces $V(x)$, $x \in Q_0$ the invariant c^V is defined up to a scalar.

If α is a dimension vector, then $\sigma = \langle \alpha, \cdot \rangle$ is a function $\mathbb{Z}^{Q_0} \to \mathbb{Z}$, but can again be identified with an element in $\mathbb{Z}^{Q_0}$. So we can view σ as a the weight defined by

$$\sigma(x) = \alpha(x) - \sum_{\substack{a \in Q_1 \\ ha = x}} \alpha(ta).$$

Similarly, $\tau = \langle \cdot, \beta\rangle$ is the weight given by

$$\tau(x) = \beta(x) - \sum_{\substack{a\in Q_1\\ ta=x}} \beta(ha).$$

The semi-invariants c^V and c_W enjoy many nice properties.

Proposition 10.6.1. *Suppose that α and β are dimension vectors with $\langle \alpha, \beta\rangle = 0$.*

(1) *The function c^V is a semi-invariant of weight $\langle \alpha, \cdot\rangle$, so*

$$c^V \in \mathrm{SI}(Q, \beta)_{\langle \alpha,\cdot\rangle},$$

(2) *The function c_W is a semi-invariant of weight $-\langle \cdot, \beta\rangle$, so*

$$c^V \in SI(Q, \alpha)_{-\langle \cdot,\beta\rangle}.$$

Proof. If $A = (A(x) \mid x \in Q_0) \in \mathrm{GL}_\beta$, then we have $(A \cdot c^V)(W) = c^V(A^{-1} \cdot W)$. On $\bigoplus_{x\in Q_0} \mathrm{Hom}(V(x), W(x))$, A^{-1} acts by a block diagonal matrix B that has $\alpha(x)$ blocks of the matrix $A^{-1}(x)$ for $x \in Q_0$. We have $\det(B) = \prod_{x\in Q_0} \det(A(x))^{-\alpha(x)}$. On $\bigoplus_{a\in Q_1} \mathrm{Hom}(V(ta), W(ha))$, A acts by a block diagonal matrix C that has $\alpha(ta)$ blocks of the matrix $A^{-1}(ha)$ for $a \in Q_1$. We have $\det(C) = \prod_{a\in Q_1} \det(A(ha))^{-\alpha(ta)}$. It follows that

$$(A \cdot c^V)(W) = \det d^V_{A^{-1}\cdot W} = \det(Cd^V_W B^{-1}) = \det(C) \det(B)^{-1} c^V(W).$$

This shows that c^V has weight $\langle \alpha, \cdot\rangle$. The second part is proved in a similar fashion. ☐

Proposition 10.6.2. *Suppose that the dimension vectors α, β satisfy $\langle \alpha, \beta\rangle = 0$.*

(1) *Consider the semi-invariant $c^V \in \mathrm{SI}(Q, \beta)_{\langle \alpha,\cdot\rangle}$ for some $V \in \mathrm{Rep}_\alpha(Q)$. Suppose that we have an exact sequence*

$$0 \longrightarrow V' \longrightarrow V \longrightarrow V'' \longrightarrow 0,$$

where V' and V'' are representations of dimension α' and α'' respectively. If $\langle \alpha', \beta\rangle > 0$ we have $c^V = 0$. And if $\langle \alpha', \beta\rangle = 0$ then we have $c^V = c^{V'} c^{V''}$.

(2) *Consider the semi-invariant $c_W \in \mathrm{SI}(Q, \alpha)_{-\langle \cdot,\beta\rangle}$ for some $W \in \mathrm{Rep}_\beta(Q)$. Suppose that we have an exact sequence*

$$0 \longrightarrow W' \longrightarrow W \longrightarrow W'' \longrightarrow 0,$$

where W' and W'' are representations of dimension β' and β', respectively. If $\langle \alpha, \beta'\rangle < 0$ we have $c_W = 0$, and if $\langle \alpha, \beta'\rangle = 0$, then we have $c_W = c_{W'} c_{W''}$.

Proof. We will only prove part (1). The proof of part (2) will be similar. For every x we can extend a basis $e_1, \dots, e_{\alpha'(x)}$ of $V'(x)$ to a basis $e_1, \dots, e_{\alpha(x)}$ of $V(x)$. We can identify the space spanned by $e_{\alpha'(x)+1}, \dots, e_{\alpha(x)}$ with $V(x)/V'(x)$. So we identify $V(x)$ with the direct sum $V'(x) \oplus V''(x)$. Choose $W \in \operatorname{Rep}_\beta(Q)$ arbitrarily. Define

$$X' = \bigoplus_{x \in Q_0} \operatorname{Hom}(V'(x), W(x)), \quad X'' = \bigoplus_{x \in Q_0} \operatorname{Hom}(V''(x), W(x))$$

and $X = X' \oplus X''$. Similarly define

$$Y' = \bigoplus_{a \in Q_0} \operatorname{Hom}(V'(ta), W(ha)), \quad X'' = \bigoplus_{a \in Q_1} \operatorname{Hom}(V''(ta), W(ha))$$

and $Y = Y' \oplus Y''$. The map d_W^V maps X' into Y'. For every $a \in Q_1$, $V(a)$ maps $V'(ta)$ into $V'(ha)$. This implies that d_W^V maps X' into Y'. We have $\dim X' - \dim Y' = \langle \alpha', \beta \rangle$. If $\langle \alpha', \beta \rangle > 0$, then d_W^V cannot be injective and $c^V(W) = \det(d_W^V) = 0$. If $\langle \alpha', \beta \rangle = 0$, then d_W^V is of the form

$$\begin{pmatrix} d_W^{V'} & \star \\ 0 & d_W^{V''} \end{pmatrix},$$

where $d_W^{V'}$ and $d_W^{V''}$ are square matrices. It follows that $c^V(W) = \det(d_W^V) = \det(d_W^{V'}) \det(d_W^{V''}) = c^{V'}(W) c^{V''}(W)$. □

Corollary 10.6.3.

(1) *If $V = V' \oplus V''$, then c^V is either a product $c^{V'} c^{V''}$ or is identically zero.*

(2) *If $W = W' \oplus W''$, then c_W is either a product $c_{W'} c_{W''}$ or is identically zero.*

10.7. The Main Theorem and Saturation Theorem

Throughout this section, Q is an acyclic quiver. The main result of this chapter says that the rings of semi-invariants are spanned by Schofield invariants.

Theorem 10.7.1. *Let γ be a dimension vector.*

(1) *If $\operatorname{SI}(Q, \gamma)_\sigma$ is nonzero for some weight σ, then $\sigma(\gamma) = 0$.*

(2) *If $\operatorname{SI}(Q, \gamma)_\sigma$ is nonzero, then $\sigma = \langle \alpha, \cdot \rangle$ for some dimension vector α and $\operatorname{SI}(Q, \gamma)_\sigma$ is spanned by semi-invariants c^V, $V \in \operatorname{Rep}_\alpha(Q)$.*

(3) *If $\operatorname{SI}(Q, \gamma)_\sigma$ is nonzero, then $\sigma = -\langle \cdot, \beta \rangle$ for some dimension vector β and $\operatorname{SI}(Q, \gamma)_\sigma$ is spanned by semi-invariants c_W, $W \in \operatorname{Rep}_\beta(Q)$.*

Combining (1) with (2) and (3) we see that $\langle \alpha, \gamma \rangle = \langle \gamma, \beta \rangle = 0$. Before we prove the theorem we give some consequences.

Corollary 10.7.2 (Reciprocity). *Let α and β be the dimension vectors satisfying $\langle \alpha, \beta \rangle = 0$. Then*

$$\dim \mathrm{SI}(Q, \beta)_{\langle \alpha, \cdot \rangle} = \dim \mathrm{SI}(Q, \alpha)_{-\langle \cdot, \beta \rangle}.$$

Proof. Let $V_1, \dots, V_s$ be the representations of dimension α such that $c^{V_1}, \dots, c^{V_s}$ form a basis of $\mathrm{SI}(Q, \beta)_{\langle \alpha, \cdot \rangle}$. These are linearly independent polynomial functions on $\mathrm{Rep}_\alpha(Q)$, so there exist s representations $W_1, \dots, W_s$ from $\mathrm{Rep}_\beta(Q)$ such that the determinant $\det(c^{V_i}(W_j))_{1 \le i,j \le s}$ is nonzero. Because $c^{V_i}(W_j) = c_{W_j}(V_i)$ for all i, j, we have that $c_{W_1}, \dots, c_{W_s} \in \mathrm{SI}(Q, \alpha)_{-\langle \cdot, \beta \rangle}$ are linearly independent. This proves

$$\dim \mathrm{SI}(Q, \beta)_{\langle \alpha, \cdot \rangle} \le \dim \mathrm{SI}(Q, \alpha)_{-\langle \cdot, \beta \rangle}.$$

The other inequality is shown in the same way. □

The following corollary follows from the main theorem 10.7.1 and Corollary 10.6.3.

Corollary 10.7.3. *The ring of semi-invariants $\mathrm{SI}(Q, \beta)$ is generated by invariants of the form c^V where V is an indecomposable representation of Q.*

Example 10.7.4. Consider again the Kronecker quiver $\theta(2)$:

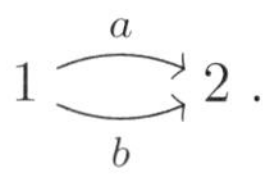

The positive roots, which are the dimension vectors of indecomposables are $(k, k-1)$, $(k, k-1)$ and (k, k) for $k \ge 1$. We have $\langle (k, k-1), \cdot \rangle = (k, -k-1)$, $\langle (k-1, k), \cdot \rangle = (k-1, 2-k)$ and $\langle (k, k), \cdot \rangle = (k, -k)$. Suppose that $\beta = (n, n)$. If $\langle \alpha, \beta \rangle = 0$ for some positive root α, then we must have $\alpha = (k, -k)$ for some $k \ge 0$. This shows that $\mathrm{SI}(Q, \beta)$ is generated by all c^V where V is an indecomposable representation of dimension (k, k) for some positive integer k. Let R be the subalgebra of $\mathrm{SI}(Q, \beta)$ generated by all c^W where W is a representation of dimension $(1, 1)$. By induction on k we show that $c^V \in R$ if V is a representation of dimension $(1, 1)$. The case $k = 1$ is clear. Suppose that V is a representation of dimension (k, k) with $k \ge 2$. Then V has a subrepresentation V' of dimension $(1, 1)$. Since $\langle (1, 1), \beta \rangle = 0$, we have $c^V = c^{V'} c^{V''}$ where $V'' = V/V'$. We have $c^{V'} \in R$ and by the induction hypothesis, we also have $c^{V''}$ lies in R. It follows that $c^V \in R$. This proves that $R \subseteq \mathrm{SI}(Q, \beta)$ and, therefore, $\mathrm{SI}(Q, \beta)$ is generated by all c^W with W of dimension $(1, 1)$. Suppose that V is a representation of dimension $(1, 1)$.

Then $d^V_W : W(x) \oplus W(y) \to W(y) \oplus W(y)$ is the matrix

$$\begin{pmatrix} W(a) & s1_{W(y)} \\ W(b) & t1_{W(y)} \end{pmatrix}$$

and $c^V(W) = \det(tW(a) - sW(b))$.

Example 10.7.5. More generally, let Q be the equioriented $\mathbf{A}_n$ quiver

$$1 \xrightarrow{a_1} 2 \xrightarrow{a_2} \cdots \xrightarrow{a_{r-1}} n.$$

If $V_{i,j}$ is the indecomposable representation of dimension $\alpha_{i,j} := \epsilon_i + \epsilon_{i+1} + \cdots + \epsilon_{j-1}$ for $1 \le i < j \le n+1$ then the semi-invariant $c^{V_{i,j}} \in \mathrm{SI}(Q,\beta)$ has weight $\langle \alpha_{i,j}, \cdot \rangle$ and is (up to a constant) equal to

$$\det(W(a_{j-1}a_{j-2}\cdots a_i)) = \det(W(a_{j-1})W(a_{j-2})\cdots W(a_i)).$$

Example 10.7.6. Let Q be the quiver of type $\mathbf{D}_4$. Up to isomorphism there are 12 indecomposable representation. If V is one of the 4 projective representations, then c^V is trivial:

$$\begin{array}{ccccc} 1 & \xrightarrow{a} & 4 & \xleftarrow{b} & 2 \\ & & \uparrow{\scriptstyle c} & & \\ & & 3 & & \end{array}.$$

If β is a dimension vector with $\beta(1) = \beta(4)$ and V is the indecomposable representation of dimension

$$\begin{matrix} 1 & 0 & 0 \\ & 0 & \end{matrix} \quad \text{and weight} \quad \begin{matrix} 1 & -1 & 0 \\ & 0 & \end{matrix}$$

then $c^V(W)$ is up to a constant equal to $\det(W(a))$. By symmetry we also get the semi-invariants $\det(W(b))$ and $\det(W(c))$.

If β is a dimension vector with $\beta(1) + \beta(2) = \beta(4)$ and V is an indecomposable of dimension

$$\begin{matrix} 1 & 1 & 1 \\ & 0 & \end{matrix} \quad \text{and weight} \quad \begin{matrix} 1 & -1 & 1 \\ & 0 & \end{matrix},$$

then $c^V(W)$ is up to a constant equal to $\det(W(a)\ W(b))$. Similarly we get the semi-invariants $\det(W(a)\ W(c))$ and $\det(W(b)\ W(c))$.

If β is a dimension vector with $\beta(1) + \beta(2) + \beta(3) = \beta(4)$ and V is the indecomposable of dimension

$$\begin{matrix} 1 & 2 & 1 \\ & 1 & \end{matrix} \quad \text{and weight} \quad \begin{matrix} 1 & -1 & 1 \\ & 1 & \end{matrix},$$

then $c^V(W)$ is up to a constant equal to $\det(W(a)\ W(b)\ W(c))$.

Finally, if β is a dimension vector with $\beta(1)+\beta(2)+\beta(3)=2\beta(4)$ and V is the indecomposable of dimension

$$\begin{matrix} 1 & 1 & 1 \\ & 1 & \end{matrix} \quad \text{and weight} \quad \begin{matrix} 1 & -2 & 1 \\ & 1 & \end{matrix},$$

then $c^V(W)$ is up to a constant equal to

$$\det\begin{pmatrix} V(a) & 0 & V(c) \\ 0 & V(b) & V(c) \end{pmatrix}.$$

Since we have exhausted all positive roots, the semi-invariants described before generate the ring $\mathrm{SI}(Q,\beta)$ for all β.

Suppose that β is a dimension vector. We define the cone of weights

$$\Sigma(Q,\beta)=\{\sigma\in\mathbb{Z}^{Q_0}\mid \mathrm{SI}(Q,\beta)_\sigma\neq 0\}. \tag{10.12}$$

For dimension vectors α and β we define: Let Q be a quiver, α, β two dimension vectors. Notice that the set

$$Z^t(\alpha,\beta):=\{(V,W)\in\mathrm{Rep}_\alpha(Q)\times\mathrm{Rep}_\beta(Q)\mid \dim\mathrm{Hom}_Q(V,W)\geq t\}$$

is closed; in fact, it is defined by the determinantal condition for the matrix d^V_W. Thus on a nonempty open subset of $\mathrm{Rep}_\alpha(Q)\times\mathrm{Rep}_\beta(Q)$ the dimension of $\mathrm{Hom}_Q(V,W)$ is constant. Since

$$\dim\mathrm{Hom}_Q(V,W)-\dim\mathrm{Ext}_Q(V,W)=\langle\alpha,\beta\rangle,$$

the dimension of $\mathrm{Ext}_Q(V,W)$ is also constant on this open set. We define $\hom_Q(\alpha,\beta)$ and $\mathrm{ext}_Q(\alpha,\beta)$ to be the generic values of $\dim\mathrm{Hom}_Q(V,W)$ (resp. $\dim\mathrm{Ext}_Q(V,W)$) on $\mathrm{Rep}_\alpha(Q)\times\mathrm{Rep}_\beta(Q)$. We refer to them as **generic hom** and **generic ext**, respectively.

We say that a dimension vector β is a **generic subdimension vector** of α (notation: $\beta\hookrightarrow\alpha$) if every representation from $\mathrm{Rep}_\alpha(Q)$ has a subrepresentation of dimension vector β. Similarly, we say that a dimension vector β is a **generic factor-dimension vector** of α (notation: $\alpha\twoheadrightarrow\beta$) if every representation of dimension vector α has a factor representation of dimension vector β. Note that

$$\beta\hookrightarrow\alpha \quad\Leftrightarrow\quad \alpha\twoheadrightarrow\alpha-\beta.$$

The connection between these notions is expressed by the following result.

Theorem 10.7.7 (Schofield). *Let α, β be two dimension vectors for the quiver Q without oriented cycles.*

(1) $\mathrm{ext}_Q(\alpha,\beta)=0$ *if and only if* $\alpha\hookrightarrow\alpha+\beta$,

(2) $\mathrm{ext}_Q(\alpha,\beta)\neq 0$ *if and only if* $\beta\twoheadrightarrow\gamma$ *and* $\langle\alpha,\gamma\rangle<0$ *for some dimension vector* γ.

We will prove Theorem 10.7.7 in the Section 10.12.

The main consequence of Theorem 10.7.1 is the explicit description of $\Sigma(Q,\beta)$ by means of one homogeneneous linear equality and homogeneous linear inequalities.

Theorem 10.7.8. *The set $\Sigma(Q,\beta)$ is a rational polyhedral cone described as follows:*

$$\Sigma(Q,\beta) = \{\sigma = \langle \alpha, \cdot\rangle \mid \sigma(\beta) = 0 \text{ and } \forall \delta \hookrightarrow \beta,\ \sigma(\delta) \leq 0\}.$$

In particular $\Sigma(Q,\beta)$ is saturated, i.e., $n\sigma \in \Sigma(Q,\beta)$ for some positive integer n implies $\sigma \in \Sigma(Q,\beta)$.

Proof of Theorem 10.7.8. Suppose that $\mathrm{SI}(Q,\beta)_\sigma \neq 0$. Then there exist a representation $W \in \mathrm{Rep}_\alpha(Q)$ that is σ-semi-stable. It follows that $\sigma(\beta) = 0$. If $\delta \hookrightarrow \beta$, then W has a subrepresentation of dimension δ. Because W is σ-semi-stable, we have $\sigma(\delta) \leq 0$.

Conversely, suppose that $\sigma(\beta) = 0$ and $\sigma(\delta) \leq 0$ for all $\delta \hookrightarrow \beta$. If we take a general representation $W \in \mathrm{Rep}_\beta(Q)$, then W is σ-semi-stable. This implies that $\mathrm{SI}(Q,\beta)_{m\sigma} \neq 0$ for some $m \geq 0$. There exists a representation V such that $c^V \in \mathrm{SI}(Q,\beta)_{m\sigma}$ is nonzero. Since $\langle \underline{\dim}\, V, \cdot\rangle = m\sigma$ is divisible by m, so is $\underline{\dim}\, V$. We can write $\underline{\dim}\, V = m\alpha$. Now we have $\sigma = \langle \alpha, \cdot \rangle$. If we set $\gamma = \beta - \alpha$, then we have $-\sigma(\gamma) = \sigma(\beta - \gamma) = \sigma(\delta) \leq 0$ for all γ with $\beta \twoheadrightarrow \gamma$. So $\langle \alpha, \gamma\rangle = \sigma(\gamma) \geq 0$ for all γ with $\beta \twoheadrightarrow \gamma$. By Theorem 10.7.7 this implies that $\mathrm{ext}_Q(\alpha,\beta) = 0$. Choose $V \in \mathrm{Rep}_\beta(Q)$ and $Z \in \mathrm{Rep}_\alpha(Q)$ such that $\mathrm{Ext}_Q(Z,W) = 0$. Then d^Z_W is an isomorphism, and $c^Z(W) \neq 0$. Therefore $\mathrm{SI}(Q,\beta)_\sigma$ is nonzero because it contains c^Z, and $\sigma \in \Sigma(Q,\beta)$. □

Suppose that $\lambda = (\lambda_1, \ldots, \lambda_n)$, $\mu = (\mu_1, \ldots, \mu_n)$ and $\nu = (\nu_1, \ldots, \nu_n)$ are partitions with $\lambda_1 \geq \cdots \geq \lambda_n \geq 0$, $\mu_1 \geq \cdots \geq \mu_n \geq 0$ and $\nu_1 \geq \cdots \geq \nu_n \geq 0$. If W is an n-dimensional vector space, then

$$\dim(\mathbb{S}_\lambda W \otimes \mathbb{S}_\mu W \otimes \mathbb{S}_\nu W^\star)^{\mathrm{GL}(W)}$$

is the Littlewood-Richardson coefficient $c^\nu_{\lambda,\mu}$. Choose $m = \lambda_1 + \mu_1$ and let $\overline{\nu} = (m - \lambda_n, m - \lambda_{n-1}, \ldots, m - \lambda_1)$. Then we have

$$c^\nu_{\lambda,\mu} = \dim(\mathbb{S}_\lambda W \otimes \mathbb{S}_\mu W \otimes \mathbb{S}_\nu W^\star)^{\mathrm{SL}(W)} = \dim(\mathbb{S}_\lambda W \otimes \mathbb{S}_\mu W \otimes \mathbb{S}_{\overline{\nu}} W)^{\mathrm{SL}(W)}.$$

By Example 10.5.4, we can build quiver $Q = T_{n,n,n}$, a dimension vector β and a weight σ such that

$$\mathrm{SI}(Q,\beta)_\sigma \cong (\mathbb{S}_\lambda W \otimes \mathbb{S}_\mu W \otimes \mathbb{S}_{\overline{\nu}} W)^{\mathrm{SL}(W)}.$$

If $\lambda = (\lambda_1, \ldots, \lambda_r)$ is a partition, then we define $k\lambda = (k\lambda_1, \ldots, k\lambda_r)$. The way λ, μ, ν and σ are related, it follows that

$$\mathrm{SI}(Q,\beta)_{k\sigma} \cong (\mathbb{S}_{k\lambda} W \otimes \mathbb{S}_{k\mu} W \otimes \mathbb{S}_{k\overline{\nu}} W)^{\mathrm{SL}(W)}.$$

Corollary 10.7.9 (Saturation for LR coefficients). *Let λ, μ and ν are partitions with $|\nu| = |\lambda| + |\mu|$. If $c^{k\nu}_{k\lambda,k\mu} \neq 0$, then $c^{\nu}_{\lambda,\mu} \neq 0$.*

Proof. Suppose that $c^{k\nu}_{k\lambda,k\mu} \neq 0$. Then we have $\mathrm{SI}(Q,\beta)_{k\sigma} \neq 0$ and $k\sigma \in \Sigma(Q,\alpha)$. so $\sigma \in \Sigma(Q,\alpha)$ and $c^{\nu}_{\lambda,\mu} = \dim \mathrm{SI}(Q,\alpha)_\sigma \neq 0$. □

10.8. Proof of the Main Theorem

Since the quiver is acyclic, we can assume without loss of generality that $Q_0 = \{1,2,\dots,n\}$ and $ha < ta$ for every $a \in Q_1$. Let $b_{i,j}$ be the number of arrows from j to i, and $p_{i,j}$ be the number of paths from j to i.

Let σ be a weight which is expressed in terms of the Euler form as $\sigma = \langle \alpha, \cdot \rangle$. The components of σ and α are expressed through each other by the formulas

$$\sigma(i) = \alpha(i) - \sum_{j>i} b_{i,j}\alpha(j), \tag{10.13}$$

$$\alpha(i) = \sigma(i) + \sum_{j>i} p_{i,j}\sigma(j). \tag{10.14}$$

The method of proof of the theorem is reduction to the case where Q is the generalized Kronecker quiver with vertices s_+ and s_-, and m arrows from s_+ to s_-, and β and σ are given by $\beta(s_+) = \beta(s_-) = n$, $\sigma(s_+) = 1$ and $\sigma(s_-) = -1$. We will apply Classical Invariant Theory with an adjustment to accomodate the definition of the semi-invariants c^V.

Let us start with a quiver Q, a dimension vector β and a weight σ. We proceed in three steps. The first step reduces to the case when Q is a quiver with exactly one source s_+, one sink s_- and the weight σ satisfies $\sigma(s_+) = 1$, $\sigma(s_-) = -1$ and it is zero on all other vertices of Q.

In the second step we reduce to the case when there are no vertices x with $\sigma(x) = 0$. The only case left is the quiver $\theta(m)$ with the weight τ. In the third step we deal with this case.

Step 1. Construct a quiver $Q(\sigma)$ as follows:

$$Q(\sigma)_0 = Q_0 \cup \{s_+, s_-\},$$

$$Q(\sigma)_1 = Q_1 \cup Q_- \cup Q_+,$$

where Q_+ contains exactly $\max\{\sigma(x_i), 0\}$ arrows from s_+ to i for $i = 1,2,\dots,n$, and Q_- consists of exactly $\max\{-\sigma(x_i), 0\}$ arrows from i to s_- for $i = 1,2,\dots,n$.

Example 10.8.1. Let Q be the quiver

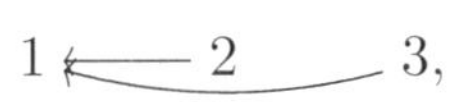

and $\sigma = (-2, 1, 1)$. Then $Q(\sigma)$ is the quiver

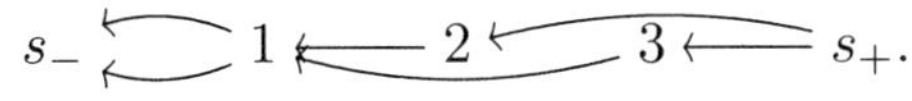

We write $\overline{Q}$ for $Q(\sigma)$. Define the weight $\overline{\sigma}$ of $\overline{Q}$ by $\overline{\sigma}(s_+) = 1$, $\overline{\sigma}(s_-) = -1$ and $\overline{\sigma}(i) = 0$ for $i = 1, \dots, n$. The dimension vector $\overline{\beta}$ is defined as follows:

$$\overline{\beta}(s_+) = \sum_{\substack{i \\ \sigma(i)>0}} \sigma(i)\beta(i), \quad \overline{\beta}(i) = \beta(i), \quad \overline{\beta}(s_-) = \sum_{\substack{i \\ \sigma(i)<0}} -\sigma(i)\beta(i).$$

Let $W \in \operatorname{Rep}_{\overline{\beta}}(\overline{Q})$. The matrices of all maps $W(a)$ with $a \in Q_+$ form a square matrix. Let $D^+(W)$ be a determinant of this block matrix. Let $D^-(W)$ be the determinant of all $W(a)$ with $a \in Q_-$. Then the correspondence $c \mapsto D^- c D^+$ gives the isomorphism of weight spaces $\operatorname{SI}(Q, \beta)_\sigma \to \operatorname{SI}(\overline{Q}, \overline{\beta})_{\overline{\sigma}}$.

Let $\overline{\alpha}$ be the dimension vector for $\overline{Q}$ such that $\overline{\sigma} = \langle \overline{\alpha}, \cdot \rangle$. Let $\overline{V}$ be a representation of $\overline{Q}$ of dimension $\overline{\alpha}$, and let $c^{\overline{V}}$ be the corresponding nonzero semi-invariant on $\operatorname{SI}(\overline{Q}, \overline{\beta})$. We have $\overline{\alpha}(s_+) = 1$,

$$\overline{\alpha}(i) = \sum_{\substack{j \\ \sigma(j)>0}} p_{i,j}\sigma(j)$$

and

$$\overline{\alpha}(s_-) = -1 + \sum_{\substack{i \\ \sigma(i)<0}} \sum_{\substack{j \\ \sigma(j)>0}} \sigma(i) p_{i,j} \sigma(j).$$

Proposition 10.8.2. *We have $c^{\overline{V}} = D^- c^V D^+$ for some $V \in \operatorname{Rep}_\alpha(Q)$.*

Proof. Notice that the weight of D^+ is σ_+ where $\sigma_+(s_+) = 1$, $\sigma_+(s_-) = 0$ and $\sigma_+(i) = \min\{0, -\sigma(i)\}$ for $i \in Q_0$. We can write $\sigma_+ = \langle \gamma_+, \cdot \rangle$ where γ_+ is given by $\gamma_+(s_+) = 1$, $\gamma_+(s_-) = 0$ and $\gamma_+(j) = 0$ for all $j \in Q_0$.

The weight of D^- is σ_- where σ_- is given by $\sigma_-(s_-) = -1$, $\sigma_-(s_+) = 0$ and $\sigma_-(i) = \max\{0, -\sigma(i)\}$ for $i \in Q_0$. We can write $\sigma_- = \langle \gamma_-, \cdot \rangle$ where $\gamma_-(s_+) = 0$,

$$\gamma_-(i) = -\sum_{\substack{j \\ \sigma(j)<0}} p_{i,j}\sigma(j)$$

and

$$\gamma_-(s_-) = -1 + \sum_{\substack{i \\ \sigma(i)>0}} \sum_{\substack{j \\ \sigma(j)<0}} p_{i,j}\sigma(i)\sigma(j).$$

One can verify that $\langle \gamma_-, \overline{\beta} \rangle = \langle \gamma_+, \overline{\beta} \rangle = 0$.

Let $\overline{V} \in \operatorname{Rep}_{\overline{\alpha}}(\overline{Q})$. Then $\overline{V}$ has an obvious submodule $\overline{V}_1 = \overline{V}|_{\overline{Q}_0 \setminus \{s_+\}}$. We have an exact sequence

$$0 \longrightarrow \overline{V}_1 \longrightarrow \overline{V} \longrightarrow \overline{V}_2 \longrightarrow 0,$$

where $\overline{V}_2$ has dimension vector γ_+. We have $c^{\overline{V}_2} = D^+$ and $c^{\overline{V}} = c^{\overline{V}_1} c^{\overline{V}_2} = c^{\overline{V}_1} D^+$.

Let M be the module defined by the exact sequence

$$0 \longrightarrow P_{s_-} \xrightarrow{\ u\ } \bigoplus_{\substack{b \\ hb=s_-}} P_{tb} \longrightarrow M \longrightarrow 0,$$

where the morphism u from P_{s_-} to a copy P_{tb} maps the trivial path $e_{s_-} \in P_{s_-}(s_-)$ to the path $b \in P_{tb}(e_{s_-})$. The dimension vector of M is γ_-, and c^M is the determinant D^-. We have

$$\begin{aligned} \dim \operatorname{Hom}_Q(M, \overline{V}_1) - \dim \operatorname{Ext}_Q(M, \overline{V}_1) &= \langle \gamma_-, \overline{\alpha} - \gamma_+ \rangle \\ &= \sigma_-(\overline{\alpha}) - \sigma_-(\gamma_+) = 1 - 0 = 1. \end{aligned}$$

So there exist a nonzero morphism $\varphi : M \to \overline{V}_1$. Let $\overline{V}_3$ be its image.

Now $\overline{V}_3$ is a submodule of $\overline{V}_1$ and $c^{\overline{V}_1} \neq 0$. By Proposition 10.6.2 we have $\langle \underline{\dim}\, \overline{V}_3, \overline{\beta} \rangle \geq 0$. We also have $c^M = D^- \neq 0$. Applying Proposition 10.6.2 to the kernel N of the map $M \to \overline{V}_3$, we get $\langle \underline{\dim}\, N, \overline{\beta} \rangle = \langle \gamma_- - \underline{\dim}\, \overline{V}_3, \overline{\beta} \rangle = -\langle \underline{\dim}\, \overline{V}_3, \overline{\beta} \rangle \geq 0$. We conclude that $\langle \underline{\dim}\, \overline{V}_3, \overline{\beta} \rangle = 0$. By Proposition 10.6.2 $c^{\overline{V}_3}$ divides the semi-invariant $c^M = D^-$. Because D^- is an irreducible semi-invariant, we must have $c^{\overline{V}_3} = D^-$, $\gamma_- = \underline{\dim}\, \overline{V}_3$ and $\overline{V}_3$ is isomorphic to M.

We have an exact sequence

$$0 \longrightarrow \overline{V}_3 \longrightarrow \overline{V}_1 \longrightarrow \overline{V}_4 \longrightarrow 0.$$

Now it is clear by the multiplicative property that $c^{\overline{V}} = c^{\overline{V}_1} D^+ = c^{\overline{V}_4} c^{\overline{V}_3} D^+ = c^{\overline{V}_4} D^- D^+$. Define a submodule $\overline{V}_5 = \overline{V}_4 \mid_{s_-}$ so we have an exact sequence

$$0 \longrightarrow \overline{V}_5 \longrightarrow \overline{V}_4 \longrightarrow \overline{V}_6 \longrightarrow 0\,.$$

Note that $\overline{V}_6$ is supported in Q. The restriction of $\overline{V}_6$ to Q will be denoted by V. We will show that the restriction of $c^{\overline{V}}$ to $\operatorname{Rep}_\beta(Q)$ is c^V.

Extend $W \in \operatorname{Rep}_\beta(Q,)$ to the module $\overline{W}$ of dimension $\overline{\beta}$ by putting

$$\overline{W}(s_+) = \bigoplus_{\substack{a \\ ta=s_+}} W(ha) \text{ and } \overline{W}(s_-) = \bigoplus_{\substack{b \\ hb=s_-}} W(tb),$$

with the maps $\overline{W}(a)$ and $\overline{W}(b)$ being the components of the identity map. Define the canonical submodule $\overline{W}_1 = \overline{W} \mid_{\{s_-\}}$. We have an exact sequence

$$0 \longrightarrow \overline{W}_1 \longrightarrow \overline{W} \longrightarrow \overline{W}_2 \longrightarrow 0 .$$

Define the submodule $\overline{W}_3 = \overline{W}_2 \mid_{\overline{Q}\setminus\{s_+\}}$ of $\overline{W}_2$. Now we have an exact sequence

$$0 \longrightarrow \overline{W}_3 \longrightarrow \overline{W}_2 \longrightarrow \overline{W}_4 \longrightarrow 0 .$$

The representation $\overline{W}_3$ has support in Q and its restriction to Q is just W. Now we have

$$c^{\overline{V}}(\overline{W}) = c^{\overline{V}_4}(\overline{W}) = c^{\overline{V}_4}(\overline{W}_1)c^{\overline{V}_4}(\overline{W}_3)c^{\overline{V}_4}(\overline{W}_4) = c^{\overline{V}_4}(\overline{W}_3)$$

because $c^{\overline{V}_4}(\overline{W}_1)$ and $c^{\overline{V}_4}(\overline{W}_4)$ are constant. Moreover,

$$c^{\overline{V}_4}(\overline{W}_3) = c^{\overline{V}_5}(\overline{W}_3)c^{\overline{V}_6}(\overline{W}_3) = c^{\overline{V}_6}(\overline{W}_3) = c^{V}(W)$$

because $c^{\overline{V}_5}(\overline{W}_4)$ is constant. This concludes the proof of the proposition. □

Step 2. Let Q, β, σ be as above and suppose that $x \in Q_0$ is a vertex with $\sigma(x) = 0$. Let $a_1, \ldots, a_s$ be the incoming arrows at vertex x and $b_1, \ldots, b_t$ be the outgoing arrows. We construct a new quiver $\overline{Q}$ from Q by omitting vertex x, and replacing the $s + t$ arrows $a_1, \ldots, a_s, b_1, \ldots, b_t$ by st arrows labeled $(ba)_{i,j}$ $(1 \leq i \leq t, 1 \leq j \leq s)$ such that $h(ba)_{i,j} = hb_i$ and $t(ba)_{i,j} = ta_j$. For example, if Q is the quiver

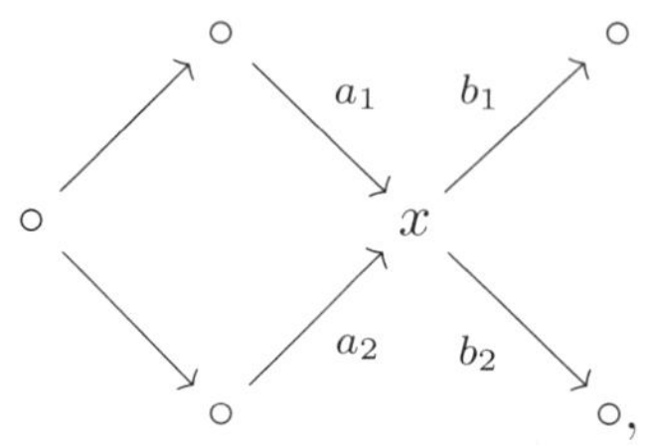

then $\overline{Q}$ is the quiver

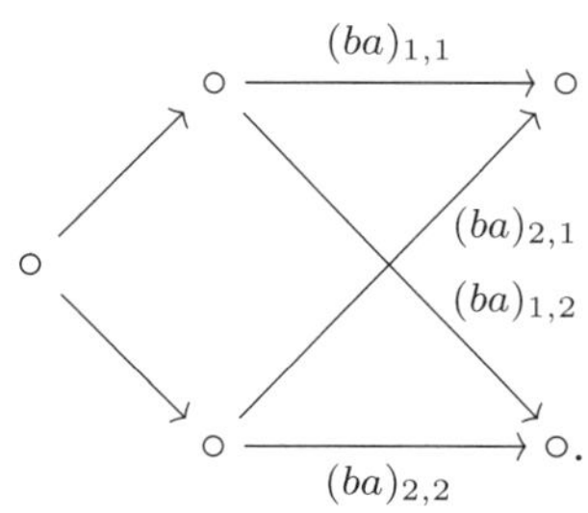

Let $\overline{\beta}$ and $\overline{\sigma}$ be the restrictions of β and σ to $\overline{Q}_0 = Q_0 \setminus \{x\}$, respectively.

We have a restriction map

$$\operatorname{res} : \operatorname{Rep}_\beta(Q) \to \operatorname{Rep}_{\overline{\beta}}(\overline{Q})$$

defined as follows. If $V \in \operatorname{Rep}_\beta(Q)$, then $\operatorname{res}(V) = V'$ where V' is defined by $V'(c) = V(c)$ for $c \in Q_1 \setminus \{a_1, \ldots, a_s, b_1, \ldots, b_t\}$ and $V'((ba)_{i,j}) = V(b_i)V(a_j)$ for all i, j. In fact, res is the quotient morphism with respect to the action of $\operatorname{GL}_{\beta(x)}$ because of the Fundamental Theorem of Invariant Theory. This quotient is equivariant with respect to the action of $\operatorname{GL}_{\overline{\beta}}(\overline{Q})$ and the induced ring homomorphism $\operatorname{res}^\star : \mathbb{C}[\operatorname{Rep}_{\overline{\beta}}(\overline{Q})] \to \mathbb{C}[\operatorname{Rep}_\beta(Q)]$ gives an isomorphism between $\operatorname{SI}(\overline{Q}, \overline{\beta})_{\overline{\sigma}}$ and $\operatorname{SI}(Q, \beta)_\sigma$.

Assuming Theorem 10.7.1. for $\operatorname{SI}(\overline{Q}, \overline{\beta})_{\overline{\sigma}}$ to be true, we need to show that every semi-invariant of type $c^{\overline{V}}$ from $\operatorname{SI}(\overline{Q}, \overline{\beta})_{\overline{\sigma}}$ pulls back to a semi-invariant of type c^V. We define a morphism

$$\operatorname{ind} : \operatorname{Rep}_{\overline{\alpha}}(\overline{Q}) \to \operatorname{Rep}_\alpha(Q)$$

as follows. If $\overline{V} \in \operatorname{Rep}_{\overline{\beta}}(\overline{Q})$, then $\operatorname{ind}(\overline{V}) = V$ where V is defined by

$$V(y) = \begin{cases} \overline{V}(y), & if\ y \neq x, \\ \bigoplus_{k=1}^s \overline{V}(ta_k), & if\ y = x. \end{cases}$$

We define linear maps $V(a)$ as follows:

$$V(a) = \begin{cases} \overline{V}(a), & if\ a \neq a_k, b_l, \\ \iota_k, & if\ a = a_k, \\ \sum_{k=1}^s \pi_k \overline{V}(ba_{k,l}), & if\ b = b_l, \end{cases}$$

where $\iota_k : V(ta_k) = \overline{V}(ta_k) \hookrightarrow V(x) = \bigoplus_{k=1}^s \overline{V}(ta_k)$ is the injection into the k-th summand, and $\pi_k : \bigoplus_{k=1}^s \overline{V}(ta_k) \to \overline{V}(ta_k)$ is the projection onto the k-th summand.

Then it is easy to check directly from the definition of semi-invariants c^V that if $\overline{W} = \operatorname{res}(W)$is a restriction of a representation W of Q of dimension β, then $c^{\overline{V}}(\overline{W}) = c^V(W)$. Notice that the functor $\operatorname{ind} : \operatorname{Rep}(\overline{Q}) \to \operatorname{Rep}(Q)$ is the left adjoint of the obvious restriction functor $\operatorname{res} : \operatorname{Rep}(Q) \to \operatorname{Rep}(\overline{Q})$, i.e., we have the natural isomorphisms

$$\operatorname{Hom}_Q(\operatorname{ind}(\overline{V}), W) = \operatorname{Hom}_{\overline{Q}}(\overline{V}, \operatorname{res}(W))$$

which explains why $c^{\bar{V}}(\bar{W})$ and $c^V(W)$ vanish simultaneously.

Step 3. It remains to deal directly with the case where Q is a generalized Kronecker quiver $\theta(m)$ with 2 vertices s_+, s_- and m arrows $a_1, a_2, \ldots, a_m : s_+ \to s_-$, and $\beta(s_+) = \beta(s_-) = n$, $\sigma(s_+) = 1$ and $\sigma(s_-) = -1$. We have

$$\operatorname{Rep}_\beta(Q) = \operatorname{Hom}(W(s_+), W(s_-))^m \cong \operatorname{Hom}(W(s_+), W(s_-)) \otimes Z,$$

where $W(s_+) \cong W(s_-) \cong \mathbb{C}^n$ and $Z \cong \mathbb{C}^m$. Using the Cauchy formula 10.3.13 we have

$$\mathbb{C}[\mathrm{Rep}_\beta(Q)] = \bigoplus_\lambda \mathbb{S}_\lambda(W(s_+)) \otimes \mathbb{S}_\lambda(W(s_-)^\star \otimes Z^\star).$$

Looking at the action of $\mathrm{GL}(W(s_+))$ we see that $\mathrm{SI}(Q,\beta)_\sigma$ contained in the summand

$$\begin{aligned}\mathbb{S}_{(1^n)}(W(s_+)) \otimes \mathbb{S}_{(1^n)}(W(s_-)^\star \otimes Z^\star) &= \bigwedge^n(W(s_+)) \otimes \bigwedge^n(W(s_-)^\star \otimes Z^\star)\\ &= \bigoplus\nolimits_\mu \bigwedge^n(W(s_+)) \otimes \mathbb{S}_\mu(W(s_-)^\star) \otimes \mathbb{S}_{\mu'}(Z^\star).\end{aligned}$$

Looking at the action of $\mathrm{GL}(W(s_-))$ we see that $\mathrm{SI}(Q,\beta)_\sigma$ can be identified with

$$\bigwedge^n W(s_+) \otimes \bigwedge^n W(s_-)^* \otimes S^n(Z^\star),$$

where $S^n(Z)$ is the n-th symmetric power.

Define a representation U with dimension vector α given by $\alpha(s_+) = 1$ and $\alpha(s_-) = n-1$ such that $U(a_1) = 0$ and $U(a_i)$ is the embedding sending 1 to the $(i-1)$'st basis vector (for $i = 2, \ldots, m$). We verify that $c^U = \det(W(a_1)) \in \mathrm{SI}(Q,\beta)_\sigma$. If $g \in \mathrm{GL}(Z)$, then we have $g \cdot c^V = c^{g \cdot V} \in \mathrm{SI}(Q,\beta)_\sigma$. This shows that the subspace of $\mathrm{SI}(Q,\beta)_\sigma$ spanned by all invariants of the form c^V where $V \in \mathrm{Rep}_\alpha(Q)$ is stable under $\mathrm{GL}(Z)$. Since $\mathrm{SI}(Q,\beta)_\sigma$, as a representation of $\mathrm{GL}(Z)$, is irreducible, we see that $\mathrm{SI}(Q,\beta)_\sigma$ is spanned by semi-invariants of the form c^V, $V \in \mathrm{Rep}_\alpha(Q)$. This concludes the proof of Theorem 10.7.1. □

Remark 10.8.3.

(1) Theorem 10.7.1 and Corollary 10.6.3. imply that the ring $\mathrm{SI}(Q,\beta)$ is generated (as a $\mathbb{C}$-algebra) by the semi-invariants c^V where V is an indecomposable representation of Q with $\langle \underline{\dim}\, V, \beta \rangle = 0$.

(2) In Theorem 10.7.8. it is enough to consider the subrepresentations $\beta' \hookrightarrow \beta$ for which general representation of dimension β' is indecomposable. Even with this restriction still a lot of inequalities $\sigma(\beta') \leq 0$ are redundant.

10.9. Semi-invariants for Dynkin Quivers

We look at the rings of semi-invariants for the Dynkin quivers. In Example 9.4.4 we saw that GL_β has a dense orbit in $\mathrm{Rep}_\beta(Q)$, and that the Sato-Kimura lemma implies that the ring of semi-invariants $\mathrm{SI}(Q,\beta)$ is a polynomial ring generated by semi-invariants. Let $W \in \mathrm{Rep}_\beta(Q)$ be a representation such that its orbit is dense. Remark 10.8.3 (1) shows that for every dimension vector β the ring $\mathrm{SI}(Q,\beta)$ is generated, as a $\mathbb{C}$-algebra, by the semi-invariants c^V where V is an indecomposable representation with a dimension vector α satisfying $\langle \alpha, \beta \rangle = 0$. Of course, we only have to consider representation V for which c^V is nonzero, which is equivalent to $\operatorname{Hom}_Q(V, W) = 0$. Let $V_1, V_2, \dots, V_r$ be the indecomposable representations with $c^{V_i} \neq 0$, and let $\alpha_1, \dots, \alpha_r$ be their dimension vectors. Let C be the monoid $\mathbb{N}\alpha_1 + \mathbb{N}\alpha_2 + \cdots + \mathbb{N}\alpha_r$. The polynomial generators of $\operatorname{SI}(Q, \beta)$ are the irreducible polynomials among the semi-invariants $c^{V_1}, \dots, c^{V_s}$. These are the semi-invariants c^{V_i} for which α_i is irreducible in the monoid C, i.e., α_i is not the sum of two smaller dimension vectors in C.

Example 10.9.1. Consider the quiver

$$\begin{array}{ccccccccc} & & & & \circ & & & & \\ & & & & \downarrow{\scriptstyle c} & & & & \\ \circ & \xrightarrow{a_1} & \circ & \xrightarrow{a_2} & \circ & \xleftarrow{b_2} & \circ & \xleftarrow{b_1} & \circ \end{array}$$

and let

$$\beta = \begin{matrix} & & n & & \\ n & 2n & 3n & 2n & n. \end{matrix}$$

Then we have

$$\langle \cdot, \beta \rangle = n\tau, \text{ where } \tau = \begin{matrix} & & -2 & & \\ -1 & -1 & 3 & -1 & -1. \end{matrix}$$

The roots α with the property $\langle \alpha, \beta \rangle = n\tau(\alpha) = 0$ are

$$\alpha_1 = \begin{matrix} & & 0 & & \\ 1 & 1 & 1 & 1 & 0, \end{matrix} \quad \alpha_2 = \begin{matrix} & & 0 & & \\ 0 & 1 & 1 & 1 & 1, \end{matrix} \quad \alpha_3 = \begin{matrix} & & 1 & & \\ 0 & 1 & 1 & 0 & 0, \end{matrix}$$

$$\alpha_4 = \begin{matrix} & & 1 & & \\ 0 & 0 & 1 & 1 & 0, \end{matrix} \quad \alpha_5 = \begin{matrix} & & 1 & & \\ 1 & 1 & 2 & 1 & 1, \end{matrix} \quad \alpha_6 = \begin{matrix} & & 1 & & \\ 1 & 2 & 2 & 1 & 0, \end{matrix}$$

$$\alpha_7 = \begin{matrix} & & 1 & & \\ 0 & 1 & 2 & 2 & 1. \end{matrix}$$

Let $V_1, V_2, \dots, V_7$ be the corresponding indecomposable representations. All the corresponding semi-invariants are nonzero. We have $\alpha_6 = \alpha_1 + \alpha_3$ and $\alpha_7 = \alpha_2 + \alpha_4$, so the ring of semi-invariants is generated by $c^{V_1}, c^{V_2}, \dots, c^{V_5}$. We have

$$c^{V_1} = \det(V(a_2a_1)\ V(b_2)), c^{V_2} = \det(V(a_2)\ V(b_2b_1)),$$

$$c^{V_3} = \det(V(a_2)\ V(c)), c^{V_4} = \det(V(b_2)\ V(c)),$$

$$c^{V_5} = \det(V(a_2a_1)\ V(b_2b_1)\ V(c)).$$

Example 10.9.2. Consider the quiver

$$1 \longrightarrow 2 \longleftarrow 3 \longleftarrow 4$$

and the dimension vector $\beta = (1, 2, 1, 2)$. We have

$$\tau = \langle \cdot, \beta \rangle = (-1, 2, -1, 1).$$

The positive roots α with $\tau(\alpha) = 0$ are

$$\alpha_1 = (1, 1, 1, 0), \alpha_2 = (0, 0, 1, 1).$$

Let V_1 and V_2 be the corresponding indecomposables. The semi-invariant c^{V_1} is nonzero, but $c^{V_2} = 0$. The ring of semi-invariants is the polynomial ring generated by c^{V_1}.

Exercises.

Exercise 10.9.1. *Find generators of the ring of semi-invariants in example* 10.9.1 *for the dimension vector*

$$\beta = \begin{array}{ccccc} & & 2n & & \\ n & 2n & 3n & 2n & n. \end{array}$$

Exercise 10.9.2. *Consider the quiver*

$$1 \longleftarrow 2 \longleftarrow 3 \longleftarrow \cdots \longleftarrow n$$

and let β be a dimension vector. Suppose that V is an indecomposable of dimension vector α, where $\alpha(k) = 1$ for $i \leq k \leq j$, and $\alpha(k) = 0$ for $k < i$ and $k > j$. Then c^V is one of the polynomial generators of $\mathrm{SI}(Q, \beta)$ *if and only if $\beta(i) = \beta(j)$ and $\beta(k) > \beta(i)$ if $i < k < j$.*

10.10. Semi-invariants for Extended Dynkin Types

Let Q be an extended Dynkin quiver and let α be the primitive isotropic root. The rings of semi-invariants for dimension vectors $m\alpha$ have a uniform description which allows us to see the simple regular representations. Let us describe this in more detail.

We will work with the distinguished orientations of the extended Dynkin quivers described in Section 7.7.3.

10.10.1. Quiver of type $\widehat{\mathbf{A}}_n$. Consider the quiver

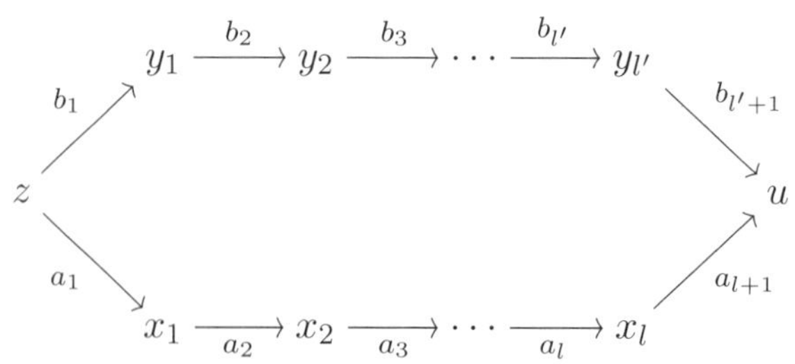

with $l + l' = n - 1$. Let α be the dimension vector

$$\begin{matrix} & 1 & \cdots & 1 & \\ 1 & & & & 1 \\ & 1 & \cdots & 1 & \end{matrix}$$

A general representation in dimension vector $m\alpha$ decomposes to a direct sum of m different indecomposable representations of dimension vector α. The ring $\mathrm{SI}(\widehat{A}_{l+1,l'+1}, m\alpha)$ is generated by the semi-invariants c^V for indecomposable representations V. But the semi-invariant c^V is nonzero only for the regular indecomposable representations V. Thus the ring $\mathrm{SI}(\widehat{A}_{l+1,l'+1}, m\alpha)$ is generated by the semi-invariants c^{E_i} $(0 \le i \le l)$, $c^{E'_j}$ $(0 \le j \le l')$ and the semi-invariants c^{V_t} where V_t are the indecomposable modules of the dimension α.

We have

$$\begin{aligned} c^{E'_i} &= \det(W(b_{i+1})) \quad 0 \le i \le l', \\ c^{E_j} &= \det(W(a_{j+1})) \quad 0 \le j \le l. \end{aligned}$$

The weight $\langle \alpha, \cdot \rangle$ is

$$\begin{matrix} & 0 & \cdots & 0 & \\ 1 & & & & -1 \\ & 0 & \cdots & 0 & \end{matrix}$$

The semi-invariants in this weight are spanned by the coefficients of the functional determinant

$$\det(SW(a) + TW(b)) = C_0 S^m + C_1 S^{m-1} T + \ldots + C_m T^m,$$

where $a = a_{l+1} \ldots a_1$, $b := b_{l'+1} \ldots b_1$. The coefficients $C_0, \ldots, C_m$ are algebraically independent because after specializing to $W(a) = 1_{W(a)}$ they specialize to the coefficients of the characteristic polynomial of $W(b)$. The ring of semi-invariants is generated by

$$c^{E_0}, \ldots, c^{E_l}, c^{E'_1}, \ldots, c^{E'_{l'}}, C_0, C_1, \ldots, C_m.$$

The products $c^{E_0} \cdots c^{E_l}$ and $c^{E'_0} \cdots c^{E'_{l'}}$ have weight $\langle \alpha, \cdot \rangle$ so they are linear combinations of $C_0, C_1, \ldots, C_m$. So we can omit two of the C_i's and still have a set of generators. The ring of semi-invariants is generated by $(l+1) + (l'+1) + (m-1) = n + m$ generators, and the generators are algebraically independent. So the ring of semi-invariants $\mathrm{SI}(\widehat{A}_{l+1,l'+1}, m\alpha)$ is a polynomial ring.

The same result can be obtained by applying the formula (10.3) to the coordinate ring of the representation space $\mathrm{Rep}_{m\alpha}(\widetilde{A}_{l+1,l'+1})$. If the weight of the semi-invariant at a vertex x_j or y_i is nonzero, then according to the Lemma 10.3.12 this semi-invariant is divisible by one of the determinants $\det W(a_j)$ or $\det W(b_i)$. If all the weights at x_j and y_i are zero, the semi-invariant comes from the semi-invariant of the Kronecker quiver in dimension vector $m\alpha = (m, m)$. However, the coordinate ring of this representation space decomposes:

$$\mathbb{C}[\mathrm{Rep}_{(m,m)}(\theta(2))] = \bigoplus_{\lambda,\mu} \mathbb{S}_\lambda W(1) \otimes \mathbb{S}_\lambda W(2)^* \otimes \mathbb{S}_\mu W(1) \otimes \mathbb{S}_\mu W(2)^*.$$

Now we can use Lemma 10.3.12 to count semi-invariants and we see that they occur only in the weights $(p, -p)$. This calculation was done in Example 10.5.2. The Hilbert function of this ring is the same as the Hilbert function of the polynomial ring on $m+1$ variables, each occurring in degree m. This is the polynomial ring generated by the coefficients $C_0, \ldots, C_m$.

10.10.2. Quiver of type $\widehat{\mathbf{D}}_n$.

Consider the quiver $\widehat{D}_n$,

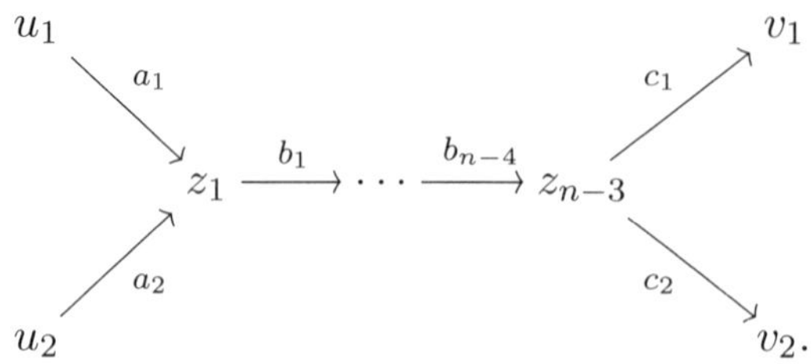

Let α be the dimension vector

$$\begin{matrix} 1 & & & & 1 \\ & 2 & \cdots & 2 & \\ 1 & & & & 1. \end{matrix}$$

The defect weight is

$$\partial = \langle \alpha, \cdot \rangle = \begin{matrix} 1 & & & & -1 \\ & 0 & \cdots & 0 & \\ 1 & & & & -1. \end{matrix}$$

It is an amusing exercise with the Littlewood-Richardson rule to see that

$$\dim \mathrm{SI}(\widehat{D}_n, m\alpha)_{\langle p\alpha, \cdot \rangle} = \binom{p+m}{m}$$

(see Exercise 10.10.1). Let us denote $C_0, \ldots, C_m$ the basis of the space of semi-invariants in weight $\langle \alpha, \cdot \rangle$.

Now three families of the simple regular nonhomogeneous modules give interesting semi-invariants.

$c^{E_0} = \det(W(a_1), W(a_2))$ and $c^{E_i} = \det(W(b_i))$for $1 \leq i \leq n-4$,

$$c^{E_{n-3}} = \det\begin{pmatrix} W(c_1) \\ W(c_2) \end{pmatrix},$$

$$\begin{aligned} c^{E_0'} &= \det(W(c_1 b_{n-4} \dots b_1 a_2)), \\ c^{E_1'} &= \det(W(c_2 b_{n-4} \dots b_1 a_1)), \\ c^{E_0''} &= \det(W(c_2 b_{n-4} \dots b_1 a_2)), \\ c^{E_1''} &= \det(W(c_1 b_{n-4} \dots b_1 a_1)). \end{aligned}$$

The ring of semi-invariants $\mathrm{SI}(\widehat{D}_n, m\alpha)$ is generated by the above semi-invariants. If $m \leq 2$ the semi-invariants $C_0, \dots, C_m$ are not needed as three products $c^{E_0}c^{E_1}\dots c^{E_{n-3}}$, $c^{E_0'}c^{E_1'}$, $c^{E_0''}c^{E_1''}$ are linearly independent so they span the weight space $\mathrm{SI}(\widehat{D}_n, m\alpha)_{\langle\alpha,\cdot\rangle}$. If $m > 2$ we need $m-2$ generators in weight $\langle\alpha,\cdot\rangle$.

For $m \geq 2$ the ring $\mathrm{SI}(\widehat{D}_n, m\alpha)$ is a polynomial ring. For $m = 1$ it is a hypersurface as there is a relation

$$c^{E_0}c^{E_1}\dots c^{E_{n-3}} + c^{E_0'}c^{E_1'} + c^{E_0''}c^{E_1''} = 0$$

in weight $\langle\alpha,\cdot\rangle$.

One could reverse the point of view and first describe this ring of semi-invariants by means of representation theory of GL_n and formula (10.3) and deduce from this the structure of simple regular modules.

10.10.3. Quiver of type $\widehat{\mathbf{E}}_6$. Consider the quiver $\widehat{E}_6$:

$$\begin{array}{ccccccccc} & & & & z_1 & & & & \\ & & & & \downarrow{\scriptstyle c_1} & & & & \\ & & & & z_2 & & & & \\ & & & & \downarrow{\scriptstyle c_2} & & & & \\ x_1 & \xrightarrow[a_1]{} & x_2 & \xrightarrow[a_2]{} & u & \xleftarrow[b_2]{} & y_2 & \xleftarrow[b_1]{} & y_1. \end{array}$$

Let α be the dimension vector

$$\begin{array}{ccccc} & & 1 & & \\ & & 2 & & \\ 1 & 2 & 3 & 2 & 1. \end{array}$$

The defect weight is

$$\partial = \langle\alpha,\cdot\rangle = \begin{array}{ccccc} & & 1 & & \\ & & 1 & & \\ 1 & 1 & -3 & 1 & 1. \end{array}$$

It is an exercise with the Littlewood-Richardson rule to see that

$$\dim \operatorname{SI}(\widehat{E}_6, m\alpha)_{\langle p\alpha,\cdot\rangle} = \binom{p+m}{m}.$$

Let $C_0, \dots, C_m$ be a basis of the space of semi-invariants in weight $\langle \alpha, \cdot\rangle$.

Now three families of the simple regular nonhomogeneous modules give interesting semi-invariants:

$$\begin{aligned}
c^{E_0} &= \det \begin{pmatrix} W(a_2) & W(b_2) & 0 \\ 0 & W(b_2) & W(c_2) \end{pmatrix}, \\
c^{E_1} &= \det \left(W(a_2a_1) \quad W(b_2b_1) \quad W(c_2c_1) \right), \\
c^{E_0'} &= \det \left(W(a_2) \quad W(b_2b_1) \right), \\
c^{E_1'} &= \det \left(W(b_2) \quad W(c_2c_1) \right), \\
c^{E_2'} &= \det \left(W(c_2) \quad W(a_2a_1) \right), \\
c^{E_0''} &= \det \left(W(c_2) \quad W(b_2b_1) \right), \\
c^{E_1''} &= \det \left(W(a_2) \quad W(c_2c_1) \right), \\
c^{E_2''} &= \det \left(W(b_2) \quad W(a_2a_1) \right)).
\end{aligned}$$

The ring of semi-invariants $\operatorname{SI}(\widehat{E}_6, m\alpha)$ is generated by the above semi-invariants. If $m \leq 2$ the semi-invariants $C_0, \dots, C_m$ are not needed as three products $c^{E_0}c^{E_1}$, $c^{E_0'}c^{E_1'}c^{E_2'}$, $c^{E_0''}c^{E_1''}c^{E_2''}$ are linearly independent so they span the weight space $\operatorname{SI}(\widehat{E}_6, m\alpha)_{\langle\alpha,\cdot\rangle}$. If $m > 2$, we need $m-2$ generators in weight $\langle \alpha, \cdot\rangle$.

For $m \geq 2$ the ring $\operatorname{SI}(\widehat{E}_6, m\alpha)$ is a polynomial ring. For $m = 1$ it is a hypersurface as there is a relation

$$c^{E_0}c^{E_1} + c^{E_0'}c^{E_1'}c^{E_2'} + c^{E_0''}c^{E_1''}c^{E_2''} = 0$$

in weight $\langle \alpha, \cdot\rangle$.

10.10.4. Quiver of type $\widehat{\mathbf{E}}_7$. Consider the quiver

$$\begin{array}{ccccccccccccc}
 & & & & & & z & & & & & & \\
 & & & & & & \downarrow c & & & & & & \\
x_1 & \xrightarrow[a_1]{} & x_2 & \xrightarrow[a_2]{} & x_3 & \xrightarrow[a_3]{} & u & \xleftarrow[b_3]{} & y_3 & \xleftarrow[b_2]{} & y_2 & \xleftarrow[b_1]{} & y_1.
\end{array}$$

Let α be the dimension vector

$$\begin{array}{ccccccc} & & & 2 & & & \\ 1 & 2 & 3 & 4 & 3 & 2 & 1. \end{array}$$

The defect weight is

$$\partial = \langle \alpha, \cdot \rangle = \begin{matrix} & & & 2 & & & \\ 1 & 1 & 1 & -4 & 1 & 1 & 1. \end{matrix}$$

Again we check by Littlewood-Richardson rule to see that

$$\dim \mathrm{SI}(\widehat{E}_7, m\alpha)_{\langle p\alpha, \cdot \rangle} = \binom{p+m}{m}.$$

Let $C_0, \ldots, C_m$ be a basis of the space of semi-invariants in weight $\langle \alpha, \cdot \rangle$.

Now three families of the simple regular nonhomogeneous modules give interesting semi-invariants.

$$\begin{aligned}
c^{E_0} &= \det \begin{pmatrix} W(a_3) & 0 & W(c) \\ 0 & W(b_3) & W(c) \end{pmatrix}, \\
c^{E_1} &= \det \left(W(a_3a_2a_1) \;\; W(b_3b_2b_1) \;\; W(c) \right), \\
c^{E_2} &= \det \left(W(a_3a_2) \;\; W(b_3b_2) \right), \\
c^{E_0'} &= \det \left(W(a_3a_2) \;\; W(c) \right), \\
c^{E_1'} &= \det \left(W(a_3) \;\; W(b_3b_2b_1) \right), \\
c^{E_2'} &= \det \left(W(b_3b_2) \;\; W(c) \right), \\
c^{E_3'} &= \det \left(W(b_3) \;\; W(a_3a_2a_1)) \right), \\
c^{E_0''} &= \det \begin{pmatrix} W(b_3b_2) & 0 & -W(a_3) & 0 \\ 0 & W(a_3a_2a_1) & W(a_3) & -W(c) \end{pmatrix}, \\
c^{E_1''} &= \det \begin{pmatrix} W(a_3a_2) & 0 & -W(b_3) & 0 \\ 0 & W(b_3b_2b_1) & W(b_3) & -W(c) \end{pmatrix}.
\end{aligned}$$

The ring of semi-invariants $\mathrm{SI}(\widehat{E}_7, m\alpha)$ is generated by the above semi-invariants. If $m \leq 2$ the semi-invariants $C_0, \ldots, C_m$ are not needed as three products $c^{E_0}c^{E_1}c^{E_2}$, $c^{E_0'}c^{E_1'}c^{E_2'}c^{E_3'}$, $c^{E_0''}c^{E_1''}$ are linearly independent so they span the weight space $\mathrm{SI}(\widehat{E}_7, m\alpha)_{\langle \alpha, \cdot \rangle}$. If $m > 2$, we need $m - 2$ generators if weight $\langle \alpha, \cdot \rangle$.

For $m \geq 2$ the ring $\mathrm{SI}(\widehat{E}_7, m\alpha)$ is a polynomial ring. For $m = 1$ it is a hypersurface as there is a relation

$$c^{E_0}c^{E_1}c^{E_2} + c^{E_0'}c^{E_1'}c^{E_2'}c^{E_3'} + c^{E_0''}c^{E_1''} = 0$$

in weight $\langle \alpha, \cdot \rangle$.

10.10.5. Quiver of type $\widehat{\mathbf{E}}_8$. Consider the quiver $\widehat{E}_8$:

$$\begin{array}{ccccccccc} & & z & & & & & & \\ & & \downarrow c & & & & & & \\ y_1 \xrightarrow[b_1]{} & y_2 \xrightarrow[b_2]{} & u & \xleftarrow[a_4]{} x_4 & \xleftarrow[a_3]{} x_3 & \xleftarrow[a_2]{} x_2 & \xleftarrow[a_1]{} x_1. \end{array}$$

Let α be the dimension vector

$$\begin{matrix} & & 3 & & & & & \\ 2 & 4 & 6 & 5 & 4 & 3 & 2 & 1. \end{matrix}$$

The defect weight is

$$\partial = \langle \alpha, \cdot \rangle = \begin{matrix} & & 3 & & & & & \\ 2 & 2 & -6 & 1 & 1 & 1 & 1 & 1. \end{matrix}$$

Again we check by Littlewood-Richardson rule to see that

$$\dim \mathrm{SI}(\widehat{E}_8, m\alpha)_{\langle p\alpha, \cdot \rangle} = \binom{p+m}{m}.$$

Let us denote $C_0, \ldots, C_m$ the basis of the space of semi-invariants in weight $\langle \alpha, \cdot \rangle$. Now three families of the simple regular nonhomogeneous modules give interesting semi-invariants: 2

$$c^{E_0} = \det \begin{pmatrix} W(a_5) & 0 & W(c) \\ 0 & W(b_2) & W(c) \end{pmatrix},$$

$$c^{E_1} = \det \left(W(a_5a_4a_3a_2a_1) \quad W(b_2b_1) \quad W(c) \right),$$

$$c^{E_2} = \det \left(W(a_5a_4a_3a_2) \quad W(b_2) \right),$$

$$c^{E_3} = \det \left(W(a_5a_4a_3) \quad W(c) \right),$$

$$c^{E_4} = \det \left(W(a_5a_4) \quad W(b_2b_1) \right),$$

$$c^{E_0'} = \det \begin{pmatrix} 0 & W(b_2) & -W(a_5a_4a_3) & 0 \\ W(b_2b_1) & W(b_2) & 0 & -W(c) \end{pmatrix},$$

$$c^{E_1'} = \det \begin{pmatrix} W(b_2) & 0 & -W(a_5a_4) & 0 \\ W(b_2) & W(a_5a_4a_3a_2a_1) & 0 & -W(c) \end{pmatrix},$$

$$c^{E_2'} = \det \begin{pmatrix} W(b_2b_1) & 0 & -W(a_5) & 0 \\ W(b_2b_1) & W(a_5a_4a_3a_2) & 0 & -W(c) \end{pmatrix},$$

$$c^{E_0''} = \det \begin{pmatrix} W(b_2) & 0 & 0 & 0 & W(c) & 0 \\ W(b_2) & 0 & 0 & W(a_5a_4) & 0 & W(c) \\ 0 & W(b_2b_1) & W(a_5a_4a_3a_2) & W(a_5a_4) & W(c) & 0 \end{pmatrix},$$

$$c^{E_1''} = \det \begin{pmatrix} W(b_2) & 0 & W(a_5) & 0 & 0 & 0 \\ W(b_2) & W(b_2b_1) & 0 & W(a_5a_4a_3) & 0 & 0 \\ 0 & 0 & W(a_5) & 0 & W(a_5a_4a_3a_2a_1) & W(c) \end{pmatrix}.$$

The ring of semi-invariants $\mathrm{SI}(\widehat{E}_8, m\alpha)$ is generated by the above semi-invariants. If $m \leq 2$ the semi-invariants $C_0, \ldots, C_m$ are not needed as three products $c^{E_0}c^{E_1}c^{E_2}$, $c^{E'_0}c^{E'_1}c^{E'_2}c^{E'_3}$, $c^{E''_0}c^{E''_1}$ are linearly independent so they span the weight space $SI(\widehat{E}_8, m\alpha)_{\langle\alpha,\cdot\rangle}$. If $m > 2$, we need $m-2$ generators if weight $\langle\alpha,\cdot\rangle$.

For $m \geq 2$ the ring $\mathrm{SI}(\widehat{E}_8, m\alpha)$ is a polynomial ring. For $m = 1$ it is a hypersurface as there is a relation

$$c^{E_0}c^{E_1}c^{E_2}c^{E_3}c^{E_4} + c^{E'_0}c^{E'_1}c^{E'_2} + c^{E''_0}c^{E''_1} = 0$$

in weight $\langle\alpha,\cdot\rangle$.

Exercises.

Exercise 10.10.1. *In the notation of Section* 10.10.2, *show that*

$$\dim \mathrm{SI}(\widehat{D}_n, m\alpha)_{\langle p\alpha,\cdot\rangle} = \binom{p+m}{m}.$$

Exercise 10.10.2. *Consider the following quiver Q of type $\widehat{\mathbf{D}}_n$:*

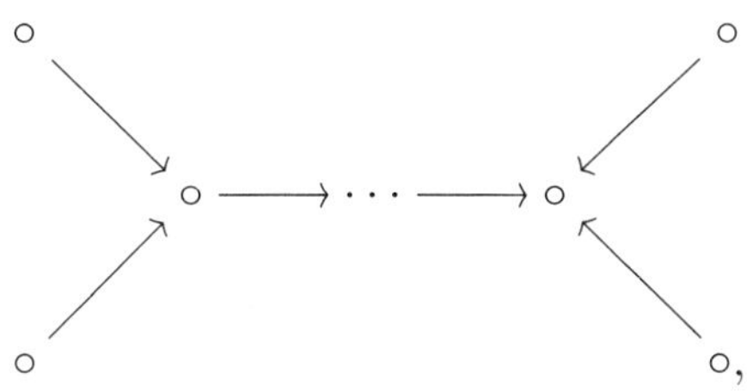

and the dimension vector

$$\begin{matrix} 1 & & & & 1 \\ & 2 & \cdots & 2 & \\ 1 & & & & 1. \end{matrix}$$

Describe explicit invariants that generate $\mathrm{SI}(Q, m\alpha)$. For which m is $\mathrm{SI}(Q, m\alpha)$ a polynomial ring?

10.11. More Examples of Rings of Semi-invariants

Let us also give some examples of calculations of the rings of semi-invariants for wild quivers.

Example 10.11.1. Let us consider the quiver $Q = \theta(2,1)$:

$$1 \underset{b}{\overset{a}{\rightrightarrows}} 2 \xrightarrow{c} 3. \tag{10.15}$$

We consider the dimension vector $\beta = (2,3,1)$. We have

$$\langle\beta,\beta\rangle = 2\cdot 2 + 3\cdot 3 + 1\cdot 1 - 2\cdot 2\cdot 3 - 3\cdot 1 = -1.$$

Let us investigate the cone $\Sigma(Q,\beta)$. The equation $\langle\alpha,\beta\rangle = 0$ gives $4\alpha(1) = 2\alpha(2)+\alpha(3)$. The subrepresentation vectors of $(2,3,1)$ include $(0,0,1)$ and $(0,1,0)$. Thus we get inequalities $\alpha(3)-\alpha(2)\le 0$ and $\alpha(2)-2\alpha(1)\le 0$. One checks easily using Theorem 10.7.8 that they define two walls of the cone $\Sigma(Q,\beta)$. The generators of the semi-group of dimension vectors in this cone are $(1,2,0)$, $(2,3,2)$ and $(3,4,4)$. The only relation between these generators is $2(2,3,2) = (1,2,0)+(3,4,4)$. The dimension vectors $(1,2,0)$ and $(3,4,4)$ are real Schur roots. Thus the weight multiplicities along extremal rays $\langle(1,2,0),\cdot\rangle$ and $\langle(3,4,4),\cdot\rangle$ are equal to 1. Let us call X, Z the semi-invariants of weights $\langle(1,2,0),\cdot\rangle$, $\langle(3,4,4),\cdot\rangle$, respectively. Consider the dimenson vector $(2,3,2)$. Restriction of Q to the subquiver supported by the vertices $\{1,2\}$ gives a general representation of the Kronecker quiver $\theta(2)$. So if, V is a general representation of dimension $(2,3,2)$ then we can choose a basis (e_1,e_2) of $V(1)$ and a basis (f_1,f_2,f_3) of $V(2)$ such that

$$V(a) = \begin{pmatrix} 1 & 0 \\ 0 & 1 \\ 0 & 0 \end{pmatrix}, \quad V(b) = \begin{pmatrix} 0 & 0 \\ 1 & 0 \\ 0 & 1 \end{pmatrix}.$$

Because V is in general position, we may assume that $V(c)$ has full rank 2. Choose $\lambda,\mu,\nu\in\mathbb{C}$, not all 0, such that $\lambda f_1+\mu f_2+\nu f_3\in V(2)$ spans the kernel of $V(c)$. The representation V has a minimal projective resolution

$$0\to P_2\oplus P_3\to P_1\oplus P_1$$

with the map being given in block form by

$$\begin{pmatrix} -b & \lambda ca+\mu cb \\ a & \nu cb \end{pmatrix}.$$

Thus the map c^V is the determinant of the map

$$W(1)\oplus W(1)\to W(2)\oplus W(3)$$

which in the block form is

$$\begin{pmatrix} W(b) & -W(a) \\ \lambda W(ca)+\mu W(cb) & \nu W(cb) \end{pmatrix}.$$

We see that the weight space is three-dimensional, with generators Y_1, Y_2, Y_3 corresponding to choices of vectors (λ,μ,ν) being $(1,0,0)$, $(0,1,0)$, $(0,0,1)$, respectively. Restricting the matrices $W(a)$, $W(b)$, $W(c)$ to the canonical forms

$$\begin{pmatrix} 1 & 0 \\ 0 & 1 \\ 0 & 0 \end{pmatrix}, \begin{pmatrix} 0 & 0 \\ 1 & 0 \\ 0 & 1 \end{pmatrix}, \begin{pmatrix} c_1 & c_2 & c_3 \end{pmatrix}$$

we see that the invariants Y_1, Y_2, Y_3 restrict to $\pm c_1, \pm c_2, \pm c_3$, so they are algebraically independent. Let us calculate the weight multiplicities $SI(Q,\beta)_\sigma$ for $\sigma = \langle\alpha,\cdot\rangle$ for $\alpha\in\Sigma(Q,\beta)$. This can be done using the formula (10.3) and Littlewood-Richardson rule (Theorem 10.3.10). For every dimension vector α in our cone we can write α uniquely as either $s(1,2,0)+m(2,3,2)$ for $s\geq 0$, or $t(3,4,4)+m(2,3,2)$ for some $t>0$. Here m is defined as the biggest positive integer for which $\alpha-m(2,3,2)$ is a dimension vector. The claim is that for α as above we have

$$\dim(\mathrm{SI}(Q,\beta)_{\langle\alpha,\cdot\rangle}) = \binom{m+2}{2}.$$

To illustrate the calculation let us do it explicitly for $\alpha = m(2,3,2)$. We use the formula (10.3) and the Cauchy formula 10.3.13, i.e., we write

$$\begin{aligned}&\mathbb{C}[\mathrm{Rep}_\beta(Q)]\\ &= \bigoplus_{\lambda,\mu,\nu} \mathbb{S}_\lambda W(1)\otimes \mathbb{S}_\lambda W(2)^*\otimes \mathbb{S}_\mu W(1)\otimes \mathbb{S}_\mu W(2)^*\otimes \mathbb{S}_\nu W(2)\otimes \mathbb{S}_\nu W(3)^*,\end{aligned}$$

where we sum over all triples of partitions (λ,μ,ν) and $W(1),W(2),W(3)$ are vector spaces of dimensions $2,3,1$, respectively. The dimension of our weight space is the multiplicity of the tensor product

$$\mathbb{S}_{(2m,2m)}W(1)\otimes \mathbb{S}_{(m,m,m)}W(2)^*\otimes \mathbb{S}_m W(3)^*$$

in the above tensor product. It becomes clear that $\nu=(m)$. Thus we are looking at the multiplicity of

$$\mathbb{S}_{(2m,2m)}W(1)\otimes \mathbb{S}_{(2m,m,m)}W(2)^*$$

in

$$\bigoplus_{\lambda,\mu} \mathbb{S}_\lambda W(1)\otimes \mathbb{S}_\lambda W(2)^*\otimes \mathbb{S}_\mu W(1)\otimes \mathbb{S}_\mu W(2)^*.$$

Writing $\lambda=(\lambda_1,\lambda_2)$ and $\mu=(\mu_1,\mu_2)$ we see that the contribution of the pair (λ,μ) can be nonzero only if $m\leq\lambda_1\leq 2m$ and $0\leq\lambda_2\leq m$. Moreover, writing $\lambda=(m+i,j)$ we see we need to have $i\leq j$. If these conditions are satisfied, the partition μ is determined (as the complement of λ to the rectangle $(2m,2m)$) and the corresponding multiplicity is one by Littlewood-Richardson rule. Thus we get our multiplicity by counting pairs $0\leq i\leq j\leq m$ which gives $\binom{m+2}{2}$. Graphing the weights from our cone in the first quadrant plane $(\alpha(2), \alpha(3))$ we get the following grid of multiplicities:

$\alpha(3)$														
12	0	0	0	0	0	0	0	0	0	0	0	0	**1**	
11	0	0	0	0	0	0	0	0	0	0	0	0	0	
10	0	0	0	0	0	0	0	0	0	0	0	**3**	0	
9	0	0	0	0	0	0	0	0	0	0	0	0	0	
8	0	0	0	0	0	0	0	0	**1**	0	**6**	0	$\underline{\mathbf{15}}$	
7	0	0	0	0	0	0	0	0	0	0	0	0	0	
6	0	0	0	0	0	0	0	**3**	0	$\underline{\mathbf{10}}$	0	**10**	0	
5	0	0	0	0	0	0	0	0	0	0	0	0	0	
4	0	0	0	0	**1**	0	$\underline{\mathbf{6}}$	0	**6**	0	**6**	0	**6**	
3	0	0	0	0	0	0	0	0	0	0	0	0	0	
2	0	0	0	$\underline{\mathbf{3}}$	0	**3**	0	**3**	0	**3**	0	**3**	0	
1	0	0	0	0	0	0	0	0	0	0	0	0	0	
0	$\underline{\mathbf{1}}$	0	**1**	0	**1**	0	**1**	0	**1**	0	**1**	0	**1**	
	0	1	2	3	4	5	6	7	8	9	10	11	12	$\alpha(2)$

The underlined multiplicities occur on the ray containing the multiples of weight $\langle (2, 3, 2), \cdot \rangle$. The extremal rays $\alpha(2) = \alpha(3)$ and $\alpha(3) = 0$ contain only the multiplicity one spaces. The apparent lack of saturation comes from not graphing the weights in three-dimensional space. The calculation of multiplicities allows to identify the ring $\mathrm{SI}(Q, \beta)$ with $\mathbb{C}[X, Y_1, Y_2, Y_3, Z)/R$ where R is a single relation between seven momomials XZ, Y_iY_j in the weight $\langle (4, 6, 4), \cdot \rangle$.

Example 10.11.2. Let $Q = \theta(2, 1)$ be the quiver as in the previous example. We consider the dimension vector $\beta = (2, 2, 1)$. As in previous example we have $\langle \beta, \beta \rangle = -1$. The orthogonality condition $\langle \alpha, \beta \rangle = 0$ reads $2\alpha(1) = \alpha(2) + \alpha(3)$. The inequalities given by subdimension vectors give $\langle \alpha, (0, 0, 1) \rangle \leq 0$, i.e., $\alpha(3) - \alpha(2) \leq 0$ and the obvious inequality $\alpha(2) \geq 0$. The set of weights in that cone is generated as a semigroup by the extremal weight. The general weight one gets in this cone is $\langle a(1, 1, 1) + b(1, 2, 0), \cdot \rangle = (a + b, -a, -2b)$. The direct calculation (similar to the one from example 10.11.1) of the multiplicity gives that this weight appears with multiplicity $\binom{a+2}{2}$. This means that there are three generators in the weight $\langle (1, 1, 1), \cdot \rangle$ (coming from the isotropic root for the extended Kronecker quiver) and one generator in the weight $\langle (1, 2, 0), \cdot \rangle$. The ring of semi-invariants is a polynomial ring on these 4 generators. The grid of weights, graphed in the plane $(\alpha(2), \alpha(3))$ is

$\alpha(3)$												
5	0	0	0	0	0	**21**	0	**21**	0	**21**	0	
4	0	0	0	0	**15**	0	**15**	0	**15**	0	**15**	
3	0	0	0	**10**	0	**10**	0	**10**	0	**10**	0	
2	0	0	**6**	0	**6**	0	**6**	0	**6**	0	**6**	
1	0	**3**	0	**3**	0	**3**	0	**3**	0	**3**	0	
0	**1**	0	**1**	0	**1**	0	**1**	0	**1**	0	**1**	
	0	1	2	3	4	5	6	7	8	9	10	$\alpha(2)$

Let us look at the examples of rings of semi-invariants for quivers with 4 vertices.

Example 10.11.3. Let us look at the quiver $Q = \theta(1, 2, 1)$:

$$1 \longrightarrow 2 \rightrightarrows 3 \longrightarrow 4.$$

and at the dimension vector $\beta = (1, 2, 2, 1)$. We have $\langle \beta, \beta \rangle = -2$. The equality $\langle \alpha, \beta \rangle = 0$ gives $\alpha(1) + 2\alpha(2) = \alpha(3) + \alpha(4)$. The inequalities $\langle \alpha, (0, 0, 0, 1) \rangle \leq 0$, $\langle \alpha, (0, 0, 1, 0) \rangle \leq 0$, $\langle \alpha, (1, 1, 2, 1) \rangle \leq 0$ give $\alpha(4) - \alpha(3) \leq 0$, $\alpha(3) - 2\alpha(2) \leq 0$, $\alpha(3) + \alpha(4) \leq 3\alpha(2)$ and we have the obvious inequality $\alpha(1) \geq 0$. Summarizing we get

$$3\alpha(2) \geq \alpha(3) + \alpha(4) \geq 2\alpha(2) \geq \alpha(3) \geq \alpha(4) \geq 0.$$

These give us the inequalities for $\Sigma(Q, \beta)$ which is a cone over a rectangle with extremal rays given by the multiples of the weights

$$\langle (0, 1, 1, 1), \cdot \rangle = (0, 1, -1, 0), \langle (2, 2, 3, 3), \cdot \rangle = (2, 0, -1, 0),$$

$$\langle (1, 1, 2, 1), \cdot \rangle = (1, 0, 0, -1), \langle (0, 1, 2, 0), \cdot \rangle = (0, 1, 0, -2).$$

It is easy to check that we get nonzero semi-invariants in each of these weights, so these are indeed the extremal rays of the cone $\Sigma(Q, \beta)$.

Exercises.

Exercise 10.11.1. *We consider the dimension vector* $\beta = (3, 2, 1)$ *for the quiver* $Q = \theta(2, 1)$ *from* (10.15) *and Example* 10.11.1. *Prove the following:*

(1) *β is an isotropic root, i.e.,* $\langle \beta, \beta \rangle = 0$.

(2) *The orthogonality condition* $\langle \alpha, \beta \rangle = 0$ *reads* $\alpha(1) = \alpha(2) + \alpha(3)$.

(3) *The inequalities given by subdimension vectors give* $\langle \alpha, (0, 0, 1) \rangle = \alpha(3) - \alpha(2) \leq 0$ *and the obvious inequality* $\alpha(2) \geq 0$.

(4) *The set of weights in the cone* $\Sigma(Q, \beta)$ *is generated as a semi-group by the extremal weight. The general weight one gets in this cone is* $\langle a(1, 1, 0) + b(2, 1, 1), \cdot \rangle = (a + 2b, -a - 3b, -a)$.

(5) *Use direct calculation (similar to the one from Example* 10.11.1*) of the multiplicity to see that this weight appears with multiplicity* $a+1$.

(6) *Prove that there are two generators in the weight* $\langle (1,1,0),\cdot\rangle$ *(coming from the isotropic root for the extended Kronecker quiver) and one generator in the weight* $\langle (2,1,1),\cdot\rangle$. *The ring of semi-invariants is a polynomial ring on these three generators.*

(7) *Prove the grid of weights, graphed in the plane* $(\alpha(2),\alpha(3))$ *is*

$\alpha(3)$							
5	0	0	0	0	0	**1**	
4	0	0	0	0	**1**	**2**	
3	0	0	0	**1**	**2**	**3**	
2	0	0	**1**	**2**	**3**	**4**	
1	0	**1**	**2**	**3**	**4**	**5**	
0	**1**	**2**	**3**	**4**	**5**	**6**	
	0	1	2	3	4	5	$\alpha(2)$

Exercise 10.11.2. *We consider the dimension vector* $\beta = (2,3,2)$ *for the quiver* $Q = \theta(2,1)$ *from* (10.15) *and Example* 10.11.1. *Prove the following:*

(1) β *is an imaginary root and* $\langle \beta,\beta\rangle = -1$.

(2) *The orthogonality condition* $\langle \alpha,\beta\rangle = 0$ *reads* $4\alpha(1) = \alpha(2) + 2\alpha(3)$.

(3) *The inequalities given by subdimension vectors give* $\langle \alpha,(0,0,1)\rangle = \alpha(3) - \alpha(2) \le 0$ *and* $\langle \alpha,(0,1,0)\rangle = \alpha(2) - 2\alpha(1) \le 0$.

(4) *The set of weights in the cone* $\Sigma(Q,\beta)$ *is generated as a semi-group by the extremal weights. The general weight one gets in this cone is* $\langle a(3,4,4) + b(1,2,1),\cdot\rangle = (3a+b,-2a,-b)$.

(5) *Use the direct calculation (similar to the one from Example* 10.11.1*) of the multiplicity to show that this weight appears with multiplicity* $\binom{b+2}{2}$. *This means that there are three generators in the weight* $\langle (1,2,1),\cdot\rangle$ *(coming from the isotropic root for the extended Kronecker quiver) and one generator in the weight* $\langle (3,4,4),\cdot\rangle$.

(6) *The ring of semi-invariants is a polynomial ring on these four generators. The grid of weights, graphed in the plane* $(\alpha(2), \alpha(3))$ *is*

$\alpha(3)$														
12	0	0	0	0	0	0	0	0	0	0	0	0	**1**	
11	0	0	0	0	0	0	0	0	0	0	0	0	0	
10	0	0	0	0	0	0	0	0	0	0	0	0	**6**	
9	0	0	0	0	0	0	0	0	0	0	**3**	0	0	
8	0	0	0	0	0	0	0	0	**1**	0	0	0	**15**	
7	0	0	0	0	0	0	0	0	0	0	**10**	0	0	
6	0	0	0	0	0	0	0	0	**6**	0	0	0	**28**	
5	0	0	0	0	0	0	**3**	0	0	0	**21**	0	0	
4	0	0	0	0	**1**	0	0	0	**15**	0	0	0	0	
3	0	0	0	0	0	0	**10**	0	0	0	0	0	0	
2	0	0	0	0	**6**	0	0	0	0	0	0	0	0	
1	0	0	**3**	0	0	0	0	0	0	0	0	0	0	
0	**1**	0	0	0	0	0	0	0	0	0	0	0	0	
	0	1	2	3	4	5	6	7	8	9	10	11	12	$\alpha(2)$

10.12. Schofield Incidence Varieties

In this section we discuss results of Schofield about generic representations, and we give the proof of Theorem 10.7.7.

We start with the part (1) of the theorem, which says that for a quiver Q and two dimension vectors β, γ we have

$$\operatorname{ext}_Q(\beta, \gamma) = 0 \Leftrightarrow \beta \hookrightarrow \beta + \gamma.$$

The proof of this result is based on analyzing a certain incidence variety.

For integers k, m with $0 \leq k \leq m$ we define the **Grassmann variety** $\operatorname{Grass}(k, m)$ as a projective variety that parametrizes k-dimensional subspaces of $\mathbb{C}^m$. If $W \in \operatorname{Grass}(k, m)$, then we can choose a complement Z in $\mathbb{C}^m$. If $\varphi \in \operatorname{Hom}(W, Z)$, then $\{(w, \varphi(w)) \mid w \in W\}$ is a k-dimensional subspace. This yields map $\iota : \operatorname{Hom}(W, Z) \hookrightarrow \operatorname{Grass}(k, m)$ which induces an isomorphism between $\operatorname{Hom}(W, X)$ as an open neigborhood of $W \in \operatorname{Grass}(k, m)$.

The tangent map

$$d\iota_0 : T_0(\operatorname{Hom}(W, X)) \to T_W(\operatorname{Grass}(k, m))$$

is an isomorphism, so $T_W(\operatorname{Grass}(k, m))$ can be identified with

$$T_0(\operatorname{Hom}(W, X)) \cong \operatorname{Hom}(W, X) \cong \operatorname{Hom}(W, \mathbb{C}^m/W).$$

The Grassmannian can be viewed as a GIT quotient of the representation space of a certain quiver (see Exercise 9.8.9).

Suppose β and γ are dimension vectors, and $\alpha = \beta + \gamma$. We define

$$\mathrm{Grass}(\beta, \alpha) = \prod_{x \in Q_0} \mathrm{Grass}(\beta(x), \alpha(x)).$$

Suppose that $V \in \mathrm{Rep}_\alpha(Q)$. We have identified $V(x)$ with $\mathbb{C}^{\alpha(x)}$, so an element $\mathcal{R} = (\mathcal{R}(x))_{x \in Q_0} \in \mathrm{Grass}(\beta, \alpha)$ is a collection of subspaces $\mathcal{R}(x) \subseteq V(x)$, $x \in Q_0$.

Consider the incidence variety

$$(10.16)\quad \begin{aligned} &Z(\beta, \alpha) \\ &= \{(V, \mathcal{R}) \in \mathrm{Rep}_\alpha(Q) \times \mathrm{Grass}(\beta, \alpha) \mid \forall a \in Q_1\ V(a)(\mathcal{R}(ta)) \subseteq \mathcal{R}(ha)\}. \end{aligned}$$

We have two natural projections:

$$\begin{array}{ccc} & Z(\beta,\alpha) & \\ {\scriptstyle p}\swarrow & & \searrow{\scriptstyle q} \\ \mathrm{Rep}_\alpha(Q) & & \mathrm{Grass}(\beta,\alpha). \end{array}$$

Proposition 10.12.1.

(1) *The projection* $p : Z(\beta, \alpha) \to \mathrm{Rep}_\alpha(Q)$ *is a proper map.*

(2) *The projection* $q : Z(\beta, \alpha) \to \mathrm{Grass}(\beta, \alpha)$ *gives* $Z(\beta, \alpha)$ *a structure of a vector bundle over* $\mathrm{Grass}(\beta, \alpha)$.

(3) *dim* $Z(\beta, \alpha) - \dim\ \mathrm{Rep}_\alpha(Q) = \langle \beta, \gamma \rangle$.

Proof.

(1) The projection p is a composition of the inclusion $Z(\beta, \alpha) \hookrightarrow \mathrm{Rep}_\alpha(Q)$ and the projection $\mathrm{Rep}_\alpha(Q) \times \mathrm{Grass}(\alpha, \beta) \to \mathrm{Rep}_\alpha(Q)$. So p is a projective morphism because $\mathrm{Grass}(\alpha, \beta)$ is a projective variety.

(2) Look at the fiber of the projection

$$q : Z(\beta, \alpha) \to \mathrm{Grass}(\beta, \alpha)$$

over the point $(\mathcal{R}(x))_{x \in Q_0} \in \mathrm{Grass}(\beta, \alpha)$. This is the set of representations $V \in \mathrm{Rep}_\alpha(Q)$ for which the family of subspaces $\mathcal{R}(x)$ forms a subrepresentation. Choosing a basis of $\mathcal{R}(x)$ and completing it to a basis of $V(x)$ for every $x \in Q_0$ allows to write all matrices $V(a)$ in a block form. The condition for $\mathcal{R}(x)$ being a subrepresentation means that for each $a \in Q_1$ the matrix of $V(a)$ in this block form has the form:

$$V(a) = \begin{pmatrix} * & * \\ 0 & * \end{pmatrix}.$$

This means that the fiber of q is an affine space of dimension

$$\sum_{a\in Q_1}(\alpha(ta)\alpha(ha)-\beta(ta)\gamma(ha)).$$

To be more precise, we can introduce the tautological $\beta(x)$-dimensional subbundles $\mathcal{R}(x)$ and the tautological $\gamma(x)$-dimensional factorbundles $\mathcal{Q}(x)$ on $\operatorname{Grass}(\beta(x),\alpha(x))$ for all $x\in Q_0$. Then the incidence variety $Z(\beta,\alpha)$ is a total space of a vector bundle $\bigoplus_{a\in Q_1}\mathcal{S}(a)$ over $\operatorname{Grass}(\beta,\alpha)$, where $\mathcal{S}(a)$ is the kernel of the natural surjection

$$\operatorname{Hom}(V(ta),V(ha))=V(ta)^*\otimes V(ha)\to\mathcal{R}(ta)^*\otimes\mathcal{Q}(ha).$$

This shows that the map q is locally trivial, and that q is indeed a vector bundle.

(3) This is just an easy calculation (see Exercise 10.12.2). □

The last missing ingredient is the analysis of the tangent space to the fiber of p. If $V\in\operatorname{Rep}_\alpha(Q)$, then we define $\operatorname{Grass}(\beta,V)$ as the fiber $p^{-1}(V)$. There is a technical difficulty that has to be overcome since $p^{-1}(V)$ may not be reduced. Suppose that

$$(\mathcal{R},V)\in\operatorname{Grass}(\beta,V)\subseteq\operatorname{Grass}(\beta,\alpha)\times\{V\}.$$

For every $x\in Q_0$ we choose a complement $\mathcal{Q}(x)$ of $\mathcal{R}(x)$ in $V(x)$. We can restrict V to the subspaces $\mathcal{R}(x)$, $x\in Q_0$, to obtain a subrepresentation that we will denote by $\mathcal{R}_V$. The quotient $V/\mathcal{R}_V$ will be denoted by $\mathcal{Q}_V$. Note that $\mathcal{Q}_V(x)$ can be identified with complement space $\mathcal{Q}(x)$ that we chose earlier.

Lemma 10.12.2. *The tangent space* $T_{(\mathcal{R},V)}\operatorname{Grass}(\beta,V)$ *at* $(\mathcal{R},V)$ *can be naturally identified with* $\operatorname{Hom}_Q(\mathcal{R}_V,\mathcal{Q}_V)$.

Proof. An open neighborhood of the point $(\mathcal{R},V)$ in $\operatorname{Grass}(\beta,\alpha)\times\{V\}$ can be identified with

$$\bigoplus_{x\in Q_0}\operatorname{Hom}(\mathcal{R}(x),\mathcal{Q}(x)).$$

An element

$$(\phi(x))_{x\in Q_0}\in\bigoplus_{x\in Q_0}\operatorname{Hom}(\mathcal{R}(x),\mathcal{Q}(x))$$

corresponds to

$$(W(x))_{x\in Q_0}\in\operatorname{Grass}(\beta,\alpha),$$

where

$$W(x)=\left\{\begin{pmatrix}v\\ \phi(x)v\end{pmatrix}\,\middle|\,v\in\mathcal{R}(x)\right\}\subseteq\mathcal{R}(x)\oplus\mathcal{Q}(x)\cong V(x).$$

The matrix of $V(a)$ is

$$\begin{pmatrix} \mathcal{R}_V(a) & \mathcal{U}(a) \\ 0 & \mathcal{Q}_V(a) \end{pmatrix},$$

where $\mathcal{U}(a) \in \operatorname{Hom}(W(ta), V(ha))$.

Locally, we calculate the equations for the subvariety $\operatorname{Grass}(\beta, V)$ of $\operatorname{Grass}(\beta, \alpha)$. The condition that $(\phi(x))_{x \in Q_0}$ lies in $\operatorname{Grass}(\beta, V)$ is that for every $a \in Q_1$ and every $v \in \mathcal{R}(ta)$ we have

$$\begin{pmatrix} \mathcal{R}_V(a) & \mathcal{U}(a) \\ 0 & \mathcal{Q}_V(a) \end{pmatrix} \begin{pmatrix} v \\ \phi(ta)v \end{pmatrix} = V(a) \begin{pmatrix} v \\ \phi(ta)v \end{pmatrix} = \begin{pmatrix} z \\ \phi(ha)z \end{pmatrix}$$

for some $z \in \mathcal{R}(ha)$. We must have

$$\mathcal{R}_V(a)v + \mathcal{U}(a)\phi(ta)v = z$$

and $\mathcal{Q}_V(a)\phi(ta)v = \phi(ha)z$. Combining both equations gives

$$\mathcal{Q}_V(a)\phi(ta)v = \phi(ha)\mathcal{R}_V(a)v + \phi(ha)\mathcal{U}(a)\phi(ta)v$$

for all $v \in \mathcal{R}(x)$. So the local equations are given by

$$\mathcal{Q}_V(a)\phi(ta) = \phi(ha)\mathcal{R}_V(a) + \phi(ha)\mathcal{U}(a)\phi(ta)$$

for all $a \in Q_1$. We get linear equations for the tangent space, by taking the linear parts of the equations defining the scheme. So the tangent space $T_{(\mathcal{Z},V)} \operatorname{Grass}(\beta, V)$ is given by

$$\Big\{\phi \in \bigoplus_{x \in Q_0} \operatorname{Hom}(\mathcal{R}(x), \mathcal{Q}(x)) \Big| \mathcal{Q}_V(a)\phi(ta) = \phi(ha)\mathcal{R}_V(a) \text{ for all } a \in Q_1\Big\}.$$

By definition, this space is $\operatorname{Hom}_Q(\mathcal{R}_V, \mathcal{Q}_V)$. □

Proposition 10.12.3. *The following conditions are equivalent:*

(1) *every representation from* $\operatorname{Rep}_\alpha(Q)$ *has a subrepresentation of dimension* β*;*

(2) *every representation from some Zariski open subset of* $\operatorname{Rep}_\alpha(Q)$ *has a subrepresentation of dimension* β*;*

(3) $\operatorname{ext}_Q(\beta, \gamma) = 0$.

Proof. The parts (1) and (2) are equivalent because the projection p is proper, so its image is a Zariski closed set.

Let us consider the differential of the map p at some point $(\mathcal{R}, V) \in Z(\beta, \alpha)$:

$$dp_{(\mathcal{R},V)} : T_{(\mathcal{R},V)} Z(\beta, \alpha) \to T_V \operatorname{Rep}_\alpha(Q).$$

The kernel of this map is $T_{(\mathcal{R},V)}\operatorname{Grass}(\beta, V) \cong \operatorname{Hom}_Q(\mathcal{R}_V, \mathcal{Q}_V)$. The cokernel of $dp_{(\mathcal{R},V)}$ has dimension

$$\begin{aligned}\dim T_V \operatorname{Rep}_\alpha(Q) &- \dim T_{(\mathcal{R},V)} Z(\beta,\alpha) + \dim \operatorname{Hom}_Q(\mathcal{R}_V, \mathcal{Q}_V) \\ &= \dim \operatorname{Rep}_\alpha(Q) - \dim Z(\beta,\alpha) + \dim \operatorname{Hom}_Q(\mathcal{R}_V, \mathcal{Q}_V) \\ &= -\langle \beta, \gamma\rangle + \dim \operatorname{Hom}_Q(\mathcal{R}_V, \mathcal{Q}_V) = \dim \operatorname{Ext}(\mathcal{R}_V, \mathcal{Q}_V).\end{aligned}$$

So the differential $dp_{(\mathcal{R},V)}$ is onto if and only if $\operatorname{Ext}_Q(\mathcal{R}_V, \mathcal{Q}_V) = 0$.

Suppose that (3) is true, i.e., $\operatorname{ext}(\beta,\gamma) = 0$. We can choose representations $\mathcal{R}$ and $\mathcal{Q}$ of dimensions β and γ, respectively, such that $\operatorname{Ext}(\mathcal{R}, \mathcal{Q}) = 0$. Let $V = \mathcal{R} \oplus \mathcal{Q}$. Then $dp_{(\mathcal{R},V)}$ is onto, and the morphism p is dominant. This implies (2).

Conversely, if (2) is true, then p is dominant and $dp_{(\mathcal{R},V)}$ is onto for some $(\mathcal{R}, V) \in Z(\beta,\alpha)$. It follows that $\operatorname{Ext}(\mathcal{R}_V, \mathcal{Q}_V) = 0$ and $\operatorname{ext}(\beta,\gamma) = 0$. □

We now start the analysis of the second incidence variety which is essential for the proof of the second part of Theorem 10.7.7. We define the rank of a homomorphism of two quiver representations to be the dimension vector of its image. The idea is to consider the rank of general homomorphism between general representations of dimensions α and β.

We define $\operatorname{Hom}_{\mathbb{C}}(\mathbb{C}^\alpha, \mathbb{C}^\beta) = \bigoplus_{x \in Q_0} \operatorname{Hom}_{\mathbb{C}}(\mathbb{C}^{\alpha(x)}, \mathbb{C}^{\beta(x)})$.

Lemma 10.12.4. *Let α and β be two dimension vectors. Then there exists a nonempty Zariski open set U of $\operatorname{Rep}_\alpha(Q) \times \operatorname{Rep}_\beta(Q)$ and a dimension vector γ such that for all $(V, W) \in U$, the value of $\dim \operatorname{Hom}_Q(V, W)$ is minimal and*

$$\{\phi \in \operatorname{Hom}_Q(V, W) \mid \operatorname{rank}(\phi) = \gamma\}$$

is open and nonempty in $\operatorname{Hom}_Q(V, W)$.

Proof. We define the variety

$$H(\alpha,\beta) \subseteq \operatorname{Hom}_{\mathbb{C}}(\mathbb{C}^\alpha, \mathbb{C}^\beta) \times \operatorname{Rep}_\alpha(Q) \times \operatorname{Rep}_\beta(Q)$$

by

$$H(\alpha,\beta) = \{(\phi, V, W) \mid \phi \in \operatorname{Hom}_Q(V, W)\}.$$

We have a projection

$$\pi : H(\alpha,\beta) \to \operatorname{Rep}_\alpha(Q) \times \operatorname{Rep}_\beta(Q).$$

The fiber $\pi^{-1}(V, W)$ is equal to $\operatorname{Hom}_Q(V, W)$. Hence there is an open subset U' of $\operatorname{Rep}_\alpha(Q) \times \operatorname{Rep}_\beta(Q)$ such that for $(V, W) \in U'$ the dimension of $\operatorname{Hom}_Q(V, W)$ is minimal.

Let δ be a dimension vector. We consider the subset

$$H_\delta(\alpha,\beta) = \{(\phi, V, W) \in H(\alpha,\beta) \mid \operatorname{rank}(\phi) = \delta\}.$$

This is a constructible subset in $H(\alpha,\beta)$. Since $\bigcup_\delta H_\delta(\alpha,\beta) = H(\alpha,\beta)$, there exists a dimension vector γ such that $H_\gamma(\alpha,\beta) \cap p^{-1}(U')$ is constructible and dense in $\pi^{-1}(U')$. Thus it contains a Zariski open subset of $\pi^{-1}(U')$. Therefore, $\pi(H_\gamma(\alpha,\beta)) \cap \pi^{-1}(U'))$ is constructible and dense in U'. So it contains a Zariski open set U of U'. This means that U and γ satisfy the conditions of the lemma. □

We say that γ is the **general rank** of a homomorphism from a general representation of dimension vector α to a general representation of dimension vector β.

Theorem 10.12.5 (Schofield). *Let α, β be dimension vectors for a quiver Q. Let γ be the general rank of a homomorphism from a general representation of dimension α to a general representation of dimension β. Then we have*

$$\alpha - \gamma \hookrightarrow \alpha \twoheadrightarrow \gamma \hookrightarrow \beta \twoheadrightarrow \beta - \gamma$$

and

$$\operatorname{ext}_Q(\alpha,\beta) = -\langle \alpha - \gamma, \beta - \gamma\rangle = \operatorname{ext}_Q(\alpha - \gamma, \beta - \gamma).$$

Proof. Since the general rank of a homomorphism from a general representation of dimension α to a general representation of dimension β is γ, we know that a general representation of dimension α has a factor of dimension vector γ and a general representation of dimension vector β has a subrepresentation of dimension vector γ. Thus $\alpha \twoheadrightarrow \gamma \hookrightarrow \beta$ and it follows that $\alpha - \gamma \hookrightarrow \alpha$ and $\beta \mapsto \beta - \gamma$.

Let $U \subseteq \operatorname{Rep}_\alpha(Q) \times \operatorname{Rep}_\beta(Q)$ be as in Lemma 10.12.4. Choose $W \in \operatorname{Rep}_\beta(Q)$ such that $(V,W) \in U$ for some V. If $U' = U \cap (\operatorname{Rep}_\alpha(Q) \times \{W\})$, then U' is open and dense subset of $\operatorname{Rep}_\alpha(Q)$ such that for all $V \in U'$, $\dim \operatorname{Hom}_Q(V,W) = \hom_Q(\alpha,\beta)$ and the general homomorphism from V to W has rank γ. Consider the variety $\operatorname{Grass}(\gamma, W) = p^{-1}(W)$ of all subrepresentations of dimension vector γ in W, where $p : Z(\gamma,\beta) \to \operatorname{Rep}_\beta(Q)$ is the projection. Since $\gamma \hookrightarrow \beta$, p is dominant (and in fact surjective by Proposition 10.12.3). If W is chosen generally, then the fiber $p^{-1}(W) = \operatorname{Grass}(\gamma, W)$ will have dimension

$$\dim \operatorname{Rep}_\beta(Q) - \dim Z(\gamma,\beta) = \langle \gamma, \beta - \gamma\rangle$$

by Proposition 10.12.1. We can view $\operatorname{Hom}_{\mathbb{C}}(\mathbb{C}^\alpha, W) \times \operatorname{Grass}(\gamma, W)$ as a vector bundle over $\operatorname{Grass}(\gamma, W)$. We define a subbundle B by

$$B = \{(\phi, \mathcal{R}) \mid \operatorname{im}(\phi) \subseteq \mathcal{R}\}.$$

The fiber over the point $\mathcal{R} \in \operatorname{Grass}(\gamma, W)$ may be identified with $\operatorname{Hom}_{\mathbb{C}}(\mathbb{C}^\alpha, \mathcal{R})$ so it has dimension $\sum_{x \in Q_0} \alpha(x)\gamma(x)$. We define an open and dense subvariety of B by

$$B^\circ = \{(\phi, \mathcal{R}) \in B \mid \operatorname{im}(\phi) = \mathcal{R}\}.$$

We have the following diagram:

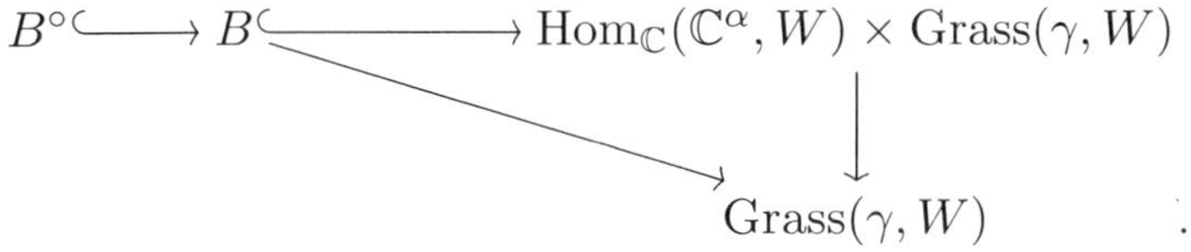

We note that B° parametrizes those elements of $\mathrm{Hom}_{\mathbb{C}}(\mathbb{C}^\alpha, W)$ whose image is a subrepresentation of W of dimension vector γ. It follows from the above discussion that

$$\dim B^\circ = \sum_{x \in Q_0} \alpha(x)\gamma(x) + \langle \gamma, \beta - \gamma \rangle.$$

We will view $\mathrm{Rep}_\alpha(Q) \times B^\circ$ as a vector bundle over B° and define a subbundle by

$$C = \{(V, (\phi, \mathcal{R})) \mid \phi : V \to W \text{ is a homomorphism of representations}\}.$$

By our description of B°, the image of ϕ is $\mathcal{R}$. The condition on V is

$$\mathcal{R}(a)\phi(ta) = \phi(ha)V(a).$$

This determines $V(a)$ uniquely up to elements in $\mathrm{Hom}(V(ta), \ker(\phi(a)))$. We have $\dim \mathrm{Hom}(V(ta), \ker(\phi(a))) = \alpha(ta)(\alpha - \gamma)(ha)$. So the dimension of a fiber above the point $(\phi, \mathcal{R}) \in B^\circ$ is given by

$$\sum_{a \in Q_1} \alpha(ta)(\alpha - \gamma)(ha).$$

Hence the dimension of C is

$$\langle \gamma, \beta - \gamma \rangle + \sum_{x \in Q_0} \alpha(x)\gamma(x) + \sum_{a \in Q_1} \alpha(ta)(\alpha - \gamma)(ha).$$

We consider the natural morphism of varieties $\theta : C \to \mathrm{Rep}_\alpha(Q)$ defined by $\theta(V, (\phi, \mathcal{R})) = V$. We have the following diagram:

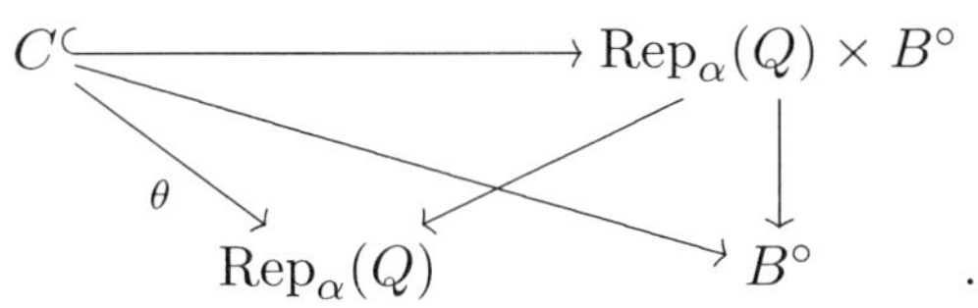

For $V \in \mathrm{Rep}_\alpha(Q)$ we have

$$\theta^{-1}(V) = \{(V, (\phi, \mathcal{R})) \mid \phi : V \to W \text{ is a homomorphism with image } \mathcal{R}\}.$$

Thus we may identify $\theta^{-1}(V)$ with

$$\{\phi \in \mathrm{Hom}_Q(V, W) \mid \mathrm{rank}(\phi) = \gamma\}.$$

Consider the set $\theta^{-1}(U')$. By assumption on W and U', we have $U' \subset \mathrm{im}(\theta)$ and hence $\theta^{-1}(U')$ is open and dense in C. Moreover, for any $V \in U'$, $\theta^{-1}(V)$

is open and dense in $\operatorname{Hom}_Q(V, W)$. Hence we can calculate the dimension of C in a different way to be

$$\dim(\operatorname{Rep}_\alpha(Q)) + \hom_Q(\alpha, \beta).$$

Thus we obtain the equality

$$\begin{aligned}\langle \gamma, \beta - \gamma \rangle + \sum_{x \in Q_0} \alpha(x)\gamma(x) + \sum_{a \in Q_1} \alpha(ta)(\alpha - \gamma)(ha) \\ = \sum_{a \in Q_1} \alpha(ta)\alpha(ha) + \hom_Q(\alpha, \beta).\end{aligned}$$

Hence

$$\langle \gamma, \beta - \gamma \rangle + \langle \alpha, \gamma \rangle = \langle \alpha, \beta \rangle + \operatorname{ext}_Q(\alpha, \beta)$$

and we conclude that

$$-\langle \alpha - \gamma, \beta - \gamma \rangle = \operatorname{ext}_Q(\alpha, \beta).$$

Let $V \in U'$ so that $\dim \operatorname{Ext}_Q(V, W) = \operatorname{ext}_Q(\alpha, \beta)$. Let $V' \hookrightarrow V$ be a subrepresentation where $\dim(V') = \alpha - \gamma$. Let W' be a quotient representation of W where $\dim(W') = \beta - \gamma$. Then by the right exactness of Ext_Q the natural map

$$\operatorname{Ext}_Q(V, W) \to \operatorname{Ext}_Q(V', W') \tag{10.17}$$

is surjective. However,

$$\dim \operatorname{Ext}_Q(V', W') \geq -\langle \alpha - \gamma, \beta - \gamma \rangle = \dim \operatorname{Ext}_Q(V, W).$$

Hence the natural map (10.17) of Ext's above is an isomorphism and

$$\dim \operatorname{Ext}_Q(V', W') = -\langle \alpha - \gamma, \beta - \gamma \rangle.$$

This implies $\operatorname{Hom}_Q(V', W') = 0$ and $\hom_Q(\alpha - \gamma, \beta - \gamma) = 0$. We conclude

$$\operatorname{ext}_Q(\alpha - \gamma, \beta - \gamma) = -\langle \alpha - \gamma, \beta - \gamma \rangle. \qquad \square$$

We record the important corollary of the above argument.

Corollary 10.12.6. *With the notation of Theorem* 10.12.5, *let V and W be representations of dimension vectors α, β, respectively, such that*

$$\dim \operatorname{Ext}_Q(V, W) = \operatorname{ext}_Q(\alpha, \beta).$$

Let V' be any subrepresentation of V of dimension vector $\alpha - \gamma$ and let W' be a factor of W of dimension vector $\beta - \gamma$. Then the natural map

$$\operatorname{Ext}_Q(V, W) \to \operatorname{Ext}_Q(V', W')$$

is an isomorphism and $\operatorname{Hom}_Q(V', W') = 0$.

Proof. This is the reasoning in the last paragraph of the proof of Theorem 10.12.5. $\square$

One way to view Theorem 10.12.5 is as an explanation for the non-vanishing of the $\operatorname{ext}_Q(\alpha,\beta)$. Indeed, we know that if $\langle\alpha,\beta\rangle < 0$, then $\operatorname{ext}_Q(\alpha,\beta)\neq 0$. Similarly, if $\alpha'\hookrightarrow\alpha$ and $\beta\twoheadrightarrow\beta'$, and $\langle\alpha',\beta'\rangle<0$, then since every representation of dimension vector α has a subrepresentation of dimension vector α' and every representation of dimension vector β has a factor of dimension β', the long exact sequence of Ext's implies that $\operatorname{ext}_Q(\alpha,\beta)\neq 0$. This suggests the idea that similar phenomenon occurs for the values of $\operatorname{ext}_Q(\alpha,\beta)$.

Now we are ready to complete the proof of Theorem 10.7.7.

Theorem 10.12.7 (Schofield). *Let α and β be dimension vectors for the quiver Q. Then*
(10.18)
$$\operatorname{ext}_Q(\alpha,\beta)=\max_{\substack{\alpha'\hookrightarrow\alpha\\ \beta\twoheadrightarrow\beta'}}\{-\langle\alpha',\beta'\rangle\}=\max_{\alpha'\hookrightarrow\alpha}\{-\langle\alpha',\beta\rangle\}=\max_{\beta\twoheadrightarrow\beta'}\{-\langle\alpha,\beta'\rangle\}.$$

Proof. Let V,W be representations of dimension vector α and β respectively and such that $\operatorname{Ext}_Q(V,W)=\operatorname{ext}_Q(\alpha,\beta)$. Let V' be a subrepresentation of V of dimension vector α' and let W' be a factor of W of dimension vector β'. Then

$$\operatorname{ext}_Q(\alpha,\beta)=\dim\operatorname{Ext}_Q(V,W)\geq\dim\operatorname{Ext}_Q(V',W')\geq-\langle\alpha',\beta'\rangle.$$

Thus $\operatorname{ext}_Q(\alpha,\beta)$ is at least as large as the remaining terms in the Theorem.

To show that the first and last term in (10.18) are equal, we want to show that $\operatorname{ext}_Q(\alpha,\beta)=-\langle\alpha,\beta'\rangle$ for some β' with $\beta\twoheadrightarrow\beta'$. We will prove this by induction on $\sum_{x\in Q_0}\beta(x)$. Theorem 10.12.5 shows that the first two terms in (10.18) are equal. We use notation of the proofs of Theorem 10.12.5 and of Corollary 10.12.6. For every subrepresentation V' of dimension vector $\alpha-\gamma$ of V and for every factor W' of dimension $\beta-\gamma$ of W the composite map

$$\operatorname{Ext}_Q(V,W)\to\operatorname{Ext}_Q(V,W')\to\operatorname{Ext}_Q(V',W')$$

is an isomorphism, while the first map is a surjection. It follows that $\operatorname{Ext}_Q(V,W)$ is isomorphic to $\operatorname{Ext}_Q(V,W')$ and so

$$\operatorname{ext}_Q(\alpha,\beta-\gamma)\leq\operatorname{ext}_Q(\alpha,\beta)=\dim\operatorname{Ext}_Q(V,W).$$

On the other hand, $\alpha-\gamma\hookrightarrow\alpha$ so we have

$$\operatorname{ext}_Q(\alpha,\beta)=\operatorname{ext}_Q(\alpha-\gamma,\beta-\gamma)\leq\operatorname{ext}_Q(\alpha,\beta-\gamma).$$

Hence $\operatorname{ext}_Q(\alpha,\beta)=\operatorname{ext}_Q(\alpha,\beta-\gamma)$. If $\gamma=0$, then $\operatorname{ext}_Q(\alpha,\beta)=-\langle\alpha,\beta\rangle$ and we are done. Otherwise, the induction hypothesis implies that there exists a dimension vector β' such that $\beta-\gamma\twoheadrightarrow\beta'$ and $\operatorname{ext}_Q(\alpha,\beta)=-\langle\alpha,\beta'\rangle$. From $\beta\twoheadrightarrow\beta-\gamma\twoheadrightarrow\beta'$ follows that $\beta\twoheadrightarrow\beta'$. This proves that the first and last term in (10.18) are equal. By symmetry, the first term and the third term are also the same. □

Exercises.

Exercise 10.12.1. *Suppose that $0 \leq k \leq n$ and consider the generalized Kronecker quiver $Q = \theta(n)$, with $Q_0 = \{1, 2\}$ and and n arrows $a_1, a_2, \ldots, a_n$ from 1 to 2. Let $\alpha = (k, 1)$, and $\sigma = (1, -k)$.*

(1) *Show that a representation $V \in \operatorname{Rep}_\alpha(Q)$ is σ-semi-stable if and only if the map*

$$\begin{pmatrix} V(a_1) \\ V(a_2) \\ \vdots \\ V(a_n) \end{pmatrix} : \mathbb{C}^k \to \mathbb{C}^n$$

is injective. Also show that in that case V is σ-stable.

(2) *Show that the variety $\operatorname{Proj}(\operatorname{SI}(Q, \alpha))$ is a projective variety that parametrizes k-dimensional subspaces of $\mathbb{C}^n$.*

Exercise 10.12.2. *Do the dimension calculation to prove part* (3) *of Proposition* 10.12.1.

Exercise 10.12.3. *Use the formula* (10.3) *to show that if Q is a tree, the semi-invariant of weight σ can occur only in one homogeneous component of $\mathbb{C}[\operatorname{Rep}_\beta(Q)]$.*

Exercise 10.12.4. *Suppose that for dimension vectors α and β and positive integers k and l we have $\operatorname{ext}_Q(k\alpha, l\beta) = kl \operatorname{ext}_Q(\alpha, \beta)$.*

10.13. Bibliographical Remarks

The basic facts about Schur functors and Littlewood-Richardson rule we review in Section 10.3 can be found in the Appendix A, Chapter 1, of the book of MacDonald [**63**]. The main results on GIT are standard. The lemma of Sato-Kimura was the essential tool in classifying prehomogeneous representations and their semi-invariants that was carried out in [**71**]. The Le Bruyn-Procesi theorem was proved in [**62**].

The determinantal semi-invariants were defined by Schofield [**72**] who also proved that they span the ring of semi-invariants for Dynkin quivers.

The proof of the Main Theorem and the Saturation Theorem follows [**25**]. One should mention the paper [**75**] of Schofield and Van den Bergh where they prove that result in the characteristic zero case using the representation theory, polarization and restitution operators. Domokos and Zubkov [**36**] found another formulation of the result on semi-invariant, and in [**27**] it was proved to be equivalent to the Main Theorem. The result was generalized to quivers with relations in two different ways in [**27**] and [**35**].

There are several interesting results on weight multiplicities of rings of semi-invariants (and for the Littlewood-Richardson coefficients in the special case of the triple flag quiver). One is the generalized Fulton conjecture (it was conjectured by Fulton in the case of Littlewood-Richardson coefficients). It says that if for some quiver Q, dimension vector β and a weight σ we have $\dim \mathrm{SI}(Q,\beta)_\sigma = 1$, then for any natural number n we have $\dim \mathrm{SI}(Q,\beta)_{n\sigma} = 1$. The generalized Fulton conjecture was proved by Belkale for the Littlewood-Richardson coefficients and in [**30**] (based on suggestion by Belkale) in the general case.

There is a similar result for the multiplicity 2. It says that if we have $\dim \mathrm{SI}(Q,\beta)_\sigma = 2$, then for any natural number n we have $\dim \mathrm{SI}(Q,\beta)_{n\sigma} = n+1$. This result was conjectured for Littlewood coefficients by King, Tollu and Toumazet. It was proved by Ikenmeyer for the Littlewood-Richardson coefficients and by Cass Sherman in [**76**] in general.

The function $n \mapsto \dim \mathrm{SI}(Q,\beta)_{n\sigma}$ was also analyzed. It was proved to be a polynomial in [**27**]. The reason is that the GIT quotient has rational singularities and the stabilizers of all the orbits are connected.

There are many structural results for the rings of semi-invariants. It follows at once from lemma of Sato-Kimura that in the case of a Dynkin quiver Q the ring $\mathrm{SI}(Q,\beta)$ is a polynomial ring. In the case of tame quiver Q it was proved in [**77**] that the ring $\mathrm{SI}(Q,\beta)$ is either a polynomial ring or a hypersurface.

There are also results on semi-invariants for some tame quivers with relations, notably in the case of gentle string algebras ([**21**]).

Another line of research is finding the bounds for the degrees of generators of the rings $\mathrm{SI}(Q,\beta)$. For the case of two vertices Derksen and Makam ([**23**], [**24**]) found the polynomial (in the number of arrows and the components of the dimension vector) bounds for the degrees of the generators.

Finally, let us mention recent results of Jiarui Fei ([**39**], [**40**]). He proved that for the triple flag quiver $T_{n,n,n}$ and the dimension vector β_n whose weight multiplicities were given by Littlewood-Richardson coefficients has an amazing cluster algebra structure associated to certain quiver with potential which was already analyzed in [**33**]. This approach allows us to find explicit generators and relations for the rings of semi-invariants $\mathrm{SI}(T_{n,n,n},\beta_n)$ for $n \leq 6$ even though the quiver $T_{n,n,n}$ is wild already for $n \geq 4$.

Chapter 11

Orthogonal Categories and Exceptional Sequences

In this chapter we investigate certain combinatorial structures emerging in quiver representations.

We start with exceptional objects. These are representations V for which $\mathrm{Hom}_Q(V,V) = \mathbb{C}, \mathrm{Ext}_Q(V,V) = 0$. They can be thought of as rigid representations and they play an important role in the sequel.

Next we discuss the canonical decomposition. It was introduced by Victor Kac in 1982. He showed that for every quiver Q and dimension vector α there is a Zariski dense open set in $\mathrm{Rep}_\alpha(Q)$ where the representation V decomposes into representations of fixed dimensions. This notion behaves well for multiples of α, so we can talk about canonical decompositions of rays of dimension vectors. For Dynkin quivers this gives an interesting triangulation of the simplex of rays of dimension vectors. For extended Dynkin quivers it gets more complicated as we get infinite triangulation with some vertices converging to a ray of isotropic root. For wild quivers one gets interesting fractal phenomena.

Next we go to tilting modules. Partial tilting modules are representations which form an open orbit in their dimension representation spaces. They are certain direct sums of exceptional objects. For Dynkin quivers we get a partial tilting module in every dimension vector, in the tame and wild cases the situation is more complicated.

We discuss the left and right orthogonal categories ${}^{\perp}C$ (resp $C^{\perp}$) which, for a given representation C, consist of modules V such that $\operatorname{Hom}_Q(V, C) = \operatorname{Ext}_Q(V, C) = 0$ (resp $\operatorname{Hom}_Q(C, V) = \operatorname{Ext}_Q(C, V) = 0$). If C is an indecomposable exceptional object, then Schofield proved that ${}^{\perp}C$ and $C^{\perp}$ are equivalent to categories of quiver representations for some quivers with $n-1$ vertices.

The third notion, that of exceptional sequence, came from Algebraic Geometry from results of Gorodentsev and Rudakov on vector bundles on projective spaces. An exceptional sequence is a certain sequence of exceptional objects which can be thought about as a frame in the Grothendieck group of the category $\operatorname{Rep}(Q)$. There are very interesting operations of mutations on exceptional sequences which lead to an action of the braid group $\mathcal{B}_n$ on the set of exceptional sequences. Crawley-Boevey proved that in the case of quiver representations the braid group action is transitive.

11.1. Schur Representations

In this section we will study Schur representations and exceptional objects. Suppose that Q is a quiver and α is a dimension vector.

Lemma 11.1.1. *If $V \in \operatorname{Rep}_\alpha(Q)$, then the codimension of the orbit $\operatorname{GL}_\alpha \cdot V$ in $\operatorname{Rep}_\alpha(Q)$ is $\dim \operatorname{Ext}_Q(V, V)$.*

Proof. The stabilizer $(\operatorname{GL}_\alpha)_V$ of V can be identified with the set of invertible elements in $\operatorname{Hom}_Q(V, V)$, so $\dim \operatorname{Hom}(V, V) = \dim(\operatorname{GL}_\alpha)_V$. Consider the orbit map $\varphi : \operatorname{GL}_\alpha : \operatorname{GL}_\alpha \to \operatorname{Rep}_\alpha(V)$ defined by $\varphi(A) = A \cdot V$. Then φ is a morphism whose fibers are cosets of $(\operatorname{GL}_\alpha)_V$ and have dimension $\dim \operatorname{Hom}_Q(V,V)$. So the dimension of the orbit is $\dim \operatorname{GL}_\alpha - \dim \operatorname{Hom}_Q(V, V)$ and the codimension of the orbit is

$$\begin{aligned}\dim \operatorname{Rep}_\alpha(Q) &- \dim \operatorname{GL}_\alpha + \dim \operatorname{Hom}_Q(V, V)\\ &= -\langle \alpha, \alpha \rangle + \dim \operatorname{Hom}_Q(V, V) = \dim \operatorname{Ext}_Q(V, V).\end{aligned}$$

□

Definition 11.1.2. A representation V of Q is called a **Schur representation** if $\operatorname{Hom}_Q(V, V) = \mathbb{C}$. Such a representation is indecomposable. A dimension vector α is called a **Schur root** if there exists a Schur representation of dimension α. If V is a Schur representation with $\operatorname{Ext}_Q(V, V) = 0$, then it is called an **exceptional object**.

Note that if V is an exceptional object of dimension α, then $\langle \alpha, \alpha \rangle = \dim \operatorname{Hom}_Q(V, V) - \dim \operatorname{Ext}_Q(V, V) = 1$ and α must be a real root.

Lemma 11.1.3. *Suppose that α is a Schur root and let $U \subseteq \operatorname{Rep}_\alpha(Q)$ be the set of all Schur representations of dimension α.*

(1) *The set U is open and dense in* $\operatorname{Rep}_\alpha(Q)$.

(2) *If α is a real root, then there is a unique indecomposable representation V of dimension α. This representation V is an exceptional object and its orbit is dense in* $\operatorname{Rep}_\alpha(Q)$.

(3) *If α is an imaginary root, then there exist infinitely many pairwise nonisomorphic Schur representations of dimension α.*

Proof. We have

$$U = \{V \in \operatorname{Rep}_\alpha(Q) \mid \dim \operatorname{Hom}_Q(V, V) \leq 1\}.$$

The function $V \mapsto \dim \operatorname{Hom}_Q(V, V)$ is upper semi-continuous, which means that U is open. Since U is nonempty, it is also dense. This proves (1). If V is a representation of dimension α, then we have

$$\langle \alpha, \alpha \rangle = \dim \operatorname{Hom}_Q(V, V) - \dim \operatorname{Ext}_Q(V, V).$$

Let $V \in U$ be a Schur representation. Then $\dim \operatorname{Hom}_Q(V, V) = 1$ and $\langle \alpha, \alpha \rangle = 1$ so it follows that $\dim \operatorname{Ext}_Q(V, V) = 0$. This proves that V has a dense orbit. Since α is real there is only one indecomposable representation and that must be V. This proves (2). If α is imaginary, then $\langle \alpha, \alpha \rangle \leq 0$ and $\dim \operatorname{Ext}_Q(V, V) > 0$ for every $V \in U$. So U is the union of orbits of Schur representations, but none of the orbits are dense in U. So U contains infinitely many orbits and there are infinitely many pairwise nonisomorphic Schur representations of dimension α. This proves (3). $\square$

Example 11.1.4. Consider the quiver

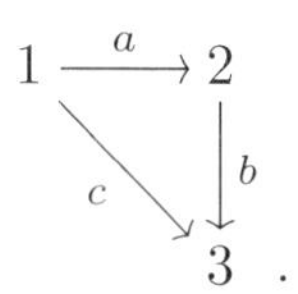

A representation V of dimension $\alpha = (1, 1, 2)$ is given by

$$V(a) = 1, \quad V(b) = \begin{pmatrix} 1 \\ 0 \end{pmatrix}, \quad V(c) = \begin{pmatrix} 0 \\ 1 \end{pmatrix}.$$

From $\operatorname{Hom}_Q(V, V) = \mathbb{C}$ it follows that V is an indecomposable Schur representation. We have $\langle \alpha, \alpha \rangle = 1$, so α is a real Schur root.

A representation W of dimension $\beta = (1, 2, 1)$ is given by

$$W(a) = \begin{pmatrix} 1 \\ 0 \end{pmatrix}, \quad W(b) = \begin{pmatrix} 0 \\ 1 \end{pmatrix}, \quad W(c) = 1.$$

One can verify that W is indecomposable, but $\dim \operatorname{Hom}_Q(W, W) = 2$. Since $\langle \beta, \beta \rangle = 1$ we have that β is a real root. Since W is not a Schur representation, α is not a Schur root.

Consider the representation Z of dimension $\gamma = (1,1,1)$ given by $Z(a) = Z(b) = Z(c) = 1$. Then $\operatorname{Hom}_Q(Z,Z) = \mathbb{C}$ so Z is a Schur representation. Because $\langle \gamma, \gamma \rangle = 0$ we have that γ is an imaginary root. So γ is an imaginary Schur root.

Finally, let R be the representation of dimension $\delta = (2,2,2)$ given by

$$R(a) = R(b) = \begin{pmatrix} 1 & 0 \\ 0 & 1 \end{pmatrix}, \quad R(c) = \begin{pmatrix} 1 & 1 \\ 0 & 1 \end{pmatrix}.$$

One can verify that R is indecomposable and $\langle \delta, \delta \rangle = 0$. So δ is an imaginary root. Let $U \subseteq \operatorname{Rep}_\delta(Q)$ be the open set of all representations T for which $T(a)$, $T(b)$ and $T(c)$ are invertible, and $T(c)^{-1}T(b)T(a)$ has two distinct eigenvalues. Suppose $T \in U$. We can choose a basis of eigenvectors t_1, t_2 of $T(1) = \mathbb{C}^2$. The vectors $t_i \in T(1)$, $V(a)t_i \in T(2)$ and $V(b)V(a)t_i \in T(3)$ span a subrepresentation T_i of dimension $(1,1,1)$ and we have $T = T_1 \oplus T_2$. So all representations in U are decomposable. There is no open set of indecomposable representations, so δ is not a Schur root.

11.2. The Canonical Decomposition

For a fixed dimension vector α we can study the decomposition of a representation V of dimension α into indecomposable representations. For V within an open dense set of $\operatorname{Rep}_\alpha(Q)$, the dimension vectors of the indecomposable summands are independent of V. A precise description of the dimension vectors of the indecomposable summands is given by the following theorem:

Theorem 11.2.1 (Kac). *Suppose that α is a dimension vector and $\alpha_1,\dots,\alpha_s$ are Schur roots such that $\alpha = \alpha_1 + \cdots + \alpha_s$. The following statements are equivalent:*

(1) *A general representation V of dimension α has a decomposition $V = V_1 \oplus V_2 \oplus \cdots \oplus V_s$ with $\underline{\dim}\, V_i = \alpha_i$ for all i.*
(2) $\operatorname{ext}_Q(\alpha_i, \alpha_j) = 0$ *for all $i \neq j$.*

Proof. Define

$$X = \mathrm{GL}_\alpha \times \operatorname{Rep}_{\alpha_1}(Q) \times \cdots \times \operatorname{Rep}_{\alpha_s}(Q)$$

and consider the morphism $\mu : X \to \operatorname{Rep}_\alpha(Q)$ defined by

$$(A, V_1, V_2, \dots, V_s) \mapsto A \cdot (V_1 \oplus \cdots \oplus V_s).$$

Let Z be the closure of the image of μ. Let us choose a general element $W = (A, V_1, \dots, V_s)$ and let $Y = \mu^{-1}\mu(W)$. Then we have

$$\dim Z = \dim X - \dim Y.$$

To compute $\dim Y$, let p be the restriction of the projection $X \to \operatorname{Rep}_{\alpha_1}(Q) \times \cdots \times \operatorname{Rep}_{\alpha_s}(Q)$ to Y. Because of the Krull-Remak-Schmidt theorem, $p(Y)$

consists of all tuples $(V_1',\dots,V_s')$ such that, after rearranging, V_i' is isomorphic to V_i for all i. We have

$$\dim p(Y)=\sum_{i=1}^{s}\dim \mathrm{GL}_{\alpha_i}\cdot V_i=\sum_{i=1}^{s}(\dim \mathrm{GL}_{\alpha_i}-\dim \mathrm{Hom}_Q(V_i,V_i)).$$

For $(V_1',\dots,V_s')$ in $p(Y)$, the fiber $F=p^{-1}(V_1',\dots,V_s')$ is isomorphic to the stabilizer of $V'=V_1'\oplus\cdots\oplus V_s'$ in GL_α which is $\mathrm{Hom}_Q(V',V')^\star$. In particular, this fiber has dimension

$$\dim F=\mathrm{Hom}_Q(V',V')^\star=\dim \mathrm{Hom}_Q(V',V')=\dim \mathrm{Hom}_Q(V,V).$$

So we have

$$\begin{aligned}\dim Y&=\dim p(Y)+\dim \mathrm{Hom}_Q(V,V)\\&=\sum_{i=1}^{s}(\dim \mathrm{GL}_{\alpha_i}-\dim \mathrm{Hom}_Q(V_i,V_i))+\dim \mathrm{Hom}_Q(V,V)\\&=\sum_{i=1}^{s}\dim \mathrm{GL}_{\alpha_i}+\sum_{i\neq j}\dim \mathrm{Hom}_Q(V_i,V_j).\end{aligned}$$

So the codimension of Z in $\mathrm{Rep}_\alpha(Q)$ is

$$\begin{aligned}\dim \mathrm{Rep}_\alpha(Q)-\dim Z&=\dim \mathrm{Rep}_\alpha(Q)-\dim X+\dim Y\\&=\dim \mathrm{Rep}_\alpha(Q)-\dim \mathrm{GL}_\alpha-\sum_{i=1}^{s}\dim \mathrm{Rep}_{\alpha_i}(Q)+\sum_{i=1}^{s}\dim \mathrm{GL}_{\alpha_i}\\&+\sum_{i\neq j}\dim \mathrm{Hom}_Q(V_i,V_j)=-\langle\alpha,\alpha\rangle+\sum_{i=1}^{s}\langle\alpha_i,\alpha_i\rangle+\sum_{i\neq j}\dim \mathrm{Hom}_Q(V_i,V_j)\\&=\sum_{i\neq j}(-\langle\alpha_i,\alpha_j\rangle+\dim \mathrm{Hom}_Q(V_i,V_j))\\&=\sum_{i\neq j}\dim \mathrm{Ext}_Q(V_i,V_j)=\sum_{i\neq j}\mathrm{ext}(\alpha_i,\alpha_j).\end{aligned}$$

□

Definition 11.2.2. If $\alpha=\alpha_1+\cdots+\alpha_s$ where $\alpha_1,\dots,\alpha_s$ are Schur roots, and a general representation of dimension α has a decomposition $V_1\oplus\cdots\oplus V_s$ where $\underline{\dim}\, V=\alpha_i$ for all i, then we write

$$\alpha=\alpha_1\oplus\alpha_2\oplus\cdots\oplus\alpha_s$$

and call it the **canonical decomposition** of α.

Example 11.2.3. Consider the quiver with one loop:

$$Q:\ 1 \circlearrowleft a\ .$$

The canonical decomposition of the dimension vector (n) is

$$(n) = (1) \oplus (1) \oplus \cdots \oplus (1).$$

This means that a general $n \times n$ matrix B is diagonalizable, i.e., there exists an invertible matrix A such that ABA^{-1} is diagonal.

Example 11.2.4. Consider the Kronecker quiver

$$\theta(2): \; 1 \underset{b}{\overset{a}{\rightrightarrows}} 2\,.$$

The real Schur roots are $(n-1, n)$ and $(n, n-1)$ for $n = 1, 2, \ldots$. There is only one imaginary Schur root, namely $(1,1)$. Suppose that (m_1, m_2) is a dimension vector. If $m_1 = m_2$, then the canonical decomposition is

$$(m_1, m_2) = (1,1)^{\oplus m_1}.$$

If $m_1 > m_2$, then there exists a positive integer n such that

$$\frac{n-1}{n} \leq \frac{m_2}{m_1} \leq \frac{n}{n+1}.$$

In this case, the canonical decomposition is

$$(m_1, m_2) = (n, n-1)^{\oplus(m_1 n - m_2(n+1))} \oplus (n+1, n)^{\oplus(m_2 n - m_1(n-1))}.$$

If $m_1 < m_2$, then there exists a positive integer n such that

$$\frac{n-1}{n} \leq \frac{m_1}{m_2} \leq \frac{n}{n+1}$$

and the canonical decomposition is

$$(m_1, m_2) = (n-1, n)^{\oplus(m_2 n - m_1(n+1))} \oplus (n, n+1)^{\oplus(m_1 n - m_2(n-1))}.$$

Lemma 11.2.5. *Suppose that $\alpha = \alpha_1 + \cdots + \alpha_s$ with $\operatorname{ext}_Q(\alpha_i, \alpha_j) = 0$ for $i \neq j$, and $\alpha_i = \alpha_{i,1} \oplus \cdots \oplus \alpha_{i,k_i}$ is the canonical decomposition of α_i for all i, then the canonical decomposition of α is*

$$\alpha = (\alpha_{1,1} \oplus \cdots \oplus \alpha_{1,k_1}) \oplus (\alpha_{2,1} \oplus \cdots \oplus \alpha_{2,k_2}) \oplus \cdots \oplus (\alpha_{s,1} \oplus \cdots \oplus \alpha_{s,k_s}).$$

Proof. The proof is Exercise 11.2.1. □

Corollary 11.2.6. *If $\alpha = \alpha_1 + \alpha_2 + \cdots + \alpha_s$ with $\operatorname{ext}_Q(\alpha_i, \alpha_j) = 0$ for $i \neq j$, then a general representation V of dimension α has a decomposition $V = V_1 \oplus V_2 \oplus \cdots \oplus V_s$ where V_i has dimension α_i for all i.*

Proof. By Lemma 11.2.5, a general representation V of dimension α has decomposition

$$V = (V_{1,1} \oplus \cdots \oplus V_{1,k_1}) \oplus (V_{2,1} \oplus \cdots \oplus V_{2,k_2}) \oplus \cdots \oplus (V_{s,1} \oplus \cdots \oplus V_{s,k_s}),$$

where $V_{i,j}$ has dimension $\alpha_{i,j}$. If we define $V_i = V_{i,1} \oplus \cdots \oplus V_{i,k_i}$, then we have $V = V_1 \oplus \cdots \oplus V_s$ and V_i has dimension α_i for all i. □

We say that a dimension vector α is σ-(semi-)stable if a general representation of dimension α is σ-(semi-)stable.

Proposition 11.2.7 (Schofield). *Let α be a nonzero dimension vector and set $\sigma = \langle \alpha, \cdot \rangle - \langle \cdot, \alpha \rangle$. Then α is a Schur root if and only if α is σ-stable.*

Proof. Note that $\sigma(\alpha) = 0$. Suppose that α is σ-stable. Then a general representation V of dimension α is σ-stable and indecomposable by Exercise 9.8.7. This proves that α is a Schur root.

Assume that α is a Schur root. Suppose that β is a arbitrary dimension vector with $\beta \hookrightarrow \alpha$ and $\beta \neq 0, \alpha$. We need to show that $\sigma(\beta) < 0$ and define $\gamma = \alpha - \beta$. From $\beta \hookrightarrow \alpha$ it follows that $\mathrm{ext}_Q(\beta, \gamma) = 0$. Suppose that V is a general representation. Because $\beta \hookrightarrow \alpha$, there exists a short exact sequence

$$0 \longrightarrow W \longrightarrow V \longrightarrow Z \longrightarrow 0,$$

where W and Z are of dimension β and γ, respectively. Suppose that $\psi : Z \to W$ is a morphism. Then the composition

$$V \longrightarrow\!\!\!\!\!\rightarrow Z \xrightarrow{\ \psi\ } W \hookrightarrow V$$

is a scalar multiple of the identity morphism. Since the composition clearly cannot be an isomorphism, it has to be 0, so $\psi = 0$. This proves $\mathrm{Hom}_Q(Z, W) = 0$ and $\hom_Q(\gamma, \beta) = 0$. Finally, $\mathrm{ext}_Q(\gamma, \beta) > 0$ because otherwise we would have $\mathrm{ext}_Q(\gamma, \beta) = \mathrm{ext}_Q(\beta, \gamma) = 0$ and a general representation of dimension $\alpha = \beta + \gamma$ would be decomposable. We have

$$\begin{aligned}\sigma(\beta) = \langle \alpha, \beta \rangle - \langle \beta, \alpha \rangle &= \langle \gamma, \beta \rangle - \langle \beta, \gamma \rangle \\ &= \hom_Q(\gamma, \beta) - \mathrm{ext}_Q(\gamma, \beta) - \hom_Q(\beta, \gamma) + \mathrm{ext}_Q(\beta, \gamma) \\ &= -\mathrm{ext}_Q(\gamma, \beta) < 0 - \hom_Q(\beta, \gamma).\end{aligned}$$

□

The proof actually shows that if a dimension vector α is stable for some weight σ (not necessarily the weight defined in the proposition), then α is a Schur root. So Schur roots are exactly the nonzero dimension vectors that are stable with respect to some weight.

We will now study how the canonical decomposition of α and a multiple $m\alpha$ are related. A root α is called **isotropic** if $\langle \alpha, \alpha \rangle = 0$. Recall that α is indivisible, if α is not of the form $k\alpha'$ where k is an integer ≥ 2 and α' is a dimension vector.

Lemma 11.2.8. *Suppose that α is a Schur root and m is a positive integer.*

(1) *If α is real or isotropic, then α is indivisible and we have*

$$m\alpha = \alpha^{\oplus m} = \underbrace{\alpha \oplus \alpha \oplus \cdots \oplus \alpha}_{m}.$$

(2) *If α is imaginary and nonisotropic, then $m\alpha$ is also a Schur root.*

Proof. If α is real and V is the unique Schur representation of dimension α, then we have

$$1 = \langle \alpha, \alpha \rangle = \dim \mathrm{Hom}_Q(V, V) - \dim \mathrm{Ext}_Q(V, V) = 1 - \dim \mathrm{Ext}_Q(V, V).$$

So we have $\mathrm{Ext}_Q(V, V) = 0$ and $\mathrm{ext}_Q(\alpha, \alpha) = 0$. Suppose that α is isotropic and V and V' are general representations of dimension α. Then V and V' are nonisomorphic. Let $\phi : V \to V'$ be a general map and let β be the dimension vector of the image. Since V and V' are nonisomorphic, we have $\beta \neq \alpha$. Since the image is a subrepresentation of V' we have $\beta \hookrightarrow \alpha$ and $\mathrm{ext}_Q(\beta, \alpha - \beta) = 0$. The kernel of ϕ is a subrepresentation of V and has dimension $\alpha - \beta$. Therefore, $\alpha - \beta \hookrightarrow \alpha$ and $\mathrm{ext}(\alpha - \beta, \beta) = 0$. If β is nonzero, then $\mathrm{ext}_Q(\alpha - \beta, \beta) = \mathrm{ext}_Q(\beta, \alpha - \beta) = 0$ implies that α is not a Schur root. So we conclude that $\beta = 0$, $\mathrm{Hom}_Q(V, V') = 0$, and $\mathrm{hom}_Q(\alpha, \alpha) = 0$. Now we have

$$0 = \langle \alpha, \alpha \rangle = \mathrm{hom}_Q(\alpha, \alpha) - \mathrm{ext}_Q(\alpha, \alpha) = 0 - \mathrm{ext}_Q(\alpha, \alpha).$$

So we have shown that if α is real or isotropic, then $\mathrm{ext}_Q(\alpha, \alpha) = 0$. Now it is clear that $m\alpha = \alpha^{\oplus m}$ is the canonical decomposition of m. Suppose that $\alpha = k\beta$ with $k \geq 2$. We have $0 = \mathrm{ext}_Q(\alpha, \alpha) = \mathrm{ext}_Q(k\beta, k\beta)$. From Exercise 10.12.4 it follows that $\mathrm{ext}_Q(\beta, \beta) = 0$. This implies that α has a nontrivial canonical decomposition.

Suppose that α is a nonisotropic imaginary Schur root, and let $m\alpha = \alpha_1 \oplus \cdots \oplus \alpha_s$ be the canonical decomposition. Choose generators $\tau_1, \ldots, \tau_r$ of the submodule $\{\tau \in \mathbb{Z}^{Q_0} \mid \tau(\alpha) = 0\}$. For k large enough, α is $(k\sigma + \tau_i)$-stable for all i. It follows that $m\alpha$ is semi-stable with respect to the weights $\sigma, k\sigma + \tau_1, \ldots, k\sigma + \tau_r$. In particular, $\alpha_1, \ldots, \alpha_s$ are stable with these weights, so $\sigma(\alpha_j) = 0$ and $(k\sigma + \tau_i)(\alpha_j) = 0$ for all i and j. So we have $\tau_i(\alpha_j) = 0$ for all i and j. It follows that $\alpha_1, \ldots, \alpha_s$ are rational multiples of α. If $s > 1$, then we have

$$\mathrm{hom}_Q(\alpha_1, \alpha_2) - \mathrm{ext}_Q(\alpha_1, \alpha_2) = \langle \alpha_1, \alpha_2 \rangle < 0$$

so $\mathrm{ext}_Q(\alpha_1, \alpha_2) \neq 0$. This contradicts Kac's characterization of the canonical decomposition (Theorem 11.2.1). $\square$

We want to give a more precise description in Theorem 11.2.11 of the generic hom's and ext's between the dimension vectors appearing in the canonical decomposition which will be used in Section 11.9 to obtain an efficient algorithm to compute the canonical decomposition. For the proof of the theorem we need the following two lemmas.

Lemma 11.2.9 (Happel-Ringel). *Suppose that V and W are indecomposable representations, $\phi : V \to W$ is a nonzero morphism and* $\operatorname{Ext}_Q(W,V) = 0$. *Then ϕ is injective or surjective.*

Proof. Suppose that $\phi : V \to W$ is nonzero and that it is neither injective, nor surjective. If U is the image of ϕ, then we have a factorization

$$\begin{array}{ccc} V & & \\ \downarrow\!\!\!\downarrow & \searrow^{\phi} & \\ U & \hookrightarrow & W\,. \end{array}$$

The surjective map $V \twoheadrightarrow U$ yields a surjective map $\operatorname{Ext}_Q(W/U,V) \twoheadrightarrow \operatorname{Ext}_Q(W/U,U)$. In particular, we have a lifting of extensions

$$\begin{array}{ccccccccc} 0 & \longrightarrow & V & \xrightarrow{\gamma_1} & Z & \longrightarrow & W/U & \longrightarrow & 0 \\ & & \downarrow{\scriptstyle \gamma_2} & & \downarrow{\scriptstyle \psi_2} & & \| & & \\ 0 & \longrightarrow & U & \xrightarrow[\psi_1]{} & W & \longrightarrow & W/U & \longrightarrow & 0. \end{array}$$

This yields an exact sequence

$$0 \longrightarrow V \xrightarrow{\binom{\gamma_1}{\gamma_2}} U \oplus Z \xrightarrow{(\psi_1\ -\psi_2)} W \longrightarrow 0.$$

If the exact sequence is split, then $V \oplus W$ is isomorphic to $U \oplus Z$, so by Krull-Remak-Schmidt U is isomorphic to V or W. Contradiction. So the sequence is nonsplit, and therefore $\operatorname{Ext}_Q(W,V) \neq 0$. Contradiction.

□

Lemma 11.2.10 (Schofield). *Suppose that α and β are Schur roots such that* $\operatorname{ext}_Q(\alpha,\beta) = 0$. *Then we have* $\operatorname{ext}_Q(\beta,\alpha) = 0$ *or* $\hom_Q(\beta,\alpha) = 0$. *If both α and β are imaginary, then* $\hom_Q(\beta,\alpha) = 0$.

Proof. Assume that α and β are Schur roots with $\operatorname{ext}_Q(\alpha,\beta) = 0$ and $\hom_Q(\beta,\alpha) \neq 0$. We will show that β or α is real, and that $\operatorname{ext}_Q(\beta,\alpha) = 0$.

Choose general representations V and W of dimensions α and β, respectively. So we may assume that they are both Schur representations, $\operatorname{Ext}_Q(V,W) = 0$ and $\operatorname{Hom}_Q(W,V) \neq 0$. Let $\phi \in \operatorname{Hom}_Q(W,V)$ be nonzero. Then ϕ is injective or surjective by Lemma 11.2.9. Suppose that ϕ is injective. Because $\phi : W \hookrightarrow V$ is injective, we have a surjective map $0 = \operatorname{Ext}_Q(V,W) \twoheadrightarrow \operatorname{Ext}_Q(W,W)$. So $\operatorname{Ext}_Q(W,W) = 0$ and β must be real. Since V was chosen generally, we have that $\beta \hookrightarrow \alpha$ and $\operatorname{ext}_Q(\beta,\alpha-\beta) = 0$.

We have $\operatorname{Ext}_Q(W, Z) = 0$ for some representation Z of dimension $\alpha - \beta$. So it follows $\operatorname{Ext}_Q(V, V \oplus Z) = 0$ and therefore $\operatorname{ext}_Q(\beta, \alpha) = 0$.

If, on the other hand, ϕ is surjective, then similar (dual) arguments show that α must be real, and $\operatorname{ext}_Q(\beta, \alpha) = 0$. □

Theorem 11.2.11 (Schofield). *Suppose that*

$$\alpha = \alpha_1^{d_1} \oplus \alpha_2^{d_2} \oplus \cdots \oplus \alpha_s^{d_s}$$

is the canonical decomposition, where $\alpha_1, \dots, \alpha_s$ are distinct Schur roots, and $d_1, \dots, d_s$ are positive integers (so $\operatorname{ext}_Q(\alpha_i, \alpha_j) = 0$ for $i \neq j$). Then, after rearranging, we also have $\hom_Q(\alpha_i, \alpha_j) = 0$ for all i, j with $i < j$.

Proof. Let us draw a graph with vertices $1, 2, \dots, s$ where we draw an arrow $i \to j$ if $i \neq j$ and $\hom_Q(\alpha_i, \alpha_j) \neq 0$. It suffices to show that the graph does not have oriented cycles. Suppose that the graph has an oriented cycle. Let's say that, after rearranging, we have a cycle

$$\alpha_1 \to \alpha_2 \to \cdots \alpha_r \to \alpha_1.$$

Let $V_1, \dots, V_r$ be general representations of dimensions $\alpha_1, \dots, \alpha_r$, respectively. Then we have $\operatorname{Ext}_Q(V_i, V_j) = 0$ for $i \neq j$. Let $\varphi_i : V_i \to V_{i+1}$ $(1 \leq i \leq r-1)$ and $\varphi_r : V_r \to V_1$ be nonzero morphisms. For each i, φ_i is either injective or surjective (but not both since the roots are distinct). They clearly cannot all be injective, otherwise there would be a proper inclusion from V_1 into itself. For a similar reason, the maps cannot all be surjective. Without loss of generality we may assume φ_1 is surjective. Let $k \geq 2$ be minimal such that φ_k is injective. Then φ_{k-1} is surjective. The composition $\varphi_k \varphi_{k-1} : V_{k-1} \to V_{k+1}$ (with the convention $V_{r+1} = V_1$) is nonzero, and neither surjective nor injective. This contradicts Lemma 11.2.9. □

Example 11.2.12. Consider the quiver

$$1 \longrightarrow 2 \longrightarrow 3 \longrightarrow 4 \longrightarrow 5 \longrightarrow 6 \longrightarrow 7 \longrightarrow 8\,.$$

An example of a canonical decomposition (using column vectors) is:

$$\begin{pmatrix}1\\2\\5\\5\\1\\2\\2\\1\end{pmatrix} = \begin{pmatrix}1\\1\\1\\1\\1\\1\\1\\1\end{pmatrix} \oplus \begin{pmatrix}0\\1\\1\\1\\0\\0\\0\\0\end{pmatrix} \oplus \begin{pmatrix}0\\0\\0\\0\\0\\1\\1\\0\end{pmatrix} \oplus \begin{pmatrix}0\\0\\1\\1\\0\\0\\0\\0\end{pmatrix}^{\oplus 3}.$$

Exercises.

Exercise 11.2.1. *Prove Lemma* 11.2.5.

Exercise 11.2.2. *As in Example* 11.2.12, *find the canonical decomposition of the dimension vector*

$$\alpha = \begin{pmatrix} 1 \\ 2 \\ 5 \\ 5 \\ 1 \\ 2 \\ 2 \\ 1 \end{pmatrix}$$

but now for the quiver

$$1 \longrightarrow 2 \longleftarrow 3 \longrightarrow 4 \longleftarrow 5 \longrightarrow 6 \longleftarrow 7 \longrightarrow 8.$$

Exercise 11.2.3. *Suppose that*

$$\alpha = \alpha_1^{\oplus d_1} \oplus \alpha_2^{\oplus d_2} \oplus \cdots \oplus \alpha_s^{\oplus d_s}$$

is a canonical decomposition where $\alpha_1, \dots, \alpha_s$ *are distinct Schur roots, and* $d_1, \dots, d_s$ *are positive integers.*

(1) *Show that* $\alpha_1, \dots, \alpha_s$ *are linearly independent over* $\mathbb{Q}$. *(Hint: Use Theorem* 11.2.11 *and consider the matrix* $(\langle \alpha_i, \alpha_j \rangle)_{i,j}$.*)*

(2) *Suppose that* $\alpha_1, \dots, \alpha_s$ *are real Schur roots. Show that*

$$(\mathbb{Q}\alpha_1 + \cdots + \mathbb{Q}\alpha_s) \cap \mathbb{Z}^{Q_0} = \mathbb{Z}\alpha_1 + \mathbb{Z}\alpha_2 + \cdots + \mathbb{Z}\alpha_s.$$

Exercise 11.2.4. *Suppose that*

$$\alpha = \alpha_1 \oplus \alpha_2 \oplus \cdots \oplus \alpha_s$$

is a canonical decomposition. Show that the orbit $\mathrm{GL}(\alpha) \cdot V \subseteq \mathrm{Rep}_Q(\alpha)$ *of a general representation* V *of dimension* α *has codimension*

$$\sum_{i=1}^{s} (1 - \langle \alpha_i, \alpha_i \rangle).$$

Exercise 11.2.5. *Let* Q *be a quiver with* $Q_0 = \{1, 2, \dots, n\}$ *and let* $\epsilon_1, \dots, \epsilon_n$ *be the standard basis vectors of* $\mathbb{Z}^n = \mathbb{Z}^{Q_0}$.

(1) *Show that* β *and* γ *are distinct imaginary Schur roots appearing in the canonical decomposition of* α. *Show that* $\langle \beta, \gamma \rangle = \langle \gamma, \beta \rangle = 0$.

(2) *Consider the Cartan matrix* $C = ((\epsilon_i, \epsilon_j))_{i,j=1}^n$ *and let* (a, b, c) *be its signature, i.e.,* C *has* a *positive eigenvalues,* b *negative eigenvalues and* c *eigenvalues equal to* 0. *Show that the number of distinct*

imaginary Schur roots appearing in the canonical decomposition of α *is at most* $b + c$.

(3) *Suppose that the quiver* Q *has a subquiver* Q' *of finite type with* r *vertices. Show that every canonical decomposition has at most* $n - r$ *distinct imaginary Schur roots.*

(4) *Show that for a quiver with 2 vertices, every canonical decomposition has at most 1 distinct imaginary Schur root.*

(5) *Show that for the triple flag quiver, every canonical decomposition has at most 1 distinct imaginary Schur root.*

11.3. Tilting Modules

Tilting modules can be used to relate the representation theory of two algebras. We used tilting modules in Section 3.4 to prove that certain algebras are Morita equivalent. A direct summand of a tilting module is a partial tilting module. We will see that for a path algebra the partial tilting modules correspond to representations that have a dense orbit in the representation space.

For an object $T \in \operatorname{Rep}(Q)$ we denote by $\operatorname{add}(T)$ the full subcategory of $\operatorname{Rep}(Q)$ consisting of objects that are direct summands of modules T^m (m arbitrary).

Definition 11.3.1. A representation T of a quiver Q is called a **tilting module** if the following conditions are satisfied:

(1) $\operatorname{Ext}_Q(T, T) = 0$.

(2) There is a short exact sequence

$$0 \longrightarrow \mathbb{C}Q \longrightarrow T_1 \longrightarrow T_2 \longrightarrow 0,$$

where T_1, T_2 are in $\operatorname{add}(T)$.

The second part of the definition can be reformulated as follows.

Lemma 11.3.2. *A module* T *is a tilting module if and only if:*

(1) $\operatorname{Ext}_Q(T, T) = 0$.

(2) *For each indecomposable projective module* P *there is a short exact sequence*

$$0 \longrightarrow P \longrightarrow T_1 \longrightarrow T_2 \longrightarrow 0,$$

where T_1, T_2 *are in* $\operatorname{add}(T)$.

Proof. First of all we can assume without loss of generality that the map $f : \mathbb{C}Q \to T_1$ is **left minimal** (compare definition 5.4.1). Indeed, if f is not right minimal, then its image is contained in some direct summand of T_1 which then appears also in T_2 and it can be cancelled out.

Next, note that for every $T' \in \mathrm{add}(T)$ and every map $g : \mathbb{C}Q \to T'$ there exists a map $s : T_1 \to T'$ such that $g = sf$. This follows by applying the functor $\mathrm{Hom}_Q(-, T')$ to our exact sequence. Thus f is **a left minimal approximation** of $\mathbb{C}Q$ in $\mathrm{add}(T)$.

The following facts about left minimal approximations are easy to check and we leave them as exercises to the reader:

(1) A direct sum of left minimal maps is left minimal.

(2) Every module has a left minimal approximation in $\mathrm{add}(T)$.

(3) A left minimal approximation is unique up to isomorphism. Thus a left minimal approximation commutes with direct sums.

The last fact means that the left minimal approximation $f : \mathbb{C}Q \to T_1$ is a direct sum of left minimal approximations $f_x : P_x \to T_{1,x}$. By Krull-Remak-Schmidt Theorem 1.7.4 the cokernels of the maps f_x are also in $\mathrm{add}(T)$. □

Lemma 11.3.3 (Bongartz). *Let T be a module such that $\mathrm{Ext}_Q(T, T) = 0$. Then there exists a module $X \in \mathrm{Rep}(Q)$ such that $T \oplus X$ is a tilting module.*

Proof. We prove the first part. Let $\{e_1, \ldots, e_m\}$ be a basis of $\mathrm{Ext}_Q(T, \mathbb{C}Q)$. Consider the exact sequence

$$0 \longrightarrow \mathbb{C}Q \longrightarrow X \longrightarrow T^m \longrightarrow 0$$

whose pull-back under the i-th canonical injection $T \hookrightarrow T^m$ is e_i. In the induced exact sequence

$$\begin{array}{ccccccc} 0 \longrightarrow & \mathrm{Hom}_Q(T, \mathbb{C}Q) & \longrightarrow & \mathrm{Hom}_Q(T, X) & \longrightarrow & \mathrm{Hom}_Q(T, T^m) \\ & & \swarrow^{\partial} & & & \\ & \mathrm{Ext}_Q(T, \mathbb{C}Q) & \longrightarrow & \mathrm{Ext}_Q(T, X) & \longrightarrow & \mathrm{Ext}_Q(T, T) = 0 \end{array}$$

the connecting homomorphism ∂ is surjective by construction, hence

$$\mathrm{Ext}_Q(T, X) = 0.$$

Moreover, we have exact sequences

$$0 = \mathrm{Ext}_Q(\mathbb{C}Q, T) \longrightarrow \mathrm{Ext}_Q(X, T) \longrightarrow \mathrm{Ext}_Q(T^m, T) = 0$$

and

$$0 = \mathrm{Ext}_Q(\mathbb{C}Q, X) \longrightarrow \mathrm{Ext}_Q(X, X) \longrightarrow \mathrm{Ext}_Q(T, X) = 0$$

which imply that $\mathrm{Ext}_Q(X, T) = \mathrm{Ext}_Q(X, X) = \mathrm{Ext}_Q(T \oplus X, T \oplus X) = 0$. Thus $T \oplus X$ is a tilting module. □

Corollary 11.3.4. *For a representation $T \in \operatorname{Rep}_\alpha(Q)$, the following statements are equivalent:*

(1) *T is a direct summand of a tilting module.*

(2) $\operatorname{Ext}_Q(T, T) = 0$.

(3) *The modules isomorphic to T form an open orbit in $\operatorname{Rep}_\alpha(Q)$.*

Proof. The first and second conditions have just been proved to be equivalent in Lemma 11.3.3. The modules isomorphic to T form the GL_α-orbit of T, so (2) and (3) are equivalent by Lemma 11.1.1. $\square$

A module satisfying the conditions of Corollary 11.3.4 is called a **partial tilting module**.

Proposition 11.3.5. *Let T be an exceptional module. Assume that T is not projective. Let X be a module constructed in the first part of Lemma 11.3.3. Then we have* $\operatorname{Hom}_Q(T, X) = \operatorname{Ext}_Q(T, X) = 0$.

Proof. We already proved that $\operatorname{Ext}_Q(T, X) = 0$. Applying the functor $\operatorname{Hom}_Q(T, \cdot)$ to the exact sequence defining X we get the exact sequence

$$\begin{array}{ccccc} 0 \longrightarrow \operatorname{Hom}_Q(T, \mathbb{C}Q) \longrightarrow \operatorname{Hom}_Q(T, X) & & & & \\ \downarrow & & & & \\ \operatorname{Hom}_Q(T, T^m) \longrightarrow \operatorname{Ext}_Q(T, \mathbb{C}Q) \longrightarrow 0\,. & & & & \end{array}$$

The right-hand side map in this sequence is an isomorphism from $\mathbb{C}^m$ to $\mathbb{C}^m$ because it takes the embedding into the i-th direct summand onto e_i. It remains to see that $\operatorname{Hom}_Q(T, \mathbb{C}Q) = 0$. Assume that T is indecomposable and not projective. If $\operatorname{Hom}_Q(T, \mathbb{C}Q)$ is nonzero, consider a nonzero homomorphism $f : T \to \mathbb{C}Q$. Then $\operatorname{im}(f) = P$ is projective because $\mathbb{C}Q$ is hereditary. This means that the epimorphism $f : T \to P$ splits and we get a contradiction. $\square$

Theorem 11.3.6. *Let Q be a quiver without oriented cycles and n vertices. Let T be a $\mathbb{C}Q$-module,*

$$T = T_1^{m_1} \oplus \ldots \oplus T_r^{m_r},$$

where T_i are pairwise nonisomorphic indecomposable modules. Then the following conditions are equivalent:

(1) *T is a tilting module.*

(2) $\operatorname{Ext}_Q(T, T) = 0$ *and $r = n$.*

Proof. (1) $\Rightarrow$ (2) Using Lemma 11.3.2 we see that if T is a tilting module, the dimension vectors of its summands T_i span $\mathbb{Z}^{Q_0}$, so $r \geq n$. Let us assume that $r > n$. We can renumber the summands T_i so for some j we have the relation

$$\alpha := \sum_{j=1}^{i} k_j\alpha_j = \sum_{j=i+1}^{n} k_j\alpha_j$$

with $k_1, k_2, \ldots, k_r$ being the nonnegative integers, and not all k_j's are equal to zero. But then the modules $\bigoplus_{j=1}^{i} T_j^{k_j}$ and $\bigoplus_{j=i+1}^{n} T_j^{k_j}$ have the same dimension vector and have no self-extensions, so both of them have an open orbit in this representation space $\operatorname{Rep}_\alpha(Q)$. This gives the required contradiction, so in fact $r \leq n$.

(2) $\Rightarrow$ (1) By Lemma 11.3.3 we can extend T to a tilting module $\widetilde{T} = T \oplus X$. By the implication (1) $\Rightarrow$ (2), $\widetilde{T}$ has the same number of isomorphism classes of indecomposable direct summands as T. Therefore X belongs to $\operatorname{add}(T)$ and T is a tilting module. □

A dimension vector α is called **prehomogeneous** if $\operatorname{Rep}_\alpha(Q)$ has a dense GL_α-orbit (this dense open orbit is isomorphic to GL_α/H for some algebraic subgroup H and such a variety is called a homogeneous space).

Proposition 11.3.7. *Suppose that α is a dimension vector. The following statements are equivalent:*

(1) *There exists a partial tilting object of dimension α.*

(2) *There exists a unique partial tilting object T of dimension α and $\mathrm{GL}_\alpha \cdot T$ is dense in $\operatorname{Rep}_\alpha(Q)$.*

(3) *The dimension vector α is prehomogeneous.*

(4) *If $\alpha = \alpha_1 \oplus \alpha_2 \oplus \cdots \oplus \alpha_s$ is the canonical decomposition, then $\alpha_1, \ldots, \alpha_s$ are real Schur roots.*

Proof. (1)$\Rightarrow$(2) If T is a partial tilting object, then we have $\operatorname{Ext}_Q(T,T) = 0$ and T has a dense open orbit. If T' is any other partial tilting object of dimension α, then it will also have a dense open orbit. The orbits of T and T' must intersect and therefore be the same. This shows that T is isomorphic to T'. (2)$\Rightarrow$(3) follows from the definition of a prehomogeneous dimension vector. (3)$\Rightarrow$(4) Let T be a general representation of dimension α. Then $\mathrm{GL}_\alpha \cdot T$ is the dense orbit in $\operatorname{Rep}_\alpha(Q)$. The codimension of the orbit of T is $\sum_{i=1}^{s}(1 - \langle\alpha_i, \alpha_i\rangle)$ by Exercise 11.2.4. So we have $\sum_{i=1}^{s}(1 - \langle\alpha_i, \alpha_i\rangle) = 0$ and $\langle\alpha_i, \alpha_i\rangle \leq 1$ for all i. It follows that $\langle\alpha_i, \alpha_i\rangle = 1$ for all i, so $\alpha_1, \ldots, \alpha_s$ are real roots. (4)$\Rightarrow$(1) Choose $T_i \in \operatorname{Rep}_{\alpha_i}(Q)$ in general position for all i. Since α_i is a real Schur root, T_i has a dense orbit in $\operatorname{Rep}_{\alpha_i}(Q)$ and $\operatorname{Ext}_Q(T_i, T_i) = 0$. By Kac's theorem, $\operatorname{ext}_Q(\alpha_i, \alpha_j) = 0$ for $i \neq 0$. This implies that $\operatorname{Ext}_Q(T_i, T_j) = 0$ for $i \neq j$. If we set $T = T_1 \oplus T_2 \oplus \cdots \oplus T_s$, then

T has dimension α and $\operatorname{Ext}_Q(T,T) = \sum_{i=1}^{s}\sum_{j=1}^{s}\operatorname{Ext}_Q(T_i,T_j) = 0$. This proves that T is partial tilting. □

We will visualize the dimension vectors of tilting modules for quivers with 3 vertices. If the T_1, T_2, T_3 are the direct summands of a tilting object T and $\alpha_1, \alpha_2, \alpha_3$ are the dimension vectors of T_1, T_2, T_3 respectively, Then the dimension vector of T is of the form $k_1\alpha_1 + k_2\alpha_2 + k_3\alpha_3$ with $k_1, k_2, k_3 > 0$. These are the integral points in the interior of a simplicial cone in $\mathbb{R}^3 = \mathbb{R}^{Q_0}$. If we project from $\mathbb{R}^3 \setminus \{0\}$ onto the projective space $\mathbb{P}^2$, then the cone becomes a triangle, with vertices $[\alpha_1]$, $[\alpha_2]$ and $[\alpha_3]$ and the tilting objects with summands T_1, T_2, T_3 correspond to the interior of the triangle. The boundary of the triangle corresponds to partial tilting objects. So the set dimension vectors of partial tilting objects have the structure of a (possibly infinite) simplicial complex. Note that the triangles correspond to multiplicity-free (i.e., no repeating summands) tilting modules.

Example 11.3.8. Consider the quiver

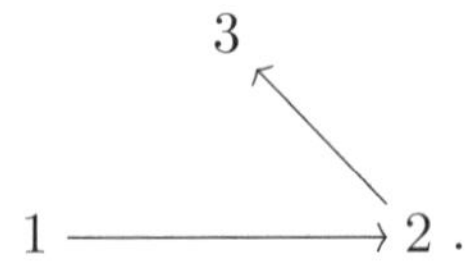

For this tame quiver there are infinitely many multiplicity-free tilting modules. The dimension vectors that are not partial tilting are of the form (k, l, k) with $k, l > 0$.

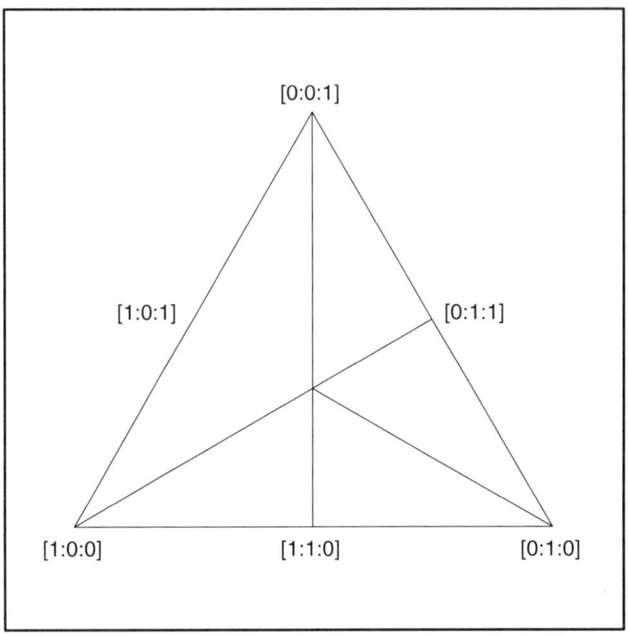

Example 11.3.9. Consider the quiver

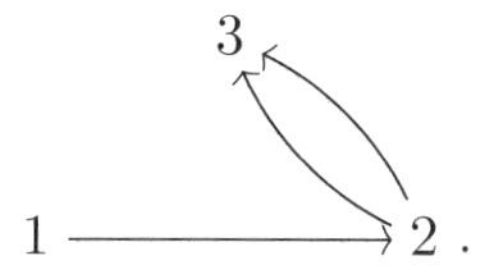

For this wild quiver, the triangulation is fractal-like. The gap in the middle corresponds to dimension vectors that are not prehomogeneous. The quadric given by $\langle \alpha, \alpha \rangle \leq 0$ is contained in this gap, but does not completely fill it.

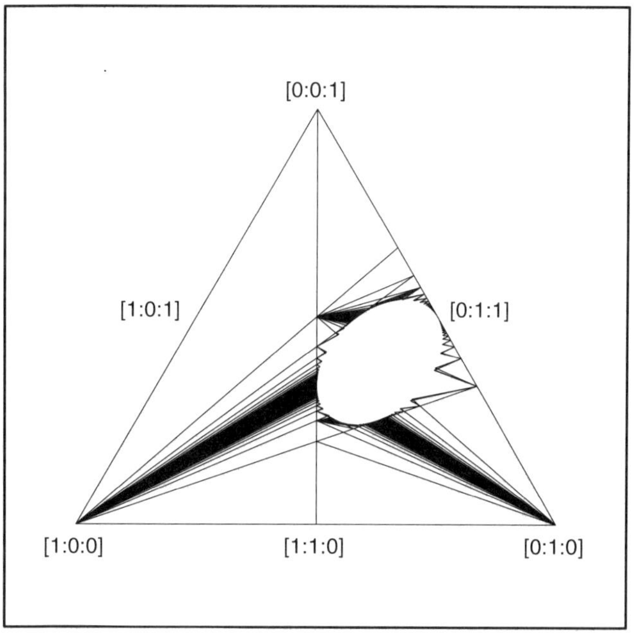

Exercises.

Exercise 11.3.1. *Let Q be a Dynkin quiver with n vertices and let β be a dimension vector. Let $T \in \operatorname{Rep}_\beta(Q)$ be the representation with the dense orbit. Prove that T is tilting if and only if $\operatorname{SI}(Q, \beta) = \mathbb{C}$.*

Exercise 11.3.2. *Let Q be a Dynkin quiver with n vertices. Prove that for dimension vector β the partial tilting module T of dimension β has exactly $n - r$ nonisomorphic indecomposable summands if and only if the ring of semi-invariants $\operatorname{SI}(Q, \beta)$ is a polynomial ring in r variables.*

11.4. Orthogonal Categories

In this section we will study the left orthogonal category and exceptional sequence. For a representation we can define a left and a right orthogonal category. We will discuss a result of Schofield which states that the left (and right) orthogonal category of an exceptional object is equivalent to the category of representations of another quiver. A partial exceptional sequence is an orthogonal sequence of exceptional objects. We will prove some basic properties of exceptional sequences.

Let Q be a quiver without oriented cycles. For representations V and W of Q, we say that $V \perp W$ if $\mathrm{Hom}_Q(V, W) = \mathrm{Ext}(V, W) = 0$. In that case we cay that V is left orthogonal to W and W is right orthogonal to V.

For a set of representations C in $\mathrm{Rep}(Q)$ we define the left **left orthogonal category** ${}^{\perp}C$ to be the full subcategory of $\mathrm{Rep}(Q)$ with the class of objects given by

$$ {}^{\perp}C = \{W \in \mathrm{Rep}(Q) \mid W \perp V \text{ for all } V \in C\}. $$

Similarly, we define the **right orthogonal category** $C^{\perp}$ to be the full subcategory of $\mathrm{Rep}(Q)$ with the class of objects given by

$$ C^{\perp} = \{W \in \mathrm{Rep}(Q) \mid V \perp W = 0 \text{ for all } V \in C\}. $$

Let us collect some properties of this notion.

Proposition 11.4.1. *Let V be an indecomposable representation from* $\mathrm{Rep}(Q)$.

(1) *Assume that V is not projective. Then we have $V^{\perp} = {}^{\perp}(\tau^{+}V)$.*
(2) *Assume that V is not injective. Then we have ${}^{\perp}V = (\tau^{-}V)^{\perp}$.*
(3) *Let U be indecomposable sincere representation (i.e., $U(x) \neq 0$ for all $x \in Q_0$). Then we get two equivalences of categories τ_{+}, τ_{-} which are inverse to each other:*

$$ {}^{\perp}U \underset{\tau_{-}}{\overset{\tau^{+}}{\rightleftarrows}} U^{\perp} . $$

Proof. Apply Theorem 6.4.1. The only thing to check is that for point (3) a projective representation cannot be in ${}^{\perp}U$. But this is assured by the sincerity of U. □

Let Q be a quiver with n vertices. Recall that a representation V of Q is exceptional if and only if $\mathrm{End}_Q(V) = K$ and $\mathrm{Ext}_Q(V, V) = 0$.

Definition 11.4.2. A sequence $\mathcal{E} = (E_1, \ldots, E_m)$ of representations is called a **partial exceptional sequence** if

(1) each E_i is an exceptional module, and
(2) for $i < j$ we have $E_j \perp E_i$.

A sequence of dimension vectors $\mathcal{E} = (\alpha_1, \dots, \alpha_m)$ is called a partial exceptional sequence if each dimension vector α_i is a Schur root and the corresponding objects $(E_1, \dots, E_m)$ form an exceptional sequence. A sequence $\mathcal{E} = (E_1, \dots, E_n)$ is a full exceptional sequence (or just an exceptional sequence) if in addition the dimension vectors of $E_1, \dots, E_n$ generate $\mathbb{Z}^{Q_0}$ as a $\mathbb{Z}$-module.

Remark 11.4.3. If $T = T_1^{m_1} \oplus \cdots \oplus T_r^{m_r}$ is a partial tilting module where $T_1, \dots, T_r$ are indecomposable and $m_1, \dots, m_r$ are positive, then in view of Theorem 11.2.11 the indecomposable summands of T can be renumbered so that $(T_1, \dots, T_r)$ is an exceptional sequence.

Example 11.4.4. Let Q be the quiver

$$1 \longrightarrow 2 .$$

Then there are three exceptional sequences in $\mathrm{Rep}(Q)$, namely $(10, 01)$, $(01, 11)$ and $(11, 10)$.

This example generalizes to an arbitrary quiver.

Example 11.4.5. Let Q be a quiver without oriented cycles with $Q_0 = \{1, 2, \dots, n\}$ such that there are no arrows from i to j if $i > j$.

(1) The the sequence $(S_1, \dots, S_n)$ of simple representations is an exceptional sequence.

(2) The sequence $(P_n, \dots, P_1)$ of indecomposable projectives is an exceptional sequence.

(3) The sequence $(I_n, \dots, I_1)$ of indecomposable injectives is an exceptional sequence.

The main point that allows us to deal with exceptional sequences is the following inductive procedure.

Theorem 11.4.6 (Schofield). *Let Q be a quiver with no oriented cycles and let V be an exceptional representation of dimension vector α. Then the category $V^\perp$ is naturally equivalent to a category of representations of a quiver $Q(\alpha)$ with $n-1$ vertices and no oriented cycles. The inverse functor from $\mathrm{Rep}(Q(\alpha))$ to $V^\perp$ is a full exact embedding into $\mathrm{Rep}(Q)$. A similar result is true for the category ${}^\perp V$.*

Before we start the proof we need one more fact.

Lemma 11.4.7. *Let $\mathcal{C}$ be an abelian category. Assume that there exists an object P in $\mathcal{C}$ which is projective and such that for every object M in $\mathcal{C}$ there is an integer $n \in \mathbb{N}$ such that P^n surjects on M. Set $B := \mathrm{End}_{\mathcal{C}}(P, P)^{\mathrm{op}}$. Then $\mathcal{C}$ is equivalent to the category of finitely presented B-modules.*

Proof. We consider the functor

$$F : \mathcal{C} \to B\text{–mod}$$

given by $F(M) = \mathrm{Hom}_{\mathcal{C}}(P, M)$. We will show that F is exact, fully faithful, i.e., for two objects M, N the map

$$F : \mathrm{Hom}_{\mathcal{C}}(M, N) \to \mathrm{Hom}_B(F(M), F(N))$$

is an isomorphism, and essentially surjective, i.e., that for each finitely presented B-module X there exists an object M in $\mathcal{C}$ such that $F(M)$ is isomorphic to X.

Exactness is obvious because P is projective. For every M in $\mathcal{C}$ we have an exact sequence

$$P^m \to P^n \to M \to 0$$

for some $m, n \in \mathbb{N}$. This means the sequence

$$\mathrm{Hom}_{\mathcal{C}}(P, P^m) \to \mathrm{Hom}_{\mathcal{C}}(P, P^n) \to \mathrm{Hom}_{\mathcal{C}}(P, M) \to 0$$

is exact. Since $\mathrm{Hom}_{\mathcal{C}}(P, P^n) = B^n$ we get a finitely presented B-module $F(M)$.

Using the above exact sequence, to check full faithfulness it is enough to do it for $N = P$. Since $F(P) = B$, this is immediate. Finally, F is essentially surjective because for each finitely presented X given by the exact sequence

$$B^m \to B^n \to X \to 0$$

we can consider the object M in $\mathcal{C}$ with the presentation

$$P^m \to P^n \to M \to 0$$

given by the same matrix as the presentation of X. $\square$

We continue with the proof of the theorem.

Proof. If V is projective, say $V = P_x$, then $V^\perp$ consists of the representations W such that $W(x) = 0$ so the statement of the theorem is true. Let us take $T := V$ and let us consider the module X constructed in the proof of Lemma 11.3.3. We have an exact sequence

$$0 \longrightarrow \mathbb{C}Q \longrightarrow X \longrightarrow T^m \longrightarrow 0\,.$$

By Proposition 11.3.5 the representation X is in $V^\perp$. Since $T \oplus X$ is a tilting module in $\mathrm{Rep}(Q)$, it has by Theorem 11.3.6 precisely n nonisomorphic indecomposable direct summands. This means X has $n-1$ nonisomorphic indecomposable direct summands. We claim that it is a projective object in $V^{\perp}$. If $W \in V^{\perp}$, then we have an exact sequence

$$\begin{array}{ccccc}
0 & & & & \\
\| & & & & \\
\mathrm{Hom}_Q(V,W) & \longrightarrow & \mathrm{Hom}_Q(X,W) & \longrightarrow & \mathrm{Hom}_Q(\mathbb{C}Q,W) \\
& & & \swarrow & \\
\mathrm{Ext}_Q(V,W) & \longrightarrow & \mathrm{Ext}_Q(X,W) & \longrightarrow & \mathrm{Ext}_Q(\mathbb{C}Q,W) \\
\| & & & & \| \\
0 & & & & 0,
\end{array}$$

where $\mathrm{Hom}_Q(V,W) = \mathrm{Ext}_Q(V,W) = 0$ because $V \perp W$ and $\mathrm{Ext}_Q(\mathbb{C}Q,W) = 0$ because $\mathbb{C}Q$ is projective. It follows that $\mathrm{Hom}_Q(X,W) = \mathrm{Hom}_Q(\mathbb{C}Q,W)$. So every homomorphism from $\mathbb{C}Q$ to W lifts to a homomorphism from X to W. For some m, $\mathbb{C}Q^m$ maps onto W. Then X^m also maps onto W. This implies $V^{\perp}$ is naturally equivalent to $\mathrm{End}_Q(X)^{op} -$ mod by Lemma 11.4.7.

Finally, $\mathrm{Ext}_Q(W_1,W_2) = \mathrm{Ext}_{V^{\perp}}(W_1,W_2)$ for $W_1, W_2 \in V^{\perp}$, as every extension of objects from $V^{\perp}$ is in $V^{\perp}$. This means that $\mathrm{Ext}_{V^{\perp}}(W_1,\cdot)$ is a right exact functor for every $W_1 \in V^{\perp}$. It follows that $\mathrm{End}_Q(X)^{op}$ is hereditary. So $\mathrm{End}_Q(X)^{\mathrm{op}}$ is Morita equivalent to some basic algebra A. By Exercise 3.3.1, A is a path algebra of some quiver Q'. So $V^{\perp}$ is naturally equivalent to the representations of Q'. This quiver has $n-1$ vertices because the representation X has $n-1$ nonisomorphic indecomposable direct summands. The quiver Q' has no oriented cycles because $\mathbb{C}Q'$ is Morita equivalent to $\mathrm{End}_Q(X)^{op}$ which is finite dimensional. □

By induction we get the following statement.

Corollary 11.4.8. *Let $\mathcal{E} = (E_1, \ldots, E_r)$ be a partial exceptional sequence. Then the category ${}^{\perp}(E_1, \ldots, E_r)$ (resp. $(E_1, \ldots, E_r)^{\perp}$) is equivalent to a category of representations of a quiver Q' (resp. Q'') with $n-r$ vertices, without oriented cycles.*

Let α be a dimension vector of an exceptional representation V. We define

$$\alpha^{\perp} = \{\beta \in \mathbb{N}^{Q_0} \ \langle \alpha, \beta \rangle = 0\}.$$

Then we have the following result.

Theorem 11.4.9. *Let Q be a quiver without oriented cycles. Let α be a real Schur root with an exceptional representation V. Let Q' be a quiver such* *that $V^\perp$ is equivalent to* $\operatorname{Rep}(Q')$. *Let α_x, $x \in Q'_0$ be the dimension vectors of the simple objects in $V^\perp$. The linear map $\Phi : \mathbb{N}^{Q'_0} \to \mathbb{N}^{Q_0}$ defined by*

$$\Phi(\beta) = \sum_{x \in Q'_0} \beta(x)\alpha_x$$

is an isometry with respect to the Euler forms: $\langle \beta, \gamma \rangle_{Q'} = \langle \Phi(\beta), \Phi(\gamma) \rangle_Q$ for all $\beta, \gamma \in \mathbb{Z}^{Q'_0}$.

Proof. Let S_x, S_y be simple objects in $V^\perp$ corresponding to the vertices $x, y \in \mathbb{Q}'_0$. We have

$$\dim \operatorname{Hom}_Q(S_x, S_y) = \dim \operatorname{Hom}_{V^\perp}(S_x, S_y) = \delta_{x,y}$$

and

$$\operatorname{Ext}_{V^\perp}(S_x, S_y) = \operatorname{Ext}_Q(S_x, S_y)$$

because an extension of two objects in $V^\perp$ is in $V^\perp$. This shows that Φ is an isometry:

$$\begin{aligned}&\langle \epsilon_x, \epsilon_y \rangle_{Q'} = \dim \operatorname{Hom}_{V^\perp}(S_x, S_y) - \dim \operatorname{Ext}_{V^\perp}(S_x, S_y)\\ &= \dim \operatorname{Hom}_Q(S_x, S_y) - \dim \operatorname{Ext}_Q(S_x, S_y) = \langle \alpha_x, \alpha_y \rangle_Q = \langle \Phi(\epsilon_x), \Phi(\epsilon_y) \rangle_Q.\end{aligned}$$

□

We proceed to construct the braid group action on the set of exceptional sequences. The starting point is provided by the completion result.

Lemma 11.4.10. *Every partial exceptional sequence*

$$(E'_1, \dots, E'_a, E''_1, \dots, E''_c)$$

can be completed to a full exceptional sequence

$$(E'_1, \dots, E'_a, E_1, \dots, E_b, E''_1, \dots, E''_c).$$

Proof. By Corollary 11.4.8 it is enough to find a complete exceptional sequence $(E_1, \dots, E_b, E''_1, \dots, E''_c)$ in ${}^\perp(E'_1, \dots, E'_a)$. Next, it suffices to find a complete exceptional sequence $(E_1, \dots, E'_b)$ in $(E''_1, \dots, E''_c)^\perp$, so we may assume $c = 0$. Now indecomposable projective, injective or simple objects give us the needed complete sequence. □

Lemma 11.4.11. *Let $\mathcal{E} = (E_1, \dots, E_n)$ and $\mathcal{F} = (F_1, \dots, F_n)$ be two complete exceptional sequences that differ in at most one place, i.e., $E_j = F_j$ for $j \neq i$. Then we have $E_i = F_i$.*

Proof. Passing to ${}^\perp(E_1, \dots, E_{i-1})$ we can assume that $i = 1$. Now passing to $(E_{i+1}, \dots, E_n)^\perp$ we can assume that Q has only one vertex. The assertion follows because there is only one exceptional representation. □

If $\mathcal{E}$ is an exceptional sequence, we denote $C(\mathcal{E})$ the smallest full subcategory of $\mathrm{Rep}(Q)$ which contains the objects from E and is closed under extensions, kernels of epimorphisms and cokernels of monomorphisms.

Lemma 11.4.12. *If $\mathcal{E} = (E_1, \ldots, E_n)$ is a complete exceptional sequence, then $C(\mathcal{E}) = \mathrm{Rep}(Q)$.*

Proof. We use induction on n. The cases $n = 0, 1$ are obvious. Now $(E_1, \ldots, E_{n-1})$ is a complete exceptional sequence for $E_n^{\perp}$ so by induction we have $C(E_1, \ldots, E_{n-1}) = E_n^{\perp}$. Thus $C(\mathcal{E})$ contains E_n and $E_n^{\perp}$.

Let us assume that E_n is not projective. Then by Proposition 11.3.5 there is a tilting module $T := E_n \oplus F$ with $F \in (E_n)^{\perp}$. For every projective module P there exists and exact sequence

$$0 \to P \to T' \to T'' \to 0$$

with $T', T'' \in \mathrm{add}(T)$. This means the projective modules are in $C(\mathcal{E})$. This implies $C(\mathcal{E}) = \mathrm{Rep}(Q)$ because every representation has a projective resolution of length ≤ 1.

If E_n is projective, say corresponding to the vertex i, then $E_n^{\perp}$ consists of representations of Q which are zero at i. Thus $C(\mathcal{E})$ contains the simples S_j $(j \neq i)$. There is also an exact sequence

$$0 \longrightarrow V \longrightarrow E_n \longrightarrow S_i \longrightarrow 0$$

with $V \in E_n^{\perp}$ so that $S_i \in C(\mathcal{E})$. Thus $C(\mathcal{E}) = \mathrm{Rep}(Q)$ as required. □

Lemma 11.4.13. *Let $\mathcal{E} = (E_1, \ldots, E_r)$ be an exceptional sequence.*

(1) $C(\mathcal{E}) = {}^{\perp}(\mathcal{E}^{\perp}) = {}^{\perp}\mathcal{F}$ *where $\mathcal{F}$ is any complete exceptional sequence in ${}^{\perp}\mathcal{E}$,*

(2) $C(\mathcal{E}) = ({}^{\perp}\mathcal{E})^{\perp} = \mathcal{G}^{\perp}$ *where $\mathcal{G}$ is any complete exceptional sequence in $\mathcal{E}^{\perp}$,*

Proof. We have $C(\mathcal{E}) \subseteq {}^{\perp}(\mathcal{E}^{\perp}) \subseteq {}^{\perp}\mathcal{F}$ since ${}^{\perp}(\mathcal{E}^{\perp})$ contains $\mathcal{E}$ and is closed under extensions, kernels of epimorphisms and cokernels of monomorphisms. Now $\mathcal{E}$ is a complete sequence for ${}^{\perp}\mathcal{F}$ so by Lemma 11.4.12 we have $C(\mathcal{E}) = {}^{\perp}\mathcal{F}$. The proof of part (2) is the same. □

The following result follows at once from the previous discussion.

Lemma 11.4.14. *Let $\mathcal{E}$ be an exceptional sequence of length r. Then the category $C(\mathcal{E})$ is equivalent to the category of representations of a quiver Q' with r vertices and no oriented cycles. Moreover, the functor* $\mathrm{Rep}(Q') \to \mathrm{Rep}(Q)$ *realizing this equivalence is exact and induces an isomorphism on both* Hom *and* Ext.

Exercises.

Exercise 11.4.1. *Find all exceptional sequences for the quivers of type* $\mathbf{A}_3$ *with all possible orientations.*

Exercise 11.4.2. *Let* Q *be the equioriented quiver of type* $\mathbf{A}_n$*:*

$$1 \longrightarrow 2 \longrightarrow \cdots \longrightarrow n\ .$$

For $1 \leq i \leq j \leq n$*, consider the indecomposable* $E_{i,j}$ *with the dimension vector* $\epsilon_i + \epsilon_{i+1} + \cdots + \epsilon_j$*. Calculate the orthogonal categories* $E_{i,j}^{\perp}$ *and* ${}^{\perp}E_{i,j}$*.*

Exercise 11.4.3. *Let* Q *be the quiver of type* D_4 *with arbitrary chosen orientation. For an indecomposable representation* E *describe the orthogonal categories* $E^{\perp}$ *and* ${}^{\perp}E$*.*

11.5. Quivers with Two Vertices

Consider *the generalized Kronecker quiver* $\theta(m)$ with two vertices $1, 2$ and m arrows $a_i : 1 \to 2$. We assume $m \geq 2$. The Euler matrix of $\theta(m)$ is

$$E(m) = \begin{pmatrix} 1 & -m \\ 0 & 1 \end{pmatrix}.$$

We start with describing the exceptional modules. Recall that the projectives P_1, P_2 have dimension vectors $(1, m)$ and $(0, 1)$. Define

$$U_i = \tau^i P_1, V_i = \tau^i P_2.$$

Proposition 11.5.1. *The modules* U_i, V_i *are exceptional objects. We have the following almost split sequences:*

$$0 \longrightarrow V_i \longrightarrow U_i^m \longrightarrow V_{i+1} \longrightarrow 0,$$

$$0 \longrightarrow U_i \longrightarrow V_{i+1}^m \longrightarrow U_{i+1} \longrightarrow 0.$$

Proof. Since the Coxeter transform takes exceptional objects to exceptional objects the first statement is obvious. The Coxeter transform takes almost split sequences to almost split sequences, so it is enough to show that the sequences above for $i = 0$ are almost split. This can be done by using reflection functors. Let us look at these sequences more closely. The first of these is

$$0 \longrightarrow P_2 \longrightarrow P_1^m \longrightarrow V_1 \longrightarrow 0.$$

The map consists of picking a basis of $\mathbb{C}^m = \mathrm{Hom}_{\theta(m)}(P_2, P_1)$. These are clearly the basis of irreducible maps, from P_2 to P_1, so our sequence is almost split. The dimension vector $\beta^{(1)}$ of $V_1 = \tau P_2$ is clearly $(m, m^2 - 1)$.

We clearly have $\mathrm{Hom}_{\theta(m)}(P_1, V_1) = \mathbb{C}^m$. Taking a basis of this space we get an exact sequence

$$0 \longrightarrow P_1 \longrightarrow V_1^m \longrightarrow U_1 \longrightarrow 0$$

and this is again an almost split sequence, so $U_1 = \tau P_1$. □

Denote $\alpha^{(i)} = \underline{\dim}\, U_i, \beta^{(i)} = \underline{\dim}\, V_i$ for $i \in \mathbb{N}$. It is easy to check

$$\langle \alpha^{(i)}, \beta^{(i)} \rangle = 0, \langle \beta^{(i+1)}, \alpha^{(i)} \rangle = 0$$

for all i. This suggests the following.

Corollary 11.5.2. *The sequences $(V_i, U_i), (U_i, V_{i+1})$ are exceptional for all $i \in \mathbb{N}$.*

Proof. Again using Coxeter transform it is enough to show that these sequences for $i = 0$ are exceptional. But this is obvious since the dimension vectors are orthogonal and the vector spaces $\mathrm{Ext}^1_{\theta(m)}(U_0, V_0) = \mathrm{Ext}^1_{\theta(m)}(V_1, U_0) = 0$ as both have projective modules as the right argument. □

Similar results are true for injective modules.

Recall that the injectives I_1, I_2 have dimension vectors $(1, 0)$ and $(m, 1)$. Define

$$W_i = \tau^{-i} I_1, Z_i = \tau^{-i} I_2.$$

Proposition 11.5.3. *The modules W_i, Z_i are exceptional objects. We have the following almost split sequences:*

$$0 \longrightarrow W_{i+1} \longrightarrow Z_i^m \longrightarrow W_i \longrightarrow 0,$$

$$0 \longrightarrow Z_{i+1} \longrightarrow W_{i+1}^m \longrightarrow Z_i \longrightarrow 0.$$

Denote $\gamma^{(i)} = \underline{\dim}\, W_i, \delta^{(i)} = \underline{\dim}\, Z_i$ for $i = 0, 1, \ldots$. It is easy to check that

$$\langle \gamma^{(i)}, \delta^{(i)} \rangle = 0, \langle \delta^{(i)}, \gamma^{(i+1)} \rangle = 0$$

for $i = 0, 1, \ldots$. This suggests the following.

Corollary 11.5.4. *The sequences $(Z_i, W_i), (W_{i+1}, Z_i)$ are exceptional for $i = 0, 1, \ldots$.*

Notice that we have one more exceptional sequence $(S_1, S_2) = (W_0, V_0)$.

Proposition 11.5.5. *The exceptional sequences listed above are all exceptional sequences for the quiver $\theta(m)$. Each exceptional object can appear exactly once as the left and as the right term in an exceptional sequence.*

Proof. Let us discuss tilting modules over $\theta(m)$. We start with pointing out that the direct sums $V_i \oplus U_i$ and $U_i \oplus V_{i+1}$ (respectively $Z_i \oplus W_i$ and $W_{i+1} \oplus Z_i$) are tilting modules. This follows from calculating Ext's for $i = 0$ in all four cases. Next, the condition $\langle \alpha, \alpha \rangle < 0$ translates to the quadratic equation $1 - mx + x^2 < 0$ for $x := \frac{\alpha_1}{\alpha_2}$. This means $\frac{m-\sqrt{m^2-4}}{2} < \frac{\alpha_1}{\alpha_2} < \frac{m-\sqrt{m^2-4}}{2}$. We claim that the dimension vectors α with open orbit occur when $\frac{\alpha_1}{\alpha_2} > \frac{m+\sqrt{m^2-4}}{2}$ or when $\frac{\alpha_1}{\alpha_2} < \frac{m-\sqrt{m^2-4}}{2}$. Clearly the condition is necessary. We will show that when $\frac{\alpha_1}{\alpha_2} < \frac{m-\sqrt{m^2-4}}{2}$ then α can be expressed as a nonnegative combination of vectors $\beta^{(i)}$ and $\alpha^{(i)}$ or of $\alpha^{(i)}$ and $\beta^{(i+1)}$. Thus a general module in dimension α will be a direct sum of copies of U_i and V_i or of U_i and V_{i+1}. $\mathrm{Rep}_\alpha(\theta(m))$ will have an open orbit.

In order to see that, let us look at the sequence of dimension vectors

$$\phi^{(1)} = \beta^{(1)}, \phi^{(2)} = \alpha^{(1)}, \phi^{(3)} = \beta^{(2)}, \phi^{(4)} = \alpha^{(2)}, \ldots .$$

Our dimension vectors satisfy the recursion

$$\phi^{(i+2)} = m\phi^{(i+1)} - \phi^{(i)}$$

with $\phi^{(1)} = (0, 1)$, $\phi^{(2)} = (1, m)$. Denoting $\phi^{(i)} = (u_i, v_i)$ we see that $\{u_i\}$ and $\{v_i\}$ satisfy the same recursion

$$u_{i+2} = mu_{i+1} - u_i, v_{i+2} = mv_{i+1} - v_i$$

with $u_1 = 0, u_2 = 1$ and $v_1 = 1, v_2 = m$ so they are the same sequence shifted by one.

So it is enough to estimate the limit $\lim_{i\to\infty} \frac{u_i}{u_{i+1}}$. Notice that the sequence $\{\frac{u_i}{u_{i+1}}\}$ is increasing because

$$\frac{u_{i+1}}{u_{i+2}} > \frac{u_i}{u_{i+1}}$$

translates to

$$\left(\frac{u_{i+1}}{u_i}\right)^2 - m\frac{u_{i+1}}{u_i} + 1 > 0$$

and we stay in this range. Our recursion in vector form can be rewritten

$$\begin{pmatrix} u_{i+1} \\ u_{i+2} \end{pmatrix} = \begin{pmatrix} 0 & 1 \\ -1 & m \end{pmatrix} \begin{pmatrix} u_i \\ u_{i+1} \end{pmatrix}.$$

So in the limit we should get an eigenvector of the matrix $\left(\begin{smallmatrix} 0 & 1 \\ -1 & m \end{smallmatrix}\right)$ which has characteristic equation $1 - mx + x^2$, so the eigenvalue is the root $\frac{m-\sqrt{m^2-4}}{2}$, and the ratio of coordinates of the eigenvector is also $\frac{m-\sqrt{m^2-4}}{2}$.

Similarly for the dimension vectors α satisfying $\frac{\alpha_1}{\alpha_2} > \frac{m+\sqrt{m^2-4}}{2}$ we use the modules W_i and Z_i.

This analysis shows that all exceptional objects in the category $\mathrm{Rep}(\theta(m))$ are U_i, V_i, W_i, Z_i. Indeed, they are the dimension vectors with the open orbit so they have to occur either in the range $\frac{\alpha_1}{\alpha_2} > \frac{m+\sqrt{m^2-4}}{2}$ or in the range $\frac{\alpha_1}{\alpha_2} < \frac{m-\sqrt{m^2-4}}{2}$. Then the general representation in such dimension vector is a multiple of one exceptional object only in indicated cases.

To conclude we observe that in the exceptional sequence (E_1, E_2) the dimension vector of one object determines by the orthogonality condition the dimension vector of the other object up to a constant multiple. Since for each exceptional object we produced an exceptional sequence with this object appearing on the left and on the right, we have all exceptional sequences. □

11.6. Two Sincerity Results

Before we move further we will need two results. Let Q be an acyclic quiver, and let α be **a prehomogeneous dimension vector**, i.e., such a dimension vector that $\mathrm{Rep}_\alpha(Q)$ has an open orbit. Let $T(\alpha)$ be the representation corresponding to the open orbit.

Lemma 11.6.1. *Let α be a sincere prehomogeneous dimension vector. Then $T(\alpha)$ is faithful.*

Proof. Let $\sum_i \lambda_i p_i$ be a nonzero combination of paths p_i from a vertex x to a vertex y, with $\lambda_i \in \mathbb{C}$ for all i and $\lambda_1 \neq 0$. We will show this element of $\mathbb{C}Q$ acts nontrivially on $T(\alpha)$. We choose a representation V of dimension vector α such that $V(p_1)$ is nonzero. We also choose V so $V(a') = 0$ for all arrows not occurring in p_1. Then $V(\sum_i \lambda_i p_i) = V(\lambda_1 p_1) \neq 0$. It follows that $V(\sum_i \lambda_i p_i)$ is nonzero on some Zariski open subset of $\mathrm{Rep}_\alpha(Q)$, so the orbit of $T(\alpha)$ has to be included in that subset. □

We have a decomposition

$$T(\alpha) = \bigoplus_{i=1}^{s} T(\alpha_i)^{n_i},$$

where $\alpha_1, \dots, \alpha_s$ are distinct real Schur roots and $n_1, \dots, n_s$ are positive integers. We assume that α is sincere. Let $\beta_1, \dots, \beta_{n-s}$ be the dimension vectors of simple objects in ${}^\perp T(\alpha)$. Then $\beta_1, \dots, \beta_{n-s}$ are real Schur roots as well. Denote by $T(\beta_j)$ the exceptional representation of dimension vector β_j. Then $T(\alpha) \in {}^\perp T(\beta_j)$ for $j = 1, \dots, n-s$. Let γ_j be a dimension vector in ${}^\perp T(\alpha)$ (identified with $\mathrm{Rep}(Q')$ for some quiver with $n-s$ vertices). We will need to know that γ_j is sincere. For this we need two lemmas.

Lemma 11.6.2. *Let α be a prehomogeneous sincere dimension vector. Let $T(\beta)$ be a simple object in ${}^\perp T(\alpha)$. We identify $T(\beta)^\perp$ with the category*

$\operatorname{Rep}(Q')$ *for some quiver* Q' *with* $n-1$ *vertices. Let* γ *be a dimension vector of* $T(\alpha)$ *considered as a representation of* Q'*. Then* γ *is sincere.*

Proof. Recall the notation of the proof of Theorem 11.4.6. The path algebra of Q' was a basic algebra Morita equivalent to $B := \operatorname{End}_Q(X)^{\mathrm{op}}$, where X was a projective generator of the category $T(\beta)^{\perp}$. We need to show that $\operatorname{Hom}_{Q'}(P, T(\alpha)) \neq 0$ for every indecomposable projective P in $T(\beta)^{\perp}$. This is equivalent to showing $\operatorname{Hom}_B(P, T(\alpha)) \neq 0$. We know that $\operatorname{Ext}_{Q'}(P, T(\beta)) = 0$, so $\operatorname{Hom}_{Q'}(P, T(\alpha)) = 0$ implies that $P \in {}^{\perp}T(\alpha)$. On the other hand by Lemma 11.6.1 we have an embedding $f : \mathbb{C}Q \to T(\alpha)^t$ for some $t \in \mathbb{N}$. For example, taking $t = \dim_Q \operatorname{Hom}_Q(\mathbb{C}Q, T(\alpha))$ and $f = (f_1, \dots, f_t)$ where $\{f_1, \dots, f_t\}$ is a basis of $\operatorname{Hom}_Q(\mathbb{C}Q, T(\alpha))$. Now recalling again the notation of proof of Theorem 11.4.6 we have an induced map $f' : X \to T(\alpha)^t$. But P is a direct summand of X and $\operatorname{Hom}_B(P, T(\alpha)) = 0$ implies that $P \subset Ker(f')$. This means $P \cap \mathbb{C}Q = 0$, so (by construction of X) P embeds in $T(\beta)^u$ for some $u \in \mathbb{N}$. But $T(\beta)$ is simple in ${}^{\perp}T(\alpha)$, so $P = T(\beta)$. This is impossible and the lemma is proved. □

11.7. The Braid Group Action on Exceptional Sequences

We will now describe the braid group action on exceptional sequences. Using this braid group one can construct many exceptional objects by applying the braid group. This action is also used in Section 11.9 in the algorithm for the canonical decomposition.

Lemma 11.7.1. *If* $\mathcal{E} = (E_1, E_2)$ *is an exceptional sequence, then there are unique representations* $R_{E_2}E_1$ *and* $L_{E_1}E_2$ *with the property that* $(E_2, R_{E_2}E_1)$ *and* $(L_{E_1}E_2, E_1)$ *are exceptional sequences in* $C(\mathcal{E})$.

Proof. This was shown in Section 11.5. Consider the exceptional sequence (E_1, E_2) with $\dim \operatorname{Ext}_Q(E_1, E_2) = m$. Consider the category $C(E_1, E_2)$. By Lemma 11.4.14 this category is equivalent to representations of a quiver with two vertices $1, 2$ and m arrows from 1 to 2. Proposition 11.5.5 lists all exceptional sequences for such quivers and the existence of representations $R_{E_2}E_1$ and $L_{E_1}E_2$ can be read off immediately. □

Lemma 11.7.2. *If* E_1 *is exceptional and not simple, then there is an exceptional sequence* (E_1, E_2) *such that* E_1 *is not simple in* $C(E_1, E_2)$.

Proof. Without loss of generality we can assume E_1 is sincere, otherwise we restrict to the support of E_1. Now we proceed by induction on n. We have $n \geq 2$ as E_1 is not simple. If $n = 2$, we have $C(E_1, E_2) = \mathbb{C}Q$ and by assumption E_1 is not simple. If $n > 2$, let E_2 be a simple object in ${}^{\perp}E_1$. Now E_1 is sincere as an object in $E_2^{\perp}$ by Lemma 11.6.2. Since $E_2^{\perp}$ is equivalent to the category of quiver representations of a quiver with $n-1$ vertices, the induction applies and we are done. □

Lemma 11.7.3. *Let $\mathcal{E} = (E_1, \ldots, E_r)$ be an exceptional sequence and let $1 \leq i < r$.*

(1) *$(E_1, \ldots, E_{i-1}, E_{i+1}, F, E_{i+2}, \ldots, E_r)$ is an exceptional sequence in $C(\mathcal{E})$ if and only if $F = R_{E_{i+1}} E_i$.*

(2) *$(E_1, \ldots, E_{i-1}, G, E_i, E_{i+2}, \ldots, E_r)$ is an exceptional sequence in $C(\mathcal{E})$ if and only if $G = L_{E_i} E_{i+1}$.*

Proof. The stated sequences are exceptional since

$$R_{E_{i+1}} E_i, L_{E_i} E_{i+1} \in C(E_i, E_{i+1}) = {}^{\perp}[(E_i, E_{i+1})^{\perp}] = [{}^{\perp}(E_i, E_{i+1})]^{\perp}$$

and $E_1, \ldots, E_{i-1} \in {}^{\perp}(E_i, E_{i+1})$, $E_{i+2}, \ldots, E_r \in (E_i, E_{i+1})^{\perp}$. The uniqueness follows from Lemmas 11.4.11 and 11.4.14 □

Definition 11.7.4. The braid group $\mathcal{B}_r$ on r strings is a group with generators $\sigma_1, \ldots, \sigma_{r-1}$ (where we think of σ_i as moving the i-th over $(i+1)$st string) with the relations

(1) $\sigma_i \sigma_j = \sigma_j \sigma_i$ when $|i-j| \geq 2$,

(2) $\sigma_i \sigma_{i+1} \sigma_i = \sigma_{i+1} \sigma_i \sigma_{i+1}$ for $1 \leq i \leq r-2$.

Let $\mathcal{E}(r, Q)$ be the set of exceptional sequences of length r in $\mathbb{C}Q-$ mod .

Lemma 11.7.5. *The assignments*

(1) $\sigma_i(E_1, \ldots, E_r) = (E_1, \ldots, E_{i-1}, E_{i+1}, R_{E_{i+1}} E_i, E_{i+2}, \ldots, E_r)$,

(2) $\sigma_i^{-1}(E_1, \ldots, E_r) = (E_1, \ldots, E_{i-1}, L_{E_i} E_{i+1}, E_i, E_{i+2}, \ldots, E_r)$

define an action of $\mathcal{B}_r$ on $\mathcal{E}(r, Q)$.

Proof. We have

$$\sigma_i^{-1} \sigma_i(E_1, \ldots, E_r) = (E_1, \ldots, E_{i-1}, F, E_{i+1}, \ldots, E_r)$$

so we have to have $F = E_i$ by Lemma 11.4.11 and the stated actions are inverse to each other. The relation $\sigma_i \sigma_j = \sigma_j \sigma_i$ for $|i-j| \geq 2$ is clear. To verify the relation $\sigma_i \sigma_{i+1} \sigma_i = \sigma_{i+1} \sigma_i \sigma_{i+1}$ note that $\sigma_i \sigma_{i+1} \sigma_i(E_1, \ldots, E_r)$ and $\sigma_{i+1} \sigma_i \sigma_{i+1}(E_1, \ldots, E_r)$ both have the form

$$(E_1, \ldots, E_{i-1}, E_{i+2}, R_{E_{i+2}} E_{i+1}, F, E_{i+3}, \ldots, E_r)$$

with $F \in C(E_1, \ldots, E_r)$ so they must be equal by Lemma 11.4.11. □

The most interesting is the case $r = n$.

Theorem 11.7.6 (Crawley-Boevey). *Let Q be a quiver with n vertices. The action of the braid group $\mathcal{B}_n$ on the set $\mathcal{E}(n, Q)$ of complete exceptional sequences is transitive.*

Proof. We prove the theorem by induction on n. For $n = 1$ there is nothing to prove. For $n = 2$ it follows from Section 11.5. Let us assume $n > 2$. Let Orb be an orbit of $\mathcal{B}_n$ acting on $\mathcal{E}(n, Q)$.

Let d be the smallest dimension of any representation in any exceptional sequence in $\mathcal{F}$. We will show that $d = 1$. So let us assume that $d > 1$. Let $\mathcal{E}$ be a complete exceptional sequence containing the representation of dimension d. If this representation occurs in place j, we can apply $\sigma_1 \dots \sigma_{j-1}$ to get an exceptional sequence with a representation of dimension d in the first place. So without loss of generality we can assume $\mathcal{E} = (E_1, \dots, E_n)$ where dimension of E_1 is equal to d. Now let (E_1, F) be an exceptional sequence from Lemma 11.7.2. This sequence extends to a complete exceptional sequence $\mathcal{F} = (E_1, F, F_3, \dots, F_n)$. Now $(E_2, \dots, E_n)$ and $(F, F_3, \dots, F_n)$ are complete exceptional sequences in ${}^{\perp}E_1$ so by induction they are in the same orbit of $\mathcal{B}_{n-1}$, and hence $\mathcal{E}$ and $\mathcal{F}$ are in the same orbit of $\mathcal{B}_n$. Now $C(E_1, F)$ contains an exceptional sequence (G_1, G_2) where G_1, G_2 are the simple objects in $C(E_1, F)$. Moreover, it follows from results of Section 11.5 that for some $k \in \mathbb{Z}$ we have $\sigma_1^k \mathcal{F} = (G_1, G_2, F_3, \dots, F_n)$. By assumption F is not a simple object in $C(E_1, F)$ so it involves both G_1 and G_2 so dim $G_1 < d$. Thus we can assume our orbit Orb contains an exceptional sequence of the form $\mathcal{E} = (S, E_2, \dots, E_n)$ where S is a simple representation. Let $\mathcal{P} = (P_1, \dots, P_n)$ be a complete exceptional sequence of projectives, with P_j being the projective cover of S. Passing to ${}^{\perp}S$ and using the induction, as before, we see that Orb contains the exceptional sequence $\mathcal{G} = (S, P_1, \dots, P_{j-1}, P_{j+1}, \dots, P_n)$. Now $\sigma_{j-1} \dots \sigma_1 \mathcal{G} = \mathcal{P}$ so we see that Orb contains $\mathcal{P}$ and we are done. □

Let us note the following corollary.

Corollary 11.7.7. *Two exceptional sequences $\mathcal{E} = (E_1, \dots, E_r)$ and $\mathcal{F} = (F_1, \dots, F_r)$ are in the same orbit of the action of the braid group $\mathcal{B}_r$ if and only if $C(E_1, \dots, E_r) = C(F_1, \dots, F_r)$.*

Exercises.

Exercise 11.7.1. *Let A_2 be the quiver of type A_2. For a given exceptional sequence find its stabilizer in the braid group B_2.*

Exercise 11.7.2. *Do the previous exercise for the quiver Q of type A_3 with arbitrary orientation.*

11.8. Examples

Example 11.8.1. Consider the equioriented quiver of type $\mathbf{A}_n$:

$$1 \longrightarrow 2 \longrightarrow \cdots \longrightarrow n\ .$$

For $1 \leq i \leq j \leq n$ consider the indecomposable module $E_{i,j}$ of dimension vector $\epsilon_i + \cdots + \epsilon_j$. Calculating the values of the Euler form we see easily that the simple objects in the category $E_{i,j}^{\perp}$ are: $E_{1,1}, E_{2,2}, \ldots, E_{i-1,i-1}, E_{i,j+1}, E_{j+2,j+2}, \ldots, E_{n,n}$ and $E_{i+1,i+1}, \ldots E_{j-1,j-1}$. The dimension vectors in two groups are orthogonal to each other (on both sides). This means the category $E_{i,j}^{\perp}$ is the product of categories of representations of equioriented quivers of type $\mathbf{A}_{n+i-j-1}$ and $\mathbf{A}_{j-i}$.

Example 11.8.2. Consider the quiver Q with three vertices and the Euler matrix

$$E = \begin{pmatrix} 1 & -1 & 0 \\ 0 & 1 & -2 \\ 0 & 0 & 1 \end{pmatrix}.$$

Consider the dimension vector $(2, 3, 3)$. It is a Schur root because the general representation will be an isomorphism on the arrow $1 \to 2$, and it will restrict to the preprojective representation on the vertices $2, 3$. Calculating the values of the Euler form we see that the simple objects in $(2, 2, 3)^{\perp}$ are $(1, 1, 2), (0, 1, 0$. But we have $\langle (1, 1, 2), (0, 1, 0) \rangle = 0$, $\langle (0, 1, 0), (1, 1, 2) \rangle = -3$. This means the right orthogonal category to $(2, 2, 3)$ is equivalent to the category of representations of the generalized Kreonecker quiver with three arrows, i.e., with the Euler matrix

$$E' = \begin{pmatrix} 1 & -3 \\ 0 & 1 \end{pmatrix}.$$

Example 11.8.3. Consider the generalized Kronecker quiver with three arrows, i.e., with the Euler matrix

$$E = \begin{pmatrix} 1 & -3 \\ 0 & 1 \end{pmatrix}.$$

Let

$$(a_0, a_1, a_2, a_3, a_4, a_5, \ldots) = (0, 1, 1, 2, 3, 5, \ldots)$$

be a Fibonacci sequence sequence defined by the usual recurrence

$$a_{n+2} = a_n + a_{n+1}.$$

It is easy to deduce from this the recurrence for the even terms of the sequence

$$a_{2n} + a_{2n+4} = 3a_{2n+2}.$$

This translates to the equation

$$\begin{pmatrix} a_{2n+2} & a_{2n} \end{pmatrix} \begin{pmatrix} 1 & -3 \\ 0 & 1 \end{pmatrix} \begin{pmatrix} a_{2n+4} \\ a_{2n+2} \end{pmatrix} = 0.$$

This means that (as proved in Section 11.5) the sequence of dimension vectors $(a_2, a_0), (a_4, a_2), (a_6, a_4), \ldots$ gives consecutive mutations of the exceptional sequences for this quiver.

For the remainder of this section we review the formula for the number of complete exceptional sequences for Dynkin quivers.

Let Q be a Dynkin quiver and $\Gamma := \Gamma(Q)$ associated graph. There is a root system Φ associated to Γ. Let $\mathcal{W}(\Gamma)$ be the Weyl group and $h(\Gamma)$ a Coxeter number of the root system Φ.

Theorem 11.8.4. *The number $e(\Gamma)$ of complete exceptional sequences in* $\mathrm{Rep}(Q)$ *equals*

$$e(\Gamma) = \frac{n!\, h(\Gamma)^n}{|\mathcal{W}(\Gamma)|}.$$

Example 11.8.5. For type $\mathbf{A}_n$ the number we get is

$$e(\mathbf{A}_n) = (n+1)^{n-1}.$$

Theorem 11.8.4 is proved type by type. We give a sketch of the proof. The details are given in [**67**]. The proof is based on the recursive formula which connects the number $e(\Gamma)$ with the numbers $e(\Gamma')$ for the subgraphs of Γ.

Definition 11.8.6. A slice is a subset Σ of vertices of $\mathrm{AR}(Q)$ such that:

(1) $T = \bigoplus_{m \in \Sigma} M$ is a sincere module.

(2) Σ is path-closed (i.e., any vertex on a path in $\mathrm{AR}(Q)$ connecting two vertices from Σ is in Σ).

(3) Let

$$0 \longrightarrow U \longrightarrow E \longrightarrow W \longrightarrow 0$$

be an almost split sequence. Assume that one of the indecomposable summands in E is in Σ. Then exactly one of the modules U, W is in Σ.

One has the following properties of slice modules.

Proposition 11.8.7. *Let Q be a Dynkin quiver and let Σ be a slice in* $\mathrm{AR}(Q)$.

(1) $T = \bigoplus_{M \in \Sigma} M$ *is a tilting module and* $\mathrm{End}(T)^{\mathrm{op}}$ *is hereditary, whose quiver is Q.*

(2) *If T is a tilting module with* $\mathrm{End}(T)^{\mathrm{op}}$ *hereditary, then* $T = \bigoplus_{M \in \Sigma} M$ *for some slice* Σ.

This proposition is proved in Section 4.2 of [**69**]. We do not include the proof here.

Lemma 11.8.8. *Let M be an indecomposable object in the τ-orbit of the projective representation P_x (i.e., $\tau^t(M) = P_x$ for some $t > 0$). Then the graph of the quiver Q' corresponding to the category $M^\perp$ is the graph of $Q|_{Q_0 \setminus x}$.*

Proof. Let M be indecomposable. If $M = P_x$, then obviously $M^\perp = \mathrm{Rep}(Q|_{Q_0\setminus x})$. If not, we can find a section Σ of $\mathrm{AR}(Q)$ such that M is a sink in Σ. Let $\Sigma = \{M_1, \dots, M_n\}$ with $M_n = M$. Let $M^\perp = \mathrm{Rep}(Q')$. Now $T := M_1 \oplus \cdots \oplus M_{n-1}$ is a partial tilting module, which is a slice for $\mathrm{AR}(Q')$ by Proposition 11.8.7. But T is also a slice for $\mathrm{End}(T)^{\mathrm{op}}$ because $\mathrm{End}(T)^{\mathrm{op}}$ is hereditary, and obviously $\mathrm{End}(T)^{\mathrm{op}}$ has the quiver which is $Q|_{Q_0\setminus\{x\}}$. So Q' is just $Q|_{Q_0\setminus\{x\}}$. □

Proposition 11.8.9. *Let $\Gamma = (\Gamma_0, \Gamma_1)$ be a Dynkin graph. For $x \in \Gamma_0$ we denote $\Gamma(x)$ the graph Γ restricted to $\Gamma_0 \setminus \{x\}$. Then*

$$e(\Gamma) = \tfrac{1}{2}h(\Gamma) \sum_{x \in \Gamma_0} e(\Gamma(x)).$$

Proof. We divide the exceptional sequences $\mathcal{E} = (E_1, \dots, E_n)$ according to the term E_1. For each indecomposable module E_1 the number of exceptional sequences starting with E_1 equals to the number of exceptional sequences in the orthogonal category $E_1^\perp$. But this category is, according to Theorem 11.4.6, equivalent to the category of representations of a quiver Q' with $n-1$ vertices. So it is enough to see what the graph Q' is for every exceptional representation of Q.. □

11.9. An Algorithm for the Canonical Decomposition

We will give an efficient algorithm for calculation the canonical decomposition. In this section we will assume that Q is an acyclic quiver with $Q_0 = \{1, 2, \dots, n\}$ and for every $a \in Q_1$ we have $ha < ta$. We first study the case where Q has 2 vertices.

Example 11.9.1.

$$Q : \circ \qquad \circ \,.$$

The Schur roots are $(1, 0)$ and $(0, 1)$ and clearly

$$\mathrm{ext}_Q((1,0),(0,1)) = \mathrm{ext}_Q((0,1),(1,0)) = 0.$$

So in this trivial case, the canonical decomposition of (p, q) is

$$(p, q) = (1, 0)^{\oplus p} \oplus (0, 1)^{\oplus q}.$$

Example 11.9.2.

$$Q : \circ \longleftarrow \circ \,.$$

The Schur roots are $(1, 0), (0, 1), (1, 1)$. It is easy to verify that

$$\operatorname{ext}_Q((1, 0), (1, 1)) = \operatorname{ext}_Q((1, 1), (1, 0)) = 0,$$

$$\operatorname{ext}_Q((0, 1), (1, 1)) = \operatorname{ext}_Q((1, 1), (0, 1)) = 0.$$

So canonical decomposition of (p, q) is as follows:

$$(p, q) = \begin{cases} (1, 1)^{\oplus q} \oplus (1, 0)^{\oplus(p-q)} & \text{if } p \geq q, \\ (0, 1)^{\oplus(q-p)} \oplus (1, 1)^{\oplus p} & \text{if } p \leq q. \end{cases}$$

Example 11.9.3. Consider the Kronecker quiver

$$Q = \theta(2) : \circ \leftleftarrows \circ \,.$$

The real Schur roots are $(n + 1, n)$ $(n \geq 0)$ and $(n, n + 1)$ $(n \geq 0)$. The only imaginary Schur root is $(1, 1)$. For all $n \geq 1$, we have an exceptional sequence

$$(n, n - 1), (n + 1, n)$$

and an exceptional sequence

$$(n, n + 1), (n - 1, n).$$

From $\operatorname{ext}_Q((n+1, n), (n, n-1)) = 0$ and $\langle (n, n-1), (n+1, n) \rangle = 1$ it follows that $\operatorname{ext}_Q((n, n - 1), (n + 1, n)) = 0$. Similarly, we have

$$\operatorname{ext}_Q((n, n + 1), (n - 1, n)) = \operatorname{ext}_Q((n - 1, n), (n, n + 1)) = 0.$$

If $p = q$, then we have a canonical decomposition

$$(p, q) = (p, q) = (1, 1)^{\oplus p}.$$

If $p < q$, then there exists an integer n such that $(n+1)/n \leq p/q \leq n/(n-1)$. Then we have a canonical decomposition

$$(p, q) = (n - 1, n)^{\oplus p'} \oplus (n, n + 1)^{\oplus q'},$$

where $p' = nq - (n + 1)p$ and $q' = np - (n - 1)q$. If $p > q$, then there exists an integer n such that $(n + 1)/n \leq q/p \leq n/(n - 1)$. We have a canonical decomposition

$$(p, q) = (n, n + 1)^{\oplus p'} \oplus (n - 1, n)^{\oplus q'},$$

where $p' = np - (n - 1)q$ and $q' = nq - (n + 1)p$.

Example 11.9.4. Let $Q = \theta(m)$ be the generalized Kronecker quiver with $Q_0 = \{1, 2\}$ and $m \geq$ arrows from 2 to 1. Define a recurrence sequence $s_0 = 1$, $s_1 = 1$ and $s_k = ms_{k-1} - s_{k-2}$ for $k \geq 1$. An easy induction shows that $s_k^2 = s_{k-1}s_{k+1} + 1$ for all k, so the sequence

$$0 = \frac{s_0}{s_1} < \frac{s_1}{s_2} < \cdots < \frac{s_2}{s_3} < \cdots$$

is increasing and it converges to the irrational number $\lambda = (m - \sqrt{m^2 - 4})/2$. The dimension vectors

$$(s_0, s_1), (s_1, s_2), (s_2, s_3), \ldots,$$

$$(s_1, s_0), (s_2, s_1), (s_3, s_2), \ldots$$

are real Schur roots. For every $n \geq 1$, the pair

$$(s_{n-1}, s_n), (s_n, s_{n+1})$$

is an exceptional sequence. As in the previous example we get

$$\operatorname{ext}_Q((s_{n-1}, s_n), (s_n, s_{n+1})) = \operatorname{ext}_Q((s_n, s_{n+1}), (s_{n-1}, s_n)) = 0.$$

Similarly, we have

$$\operatorname{ext}_Q((s_n, s_{n-1}), (s_{n+1}, s_n)) = \operatorname{ext}_Q((s_{n+1}, s_n), (s_n, s_{n-1})) = 0.$$

We will study the canonical decomposition of $(p, q) \in \mathbb{N}^2$. Suppose that $q \neq 0$ and

$$\frac{m - \sqrt{m^2 - 4}}{2} = \lambda < \frac{p}{q} < \frac{1}{\lambda} = \frac{m + \sqrt{m^2 - 4}}{2}.$$

This condition is equivalent to

$$\langle (p, q), (p, q) \rangle = (p - \lambda q)(p - \lambda^{-1} q) < 0.$$

We know that there cannot be more than 2 distinct dimension vectors in the canonical decomposition. Suppose that $(p, q) = \alpha_1^{\oplus d_1} \oplus \alpha_2^{\oplus d_2}$ is the canonical decomposition where α_1, α_2 are distinct and d_1, d_2 are positive. The roots α_1 and α_2 cannot both be imaginary by Exercise 11.2.5. Without loss of generality we may assume that α_1 is real. Then α_2 is in the right or left orthogonal category of α_1 and must also be real. But then we have

$$0 \leq d_1^2 \langle \alpha_1, \alpha_1 \rangle + d_1 d_2 (\langle \alpha_2, \alpha_1 \rangle + \langle \alpha_1, \alpha_2 \rangle) + d_2^2 \langle \alpha_2, \alpha_2 \rangle = \langle (p, q), (p, q) \rangle < 0.$$

Contradiction. So only one imaginary root appears in the canonical decomposition of (p, q). So $(p, q) = (p', q')^{\oplus k}$ is the canonical decomposition. Since (p', q') cannot be isotropic, we have $k = 1$. We conclude that (p, q) is a Schur root.

If $p < \lambda q$, then there exists an n such that

$$\frac{s_{n-1}}{s_n} \leq \frac{p}{q} \leq \frac{s_n}{s_{n+1}}.$$

In this case, the canonical decomposition of (p, q) is

$$(p, q) = (s_{n-1}, s_n)^{\oplus p'} \oplus (s_n, s_{n+1})^{\oplus q'}.$$

If $q < \lambda p$, then there exists an n such that

$$\frac{s_{n-1}}{s_n} \leq \frac{q}{p} \leq \frac{s_n}{s_{n+1}}.$$

In this case, the canonical decomposition of (p, q) is

$$(p, q) = (s_{n+1}, s_n)^{\oplus p'} \oplus (s_n, s_{n-1})^{\oplus q'}.$$

Lemma 11.9.5. *Suppose that X is a representation, Y is a Schur representation,* $\operatorname{Hom}_Q(X, Y) = \operatorname{Hom}_Q(Y, X) = 0$,

$$0 \longrightarrow X \longrightarrow Z \longrightarrow Y \longrightarrow 0 \tag{11.1}$$

is an extension such that the connecting map ∂ in the long exact sequence

$$\begin{array}{l} 0 \longrightarrow \operatorname{Hom}_Q(Y, X) \longrightarrow \operatorname{Hom}_Q(Z, X) \longrightarrow \operatorname{Hom}_Q(X, X) \\ \xrightarrow{\partial} \operatorname{Ext}_Q(Y, X) \longrightarrow \operatorname{Ext}_Q(Z, X) \longrightarrow \operatorname{Ext}_Q(X, X) \longrightarrow 0 \end{array}$$

is injective. Then Z is a Schur representation and we have $\operatorname{Hom}_Q(Z, X) = 0$.

Proof. Since the connecting map is injective, we have $\operatorname{Hom}_Q(Z, X) \cong \operatorname{Hom}_Q(Y, X) = 0$. From the exact sequence

$$0 \longrightarrow \operatorname{Hom}_Q(Y, Y) \longrightarrow \operatorname{Hom}_Q(Z, Y) \longrightarrow \operatorname{Hom}_Q(X, Y)$$

and $\operatorname{Hom}_Q(X, Y) = 0$ follows that $\operatorname{Hom}_Q(Z, Y) = \operatorname{Hom}_Q(Y, Y) \cong \mathbb{C}$. From the exact sequence

$$0 \longrightarrow \operatorname{Hom}_Q(Z, X) \longrightarrow \operatorname{Hom}_Q(Z, Z) \longrightarrow \operatorname{Hom}_Q(Z, Y)$$

and $\operatorname{Hom}_Q(Z, X) = 0$ we deduce that $\operatorname{Hom}_Q(Z, Z) = \mathbb{C}$. So Z is a Schur representation. □

Corollary 11.9.6. *If α, β are Schur roots,* $\hom_Q(\alpha, \beta) = \hom_Q(\beta, \alpha) = 0$ *and* $\operatorname{ext}_Q(\beta, \alpha) \neq 0$, *then $\alpha + \beta$ is a Schur root.*

Proof. Choose X and Y as general representations of dimension α and β, respectively. They both are Schur representations. Let

$$0 \longrightarrow X \longrightarrow Z \longrightarrow Y \longrightarrow 0$$

be a generic extension. If we apply the function $\operatorname{Hom}_Q(-, X)$, then the connection homomorphism

$$\partial : \operatorname{Hom}_Q(X, X) = \mathbb{C} \to \operatorname{Ext}_Q(Y, X)$$

is nonzero and therefore injective. From Lemma 11.9.5 it follows that Z is a Schur representation and $\alpha+\beta$ is a Schur root. $\square$

Corollary 11.9.7. *If α, β are imaginary Schur roots with $\hom_Q(\alpha,\beta) = \hom_Q(\beta,\alpha) = 0$ and $\operatorname{ext}(\beta,\alpha) \neq 0$, then $p\alpha + q\beta$ is a Schur root for all $p, q > 0$.*

Proof. We prove by induction on p and q that $p\alpha+q\beta$ is a Schur root. If $p\alpha$ and $q\beta$ are both Schur roots, then so is $p\alpha+q\beta$ by Corollary 11.9.6. If $p\alpha$ is not a Schur root, then α is isotropic and $p > 1$. By induction hypothesis we may assume that $(p-1)\alpha + q\beta$ is a Schur root. Now

$$\hom_Q(\alpha,\alpha) = \hom_Q(\alpha,\beta) = \hom_Q(\beta,\alpha) = 0,$$

therefore, we have

$$\hom_Q(\alpha,(p-1)\alpha+q\beta) = \hom_Q((p-1)\alpha+q\beta,\alpha) = 0$$

and

$$\operatorname{ext}_Q((p-1)\alpha+q\beta,\alpha) = -q\langle\beta,\alpha\rangle > 0.$$

By Corollary 11.9.6 we conclude that $p\alpha + a\beta = \alpha + ((p-1)\alpha+q\beta)$ is a Schur root. If $q\beta$ is not a Schur root, then a similar proof shows that $p\alpha+q\beta$ is a Schur root. $\square$

Corollary 11.9.8. *Suppose that α is a real Schur root and β is a Schur root such that $\hom_Q(\alpha,\beta) = \hom_Q(\beta,\alpha) = 0$. Then $\beta + t\alpha$ is a Schur root for every integer t with $0 \leq t \leq \max\{-\langle\alpha,\beta\rangle, -\langle\beta,\alpha\rangle\}$.*

Proof. Suppose that W and Y are general representations of dimension α and dimension β, respectively. By the assumptions of the Corollary, we have $\operatorname{Hom}_Q(W,Y) = \operatorname{Hom}_Q(Y,W) = 0$ and $\operatorname{Hom}_Q(Y,Y) \cong \mathbb{C}$. Suppose that $t \leq -\langle\beta,\alpha\rangle$. Let

$$0 \longrightarrow W^t \longrightarrow Z \longrightarrow Y \longrightarrow 0 \tag{11.2}$$

be a generic extension. In the long exact sequence

$$\begin{array}{ccccccc} 0 \longrightarrow & \operatorname{Hom}_Q(Y,W) & \longrightarrow & \operatorname{Hom}_Q(Z,W) & \longrightarrow & \operatorname{Hom}_Q(W^t,W) & \\ & & & \partial & & & \\ & \operatorname{Ext}_Q(Y,W) & \longrightarrow & \operatorname{Ext}_Q(Z,W) & \longrightarrow & \operatorname{Ext}_Q(W^r,W) & \longrightarrow 0 \end{array}$$

the connecting map ∂ is injective because $t \leq -\langle\beta,\alpha\rangle = \dim \operatorname{Ext}_Q(Y,W)$ and Z is a generic extension. If we apply $\operatorname{Hom}(-,W^t)$ to (11.2), then the connecting map in

$$\begin{array}{ccccccc} 0 \longrightarrow \operatorname{Hom}_Q(Y,W^t) & \longrightarrow & \operatorname{Hom}_Q(Z,W^t) & \longrightarrow & \operatorname{Hom}_Q(W^t,W^t) & & \\ & & \partial & & & & \\ \operatorname{Ext}_Q(Y,W^t) & \longrightarrow & \operatorname{Ext}_Q(Z,W^t) & \longrightarrow & \operatorname{Ext}_Q(W^t,W^t) & \longrightarrow & 0 \end{array}$$

is injective. By Lemma 11.9.5 (with $X = W^t$), Z is a Schur representation and $\beta + t\alpha$ is a Schur root.

Suppose that $t \leq -\langle \alpha, \beta \rangle$. Then α is a real Schur root, β is a Schur root and $\hom_{Q^{\mathrm{op}}}(\alpha,\beta) = \hom_{Q^{\mathrm{op}}}(\beta,\alpha) = 0$ and $t \leq -\langle \beta, \alpha \rangle_{Q^{\mathrm{op}}}$. So we apply the corollary to the opposite quiver and conclude that $\beta + t\alpha$ is a Schur root forQ^{op} and therefore it is a Schur root for Q. □

Lemma 11.9.9. *Suppose that $r \geq 3$, $\alpha_1, \alpha_2, \ldots, \alpha_r$ is a sequence of Schur roots and*

(1) $\hom_Q(\alpha_i, \alpha_{i+1}) > 0$ *for* $i = 1, \ldots, r-1$*;*

(2) $\operatorname{ext}_Q(\alpha_i, \alpha_j) = 0$ *if* $i \neq j$ *and* $(i,j) \neq (1,r)$.

Then we have $\operatorname{ext}_Q(\alpha_1, \alpha_r) = 0$ *and* $\langle \alpha_1, \alpha_r \rangle \geq 0$.

Proof. We prove the lemma by induction on r.

Suppose that conditions (1) and (2) in the lemma are satisfied, and assume $\operatorname{ext}_Q(\alpha_1,\alpha_r) \neq 0$. Because $\operatorname{ext}_Q(\alpha_r,\alpha_1) = 0$, Lemma 11.2.10 implies that $\hom_Q(\alpha_1,\alpha_r) = 0$. Choose a general representation S_i of dimension α_i for $i = 1, 2, \ldots, r$. Because $\hom_Q(\alpha_{r-1},\alpha_r) \neq 0$, we can choose a nonzero morphism $\varphi : S_{r-1} \to S_r$. Because $\operatorname{ext}_Q(\alpha_r,\alpha_{r-1}) = 0$, we have $\operatorname{Ext}_Q(S_r, S_{r-1}) = 0$ and φ is either injective or surjective by Lemma 11.2.9.

Suppose that φ is surjective. If $\psi : S_r \to S_1$ is a nonzero morphism, then $\psi\varphi : S_{r-1} \to S_1$ is nonzero, and $\hom_Q(\alpha_{r-1},\alpha_1) \neq 0$. This contradicts Theorem 11.2.11. So we get $\operatorname{Hom}_Q(S_r,S_1) = 0$ and $\hom_Q(\alpha_r,\alpha_1) = 0$. Because we already know that $\hom_Q(\alpha_1,\alpha_r) = 0$ and $\operatorname{ext}_Q(\alpha_1,\alpha_r) \neq 0$, $\alpha_1 + \alpha_r$ is a Schur root by Corollary 11.9.6. Since $\operatorname{ext}_Q(\alpha_r,\alpha_1) = 0$, we have $\alpha_r \hookrightarrow \alpha_1 + \alpha_r \twoheadrightarrow \alpha_1$. From $\hom_Q(\alpha_{r-1},\alpha_r) \neq 0$ and $\alpha_r \hookrightarrow \alpha_1 + \alpha_r$ it follows that $\hom_Q(\alpha_{r-1}, \alpha_1 + \alpha_r) \neq 0$. From $\alpha_1 + \alpha_r \twoheadrightarrow \alpha_1$ and $\hom_Q(\alpha_1,\alpha_2)$ it follows that $\hom_Q(\alpha_1+\alpha_r,\alpha_2) \neq 0$. If $2 \leq i \leq r-1$, then $\operatorname{ext}_Q(\alpha_i,\alpha_1) = \operatorname{ext}_Q(\alpha_i,\alpha_r) = 0$ implies that $\operatorname{ext}_Q(\alpha_i, \alpha_1+\alpha_r) = 0$. Similarly, we have $\operatorname{ext}_Q(\alpha_1+\alpha_r,\alpha_i) = 0$. It follows that

$$\hom_Q(\alpha_1+\alpha_r,\alpha_2), \hom_Q(\alpha_2,\alpha_3), \ldots, \hom_Q(\alpha_{r-2},\alpha_r), \hom_Q(\alpha_{r-1},\alpha_1+\alpha_r)$$

are all nonzero. This contradicts Theorem 11.2.11. We conclude that $\varphi : S_{r-1} \to S_r$ is injective.

From $\hom_Q(\alpha_{r-2},\alpha_{r-1}) \neq 0$ we deduce that $\hom_Q(\alpha_{r-2},\alpha_r) \neq 0$.

If $r = 3$, then this contradicts $\hom_Q(\alpha_1, \alpha_r) = 0$, and we can conclude that $\mathrm{ext}_Q(\alpha_1, \alpha_r) = 0$ and we are done.

If $r > 3$, then by the induction hypothesis, we can apply the lemma to the sequence of dimension vectors $\alpha_1, \alpha_2, \dots, \alpha_{r-2}, \alpha_r$ to conclude that $\mathrm{ext}_Q(\alpha_1, \alpha_r) = 0$. □

Lemma 11.9.10. *Suppose that $r \geq 3$ and $\alpha_1, \alpha_2, \dots, \alpha_r$ is a sequence of Schur roots such that:*

(1) $\hom_Q(\alpha_i, \alpha_j) = \mathrm{ext}_Q(\alpha_i, \alpha_j) = 0$ *if* $1 \leq i < j \leq r$.

(2) $\langle \alpha_j, \alpha_i \rangle \geq 0$ *if* $1 \leq i < j \leq r$, *and* $i > 1$ *or* $j < r$.

(3) $\langle \alpha_r, \alpha_1 \rangle < 0$.

We can rearrange $\alpha_1, \alpha_2, \dots, \alpha_r$ such that

(1) $\hom_Q(\alpha_i, \alpha_j) = \mathrm{ext}_Q(\alpha_i, \alpha_j) = 0$ *for all* $i < j$.

(2) $\langle \alpha_{k+1}, \alpha_k \rangle < 0$ *for some* k.

Proof. Let S be the set of all indices i with $2 \leq i \leq r-1$ for which there exists a sequence $m_1 = i < m_2 < \cdots < m_s = r$ with $\hom_Q(\alpha_{m_{k+1}}, \alpha_{m_k}) > 0$ for $k = 1, 2, \dots, s-1$. Suppose $S = \{i_1, \dots, i_t\}$ with $i_1 < i_2 < \cdots < i_t$. Choose indices $j_1 < j_2 < \cdots < j_{r-t-2}$ such that

$$\{2, 3, \dots, r-1\} \setminus S = \{j_1, \dots, j_{r-t-2}\}.$$

The sequence

$$\alpha_{i_1}, \alpha_{i_2}, \dots, \alpha_{i_t}, \alpha_1, \alpha_r, \alpha_{j_1}, \dots, \alpha_{j_{r-t-2}}$$

has the desired properties.

We only have to check (1) for all the pairs of indices (i, j) which are inverted after reordering. It suffices to show (1) for the pairs $(i_k, 1)$, (r, j_l), (i_k, j_l) where $1 \leq k \leq t$, $1 \leq l \leq r-t-2$ and $i_k > j_l$, because all other pairs of indices are not inverted after the reordering.

Suppose that $\hom_Q(\alpha_{i_k}, \alpha_1) > 0$. There exist indices $m_1 = i_k < m_2 < \cdots < m_s = r$ such that $\hom_Q(\alpha_r, \alpha_{m_{s-1}}) > 0$, $\hom_Q(\alpha_{m_{s-1}}, \alpha_{m_{s-2}}) > 0, \dots$ and $\hom_Q(\alpha_{m_2}, \alpha_{i_k}) > 0$.

By Lemma 11.9.9 we get $\langle \alpha_r, \alpha_1 \rangle \geq 0$. Contradiction.

By the definition of S, we have $\hom_Q(\alpha_r, \alpha_{j_l}) = 0$ for all l, and

$$\hom_Q(\alpha_{i_k}, \alpha_{j_l}) = 0$$

for all k, l. □

We try to find the expression $\alpha = \sum_{i=1}^{s} r_i \alpha_i$ satisfying the following properties:

P1: $r_i \geq 0$ for $i = 1, \dots, s$,

P2: each α_i is a Schur root,

P3: for each $i < j$, α_i is left orthogonal to α_j,

P4: $r_i = 1$ whenever α_i is imaginary and not isotropic,

P5: for each $i < j$ $\langle \alpha_j, \alpha_i \rangle \geq 0$.

Algorithm 11.9.11.

(1) Input: A quiver Q with n vertices, a dimension vector α.

(2) Write $\alpha = \sum_{i=1}^{s} r_i \alpha_i$ satisfying **P1, P2, P3, P4** (for example, take $s = n$ and $\alpha_i = \epsilon_i$ for all i),

(3) Omit all summands $r_i \alpha_i$ with $r_i = 0$. We may assume $r_i > 0$ for all i.

(4) If $\alpha = \sum_{i=1}^{s} r_i \alpha_i$ satisfies **P5**, terminate with output.

(5) There exists i and j with $i < j$ such that $\langle \alpha_i, \alpha_j \rangle < 0$. We take i and j such that $j - i$ is minimal. Use Lemma 11.9.10 to rearrange our roots $\alpha_i, \ldots, \alpha_j$ so they satisfy **P1, P2, P3, P4** and $\langle \alpha_{k+1}, \alpha_k \rangle < 0$ for some k,

(6) There exists i such that $\langle \alpha_{i+1}, \alpha_i \rangle < 0$. We modify the pair (α_i, α_{i+1}) (and r_i, r_{i+1}) so that the Euler form $\langle \alpha_{i+1}, \alpha_i \rangle$ becomes nonnegative. The modification is performed as follows. Write $\xi = \alpha_i, \eta = \alpha_{i+1}$. Let us write $p = r_i$, $q = r_{i+1}$. Let us also write $\zeta = p\xi + q\eta$. We will replace (ξ, η) by (ξ', η') where ξ', η' are positive linear combinations of ξ, η. Sometimes we will replace (ξ, η) by just one positive linear combination of ξ, η thus reducing the number s. The replacement is performed according to the following scheme:
  (a) ξ, η are real Schur roots. The category spanned by ξ, η is the category of representations of a quiver with two vertices, no cycles and $m := -\langle \eta, \xi \rangle$ arrows. The canonical decomposition of ζ in this category is described in Examples 11.9.1, 11.9.3 and 11.9.4.
    (i) $\langle \zeta, \zeta \rangle > 0$: There exist nonnegative combinations $\xi', \eta' \in \Gamma$ of ξ and η such that ξ' is left orthogonal to η', $\langle \eta', \xi' \rangle = m \geq 0$ and $\zeta = p'\xi' + q'\eta'$ for certain nonnegative integers p', q'. Replace (ξ, η) by (ξ', η').
    (ii) $\langle \zeta, \zeta \rangle = 0$: Replace (ξ, η) by ζ' where $\zeta = k\zeta'$ is a positive integer and $\zeta' \in \Gamma$ is indivisible.
    (iii) $\langle \zeta, \zeta \rangle < 0$: Replace (ξ, η) by ζ.
  (b) ξ is real, η is imaginary.
    (i) $p + q\langle \eta, \xi \rangle \geq 0$: Replace (ξ, η) by (ξ', η'), where $\xi' = \eta - \langle \eta, \xi \rangle \xi$ and $\eta' = \xi$. Now we have $\zeta = q\xi' + (p + q\langle \eta, \xi \rangle)\eta'$.
    (ii) $p + \langle \eta, \xi \rangle < 0$: Replace (ξ, η) by ζ.

(c) ξ is imaginary, η is real.
  (i) $q + p\langle \eta, \xi \rangle \geq 0$: Replace (ξ, η) by (ξ', η'), where $\xi' = \eta$ and $\eta' = \xi - \langle \eta, \xi \rangle \eta$. Now we have $\zeta = (q + \langle \eta, \xi \rangle)\xi' + p\eta'$.
  (ii) $q + p\langle \eta, \xi \rangle < 0$: Replace (ξ, η) by ζ.
(d) ξ, η are imaginary: Replace (ξ, η) by ζ.

(7) Keep repeating steps (3)–(6) (until the loop gets broken in step (4)).

Theorem 11.9.12. *Algorithm* 11.9.11 *terminates after finitely many steps with output* $\alpha = \sum_{i=1}^{s} r_i \alpha_i$ *satisfying* **P1**–**P5**. *The canonical decomposition of* α *is*

$$\alpha = (\alpha_1)^{\oplus r_1} \oplus \dots \oplus (\alpha_s)^{\oplus r_s}.$$

Proof. We first prove that throughout the algorithm, $\alpha = \sum_{i=1}^{s} r_i \alpha_i$ satisfies **P1**–**P4**. The only step for which this is not clear is (6). It is easy to verify that **P1** and **P4** are satisfied after step (6). We show that **P2** and **P3** are also satisfied after step (6), case by case. Note that any combination of ξ and η is left orthogonal to α_j with $j > i+1$ and right orthogonal to α_j with $j < i$. To show property **P3**, it suffices to show that η' is left orthogonal to ξ' in cases (a)(i), (b)(i) and (c)(i). From the canonical decomposition for quivers with 2 vertices, we know that ξ' and η' are Schur roots and ξ' is left orthogonal to η' in case (a)(i), ζ' is a Schur root in case (a)(ii), and ζ is a Schur root in case (a)(iii). By Corollary 11.9.7, ζ is a Schur root in case (d). By Corollary 11.9.8, ζ is a Schur root in case (b)(ii) and (c)(ii). For the same reason, η' and ξ' are Schur roots in case (b)(i) and (c)(i). In case (b)(i) we have $\operatorname{ext}_Q(\xi, \eta) = \operatorname{ext}_Q(\xi, \xi) = 0$, so it follows that $\operatorname{ext}_Q(\eta', \xi') = \operatorname{ext}_Q(\xi, \eta - \langle \eta, \xi \rangle \xi) = 0$. Therefore, $\operatorname{ext}_Q(\xi', \eta')$ or $\hom_Q(\xi, \eta')$ is equal to zero. Because $\hom_Q(\xi', \eta') - \operatorname{ext}_Q(\xi', \eta') = \langle \xi', \eta' \rangle$, we have $\hom_Q(\xi', \eta') = \operatorname{ext}_Q(\xi', \eta') = 0$. So property **P3** is satisfied after step (b)(i). Similarly, **P3** is satisfied after step (c)(i).

Every time we run through step (6), the sum $\sum_{i=1}^{s} r_i$ decreases, so the algorithm has to terminate. $\square$

Example 11.9.13. Consider the quiver

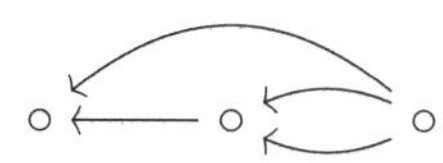

As we follow the algorithm, we keep track of the vector $r = (r_1, \dots, r_s)$, the matrix A whose rows are $\alpha_1, \dots, \alpha_s$ and the lower triangular matrix $E = (\langle \alpha_i, \alpha_j \rangle)$. Let us find the canonical decomposition of $(11, 7, 3)$:

$$r = \begin{pmatrix} 6 & 33 & 17 \end{pmatrix}, A = \begin{pmatrix} 1 & 0 & 0 \\ 0 & 1 & 0 \\ 0 & 0 & 1 \end{pmatrix}, E = \begin{pmatrix} 1 & 0 & 0 \\ -1 & 1 & 0 \\ -1 & -2 & 1 \end{pmatrix}.$$

Now $\langle \alpha_2, \alpha_1 \rangle = -1$ and $r_1 > r_2$, so we replace (α_1, α_2) by $(\alpha_1 + \alpha_2, \alpha_1)$:

$$r = \begin{pmatrix} 27 & 6 & 17 \end{pmatrix}, A = \begin{pmatrix} 0 & 1 & 0 \\ 1 & 1 & 0 \\ 0 & 0 & 1 \end{pmatrix}, E = \begin{pmatrix} 1 & 0 & 0 \\ 1 & 1 & 0 \\ -2 & -3 & 1 \end{pmatrix}.$$

Now $\langle \alpha_3, \alpha_2 \rangle = -3$ and $r_2/r_3 = \frac{6}{17}$. So α_2 and α_3 generate a generalized Kronecker quiver $\theta(3)$. The sequence s is given by $1, 3, 8, 21, \dots$. We have

$$\frac{1}{3} \leq \frac{6}{17} \leq \frac{3}{8}.$$

This gives us a decomposition $(6, 17) = (1, 3)^{\oplus 3} \oplus (3, 8)$.

$$r = \begin{pmatrix} 27 & 3 & 1 \end{pmatrix}, A = \begin{pmatrix} 0 & 1 & 0 \\ 1 & 1 & 3 \\ 3 & 3 & 8 \end{pmatrix}, E = \begin{pmatrix} 1 & 0 & 0 \\ -5 & 1 & 0 \\ -13 & 3 & 1 \end{pmatrix}.$$

We have $\langle \alpha_2, \alpha_1 \rangle = -5$. We decompose $(27, 1)$ for the quiver $\theta(5)$ and we get $(27, 3) = (5, 1)^{\oplus 3} \oplus (1, 0)^{\oplus 12}$.

$$r = \begin{pmatrix} 3 & 12 & 1 \end{pmatrix}, A = \begin{pmatrix} 1 & 6 & 3 \\ 0 & 1 & 0 \\ 3 & 3 & 8 \end{pmatrix}, E = \begin{pmatrix} 1 & 0 & 0 \\ 5 & 1 & 0 \\ -62 & -13 & 1 \end{pmatrix}.$$

We have $\langle \alpha_3, \alpha_2 \rangle = -13$. The dimension vector $\zeta = (12, 1)$ for the quiver $\theta(13)$ satisfies $\langle \zeta, \zeta \rangle = -11$. So ζ is a Schur root. So we replace (α_2, α_3) by $12\alpha_2 + \alpha_3$.

$$r = \begin{pmatrix} 3 & 1 \end{pmatrix}, A = \begin{pmatrix} 1 & 6 & 3 \\ 3 & 15 & 8 \end{pmatrix}, E = \begin{pmatrix} 1 & 0 \\ -2 & 1 \end{pmatrix}.$$

We decompose $(3, 1)$ for the Kronecker quiver $\theta(2)$ and get $(3, 1) = (2, 1) \oplus (1, 0)$. So we get

$$r = \begin{pmatrix} 1 & 1 \end{pmatrix}, A = \begin{pmatrix} 5 & 27 & 14 \\ 1 & 6 & 3 \end{pmatrix}, E = \begin{pmatrix} 1 & 0 \\ 2 & 1 \end{pmatrix}.$$

The matrix has E has no negative entries so we are done. The canonical decomposition is

$$(6, 33, 17) = (5, 27, 14) \oplus (1, 6, 3).$$

Exercises.

Exercise 11.9.1. *Show that if Q is a quiver without oriented cycles and n vertices, and the canonical decomposition of a dimension vector contains r distinct real roots and s distinct imaginary roots, then $r + 2s \leq n$.*

11.10. Bibliographical Remarks

The canonical decomposition of a dimension vector was introduced by Victor Kac in the paper [**54**]. We give the definition and main properties of this notion. In Section 9 we give a fast algorithm for calculation of canonical decomposition following [**26**]. We fix a small error from that paper that was brought to our attention by Christine Riedtmann.

The exceptional sequences were introduced by Gorodentsev, Rudakov and Bondal for studying vector bundles in algebraic geometry. In the quiver setting the results on the action of the braid group were proved by Crawley-Boevey [**16**] and Ringel [**68**], based on the results of Schofield [**74**]. The relation to orthogonal categories is also based on a paper of Schofield [**74**].

For further reading we recommend the results of Hille [**49**] on the volume of tilting modules. The applications of exceptional sequences to semi-invariants can be found in [**30**] and [**15**].

Chapter 12

Cluster Categories

Cluster categories emerged as a categorification of cluster algebras that were introduced by Fomin and Zelevinsky. The motivation came from calculations involving canonial bases in the Lie Theory, but soon similar structures were found in many other areas of mathematics.

The cluster categories can be thought about as extending canonical decomposition to dimension vectors with negative coordinates. The exceptional objects in a cluster category contain the exceptional objects in $\mathrm{Rep}(Q)$ but we have n additional exceptional objects. For Dynkin quiver the underlying combinatorial structure is the **a cluster complex** which is a triangulation of a sphere which extends the canonical decomposition. Cluster complexes are very interesting, for type $\mathbf{A}_n$ one gets the **associahedron**.

From the point of view of quiver representations the cluster category can be obtained from a derived category of quiver representations with respect to some functor. One of the main features is that it does not depend on orientation of the arrows in a quiver Q. One also has **mutation functors** which allow us to do a reflection at a vertex that is not a source or a sink.

The emergence of cluster categories and associated notion of cluster tilted algebras led to a transformation of the area of representations of Artin algebras where researches look for similar beautiful combinatorial structures as they point to the connections to other fields of mathematics.

We start with presenting a combinatorial model for the $\mathbf{A}_n$ type which uses the representation of associahedron in terms of triangulation of a regular $(n+3)$-gon. This material is due to Caldero, Chapoton and Schiffler [**14**]. Then we give a definition of clusters in terms of decorated representations following a paper of Marsh, Reineke and Zelevinsky [**64**].

Next we provide some basic information on derived category of Rep(Q) and finally we follow by giving the definition of the cluster category from the famous paper [**12**]. In the final section we give a definition and examples of cluster tilted algebras and we indicate that they appear as **Jacobian algebras** for **Quivers with Potential**. This points to our own work [**32**], [**33**] and towards further connections with Mathematical Physics.

12.1. A Combinatorial Model for Type $\mathbf{A}_n$

Consider the regular $(n+3)$-gon Π_{n+3}. We label the vertices of Π_{n+3} cyclically by the numbers $0, 1, \ldots, n+2$ in clockwise direction. We denote the line segment between i and j by $[i, j]$. So the sides of the $n+3$-gon are $[0, 1]$, $[0, 1]$, …, $[n+2, n+3]$ and $[n+3, 0]$. There are $n(n+3)/2$ diagonals. The triangle with vertices i, j, k will be denoted by $\triangle(i, j, k)$. A quadrangle with vertices i, j, k, l will be written as $\square(i, j, k, l)$. Let D be the set of diagonals.

Definition 12.1.1. An (abstract) **simplicial complex** on a finite set X is a set Δ of nonempty subsets of X such that $A \subseteq B$ and $B \in \Delta$ implies $A \in \Delta$. The k-dimensional cells are the sets $A \in \Delta$ with $k+1$ elements.

Using the $(n+3)$-gon, we will construct a simplicial complex associated to type $\mathbf{A}_n$.

Definition 12.1.2. The **associahedron** $\Delta(\mathbf{A}_n)$ is the simplicial complex on D whose cells of dimension k correspond to the collections of $k+1$ diagonals of Π_{n+3} that do not cross.

A maximal element in $\Delta(\mathbf{A}_n)$ is a triangulation of Π_{n+3} and contains n diagonals. So the associahedron $\Delta(\mathbf{A}_n)$ has dimension $n-1$ and its maximal cells correspond to triangulations of Π_{n+3} whose vertices are just the vertices of Π_{n+3}. We call such triangulations the **admissible triangulations** of Π_{n+3}.

Remark 12.1.3. The associahedron has many other interpretations. For example, its maximal cells are the ways to distribute the brackets in evaluating the product of $n+1$ elements in an algebra.

Each triangulation can be described by listing its n interior diagonals.

In Figure 1 is an example of an admissible triangulation consisting of the diagonals $[1, 7]$, $[2, 7]$, $[2, 4]$, $[4, 7]$, $[5, 7]$.

Example 12.1.4. The associahedron $\Delta(\mathbf{A}_1)$ consists of two points. Indeed, we have two triangulations of the square: one with the diagonal $[0, 2]$ and one with the diagonal $[1, 3]$.

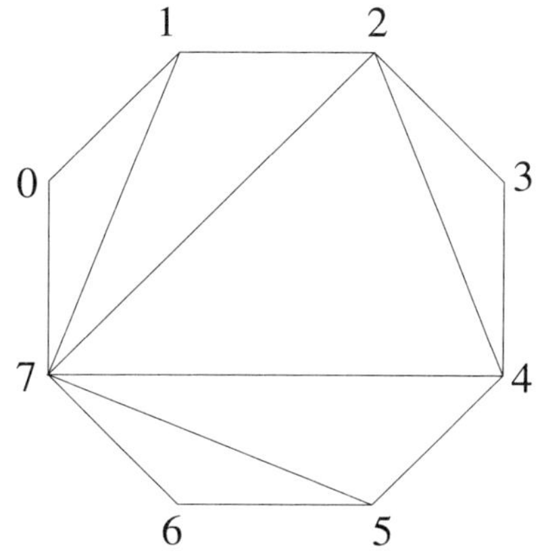

Figure 1

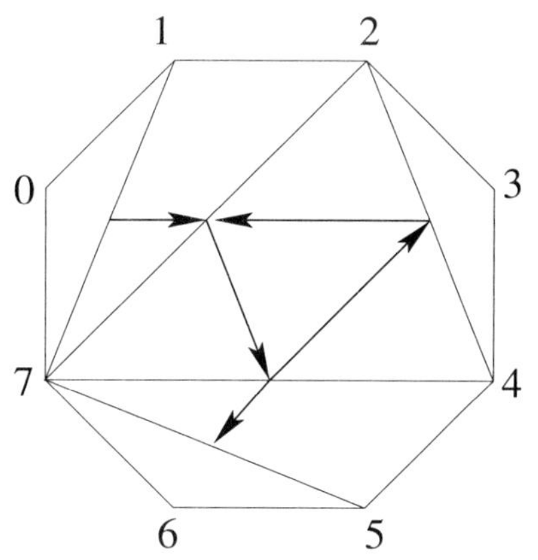

Figure 2

Example 12.1.5. $\Delta(\mathbf{A}_2)$ is a pentagon. We have five triangulations of Π_5: one with the diagonals $[0, 2]$ and $[0, 3]$ and its five rotations.

To an admissible triangulation T we will associate a quiver $Q(T)$. The vertices of $Q(T)$ are the midpoints of the interior diagonals of the admissible triangulation T. The arrows are the directed line segments connecting two midpoints of two sides of a triangle. By convention, inside the triangle $\Delta(i, j, k)$ the arrow points from the middle of $[i, j]$ to the middle of $[i, k]$ if rotating from the diagonal $[i, j]$ to $[i, k]$ moves in the clockwise direction.

In Figure 2 we see how the quiver $Q(T)$ is obtained from the triangulation of Figure 1.

Example 12.1.6. There are 14 triangulations of Π_6 (see Figure 3). The one with diagonals $[0, 2]$, $[0, 3]$, $[0, 4]$ and its 5 rotations, the one with the

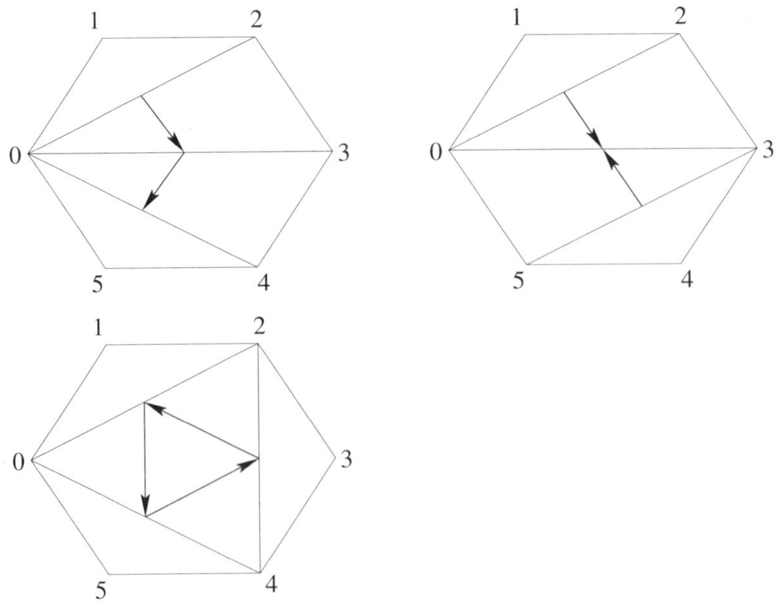

Figure 3

diagonals $[0,2]$, $[0,3]$, $[3,5]$ and its 2 rotations, the one with $[0,3]$, $[0,4]$, $[1,3]$ and its 2 rotations, and the triangulation with diagonals $[0,2]$, $[0,4]$, $[2,4]$ and its rotation.

Example 12.1.7. Consider the triangulation T of Π_{n+3} with interior diagonals $[0,2], [0,3], \ldots [0,n+1]$. The quiver $Q(T)$ has n vertices $x_2, \ldots, x_{n+1}$ with the arrows from x_i to x_{i+1}, so we get an equioriented quiver of type $\mathbf{A}_n$.

Definition 12.1.8. We call a triangle $\Delta(i,j,k)$ **internal** if all it sides are the internal diagonal. An admissible triangulation T of Π_{n+3} is **acyclic** if it does not have internal triangles.

The following lemmas are left as exercises to the reader.

Lemma 12.1.9. *Let T be an admissible triangulation of Π_{n+3}. Then T is acyclic if and only if $Q(T)$ ia a quiver of type $\mathbf{A}_n$.*

Definition 12.1.10. Two admissible triangulations of Π_{n+3} are **adjacent** if they differ by only one internal diagonal.

In Figure 4, we get the second triangulation from the first by flipping the diagonal $[2,7]$ to the diagonal $[1,4]$.

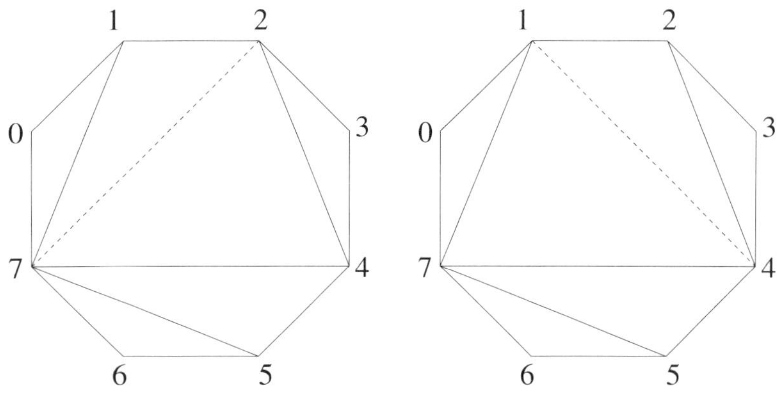

Figure 4

Lemma 12.1.11. *Two admissible triangulations T and T' of Π_{n+3} are adjacent if there exist vertices i, j, k, l (ordered cyclically) such that T' is obtained from T by flipping a diagonal $[i, k]$ to a diagonal $[j, l]$ in the quadrangle $\square(i, j, k, l)$ with vertices i, j, k, l. This means in triangulation T we will have triangles $\Delta(i, j, k)$ and $\Delta(i, k, l)$ and in T' the triangles $\Delta(i, j, l)$ and $\Delta(j, k, l)$. All other triangles in T and T' are the same.*

Let T be an acyclic triangulation of Π_{n+3}. We show how one can see the AR quiver of the quiver $Q(T)$ from our combinatorial picture. We build a new quiver $\mathrm{AR}(T)$ as follows.

Vertices of $\mathrm{AR}(T)$ are the internal diagonals of Π_{n+3} which are **not** the internal diagonals of T. To an internal diagonal $[i, j]$ that does not lie in T we can associate a dimension vector $\underline{d}_{T,[i,j]}$ such that $\underline{d}_{T,[i,j]}([k, l]) = 1$ if $[i, j]$ intersects the diagonal $[k, l]$ of T, and $\underline{d}_{T,[i,j]}([k, l]) = 0$ otherwise. In the quiver $\mathrm{AR}(T)$ we associate to $[i, j]$ the representation of $Q(T)$ with dimension vector $\underline{d}_{T,[i,j]}$.

The arrows in the quiver $\mathrm{AR}(T)$ are constructed as follows. Given a diagonal $[i, j]$ we can move it in two ways in counterclockwise direction, to $[i-1, j]$ and to $[i, j-1]$. If the new diagonal is not an outside wall or an internal diagonal in T this determines an arrow in $\mathrm{AR}(T)$.

Example 12.1.12. In Figure 5, we consider the triangulation T of Π_6 consisting of the diagonals $[0, 2]$, $[0, 3]$ and $[0, 4]$. The diagonal $[1, 4]$ corresponds to the module with dimension vector $(1, 1, 0)$ that maps to the diagonals $[0, 4]$ and $[1, 3]$. The diagonal $[0, 4]$ belongs to T, but $[1, 3]$ does not and corresponds to the module with dimension vector $(1, 0, 0)$. So there is a map in the AR quiver from the representation with dimension vector $(1, 1, 0)$ to the representation with dimension vector $(1, 0, 0)$.

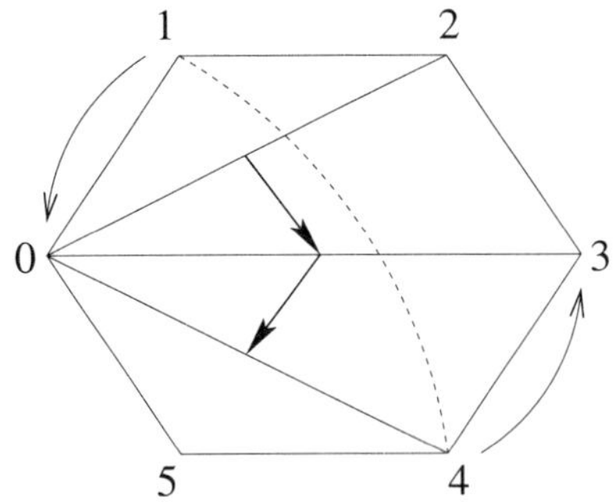

Figure 5

Proposition 12.1.13. *Let T be an admissible acyclic triangulation of Π_{n+3}. Then the quiver* $\mathrm{AR}(T)$ *is isomorphic to the AR quiver of* $Q(T)$.

We notice that for each acyclic triangulation T the graph $\mathrm{AR}(T)$ can be extended to the graph $\mathrm{AR}(\Pi_{n+3})$ that contains all the interior diagonals of Π_{n+3} with the arrows given by rotating the internal diagonals. This graph does not depend on the choice of triangulation. For a given triangulation T we can decorate the diagonal $[i, j]$ corresponding to the vertex of $Q(T)$ by the negative of the dimension of the projective corresponding to the vertex $[i, j]$.

The natural question that arises is whether for the triangulations of Π_{n+3} that are not acyclic the part of the quiver Π_{n+3} restricted to the internal diagonals of T is the Auslander-Reiten quiver of some finite dimensional algebra. For a triangulation T which is not acyclic we can also build a quiver $Q(T)$, but it will not be a Dynkin quiver. So in such case we need some relations to get a finite dimensional algebra.

Example 12.1.14. Let $n = 3$ and let T be a triangulation of the polygon Π_6 with the internal diagonals $[0, 2]$, $[0, 4]$ and $[2, 4]$. They correspond, respectively, to vertices $1, 2, 3$. The quiver $Q(T)$ is a cycle

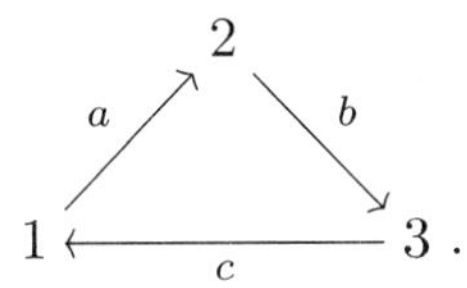

The graph $\text{AR}(\Pi_6)$ looks like this:

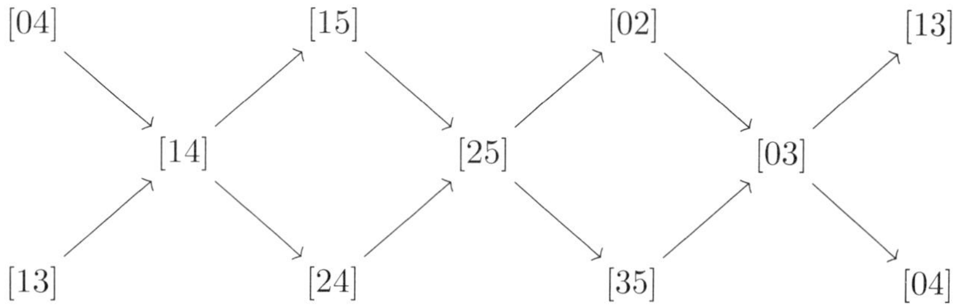

Here, the two vertices $[0,4]$ should be identified, and the same for $[1,3]$. So the graph can be embedded into a Möbius band.

The graph $\text{AR}(T)$ looks like this:

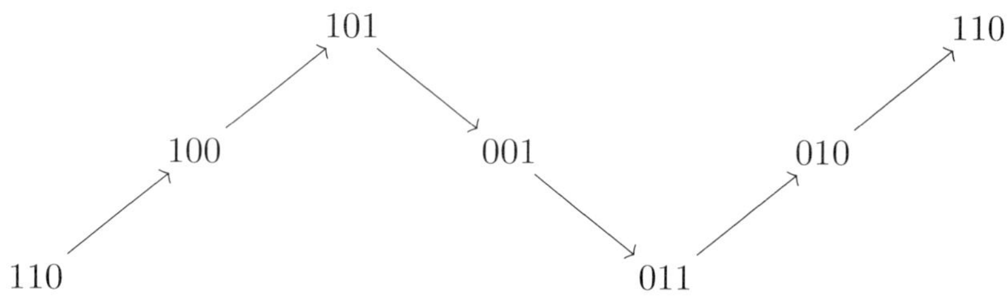

Here we identify the two vertices labeled 110. This is also the AR quiver of finite dimensional algebra

$$\mathbb{C}(Q(T))/\langle ba, cb, ac\rangle.$$

The general situation is described as follows.

Let T be an admissible triangulation of Π_{n+3}. Then the quiver $Q(T)$ has 3-cycles corresponding to internal triangles of T. Define the algebra $A(T) := \mathbb{C}(Q(T))/J(T)$ where $J(T)$ is the two sided ideal in $\mathbb{C}(Q(T))$ generated by quadratic relations $ba = cb = ac = 0$ for any oriented cycle cba in T corresponding to an internal triangle. Then we have the following.

Proposition 12.1.15. *The Auslander-Reiten quiver of $A(T)$ is equal to $\text{AR}(T)$.*

The AR quivers of adjacent triangulations are closely related as the following proposition shows.

Proposition 12.1.16. *Suppose the triangulation T_2 is obtained from T_1 by flipping the diagonal $[i,k]$ to $[j,l]$. We identify the vertex $[i,k]$ of $Q(T_1)$ with the vertex $[j,l]$ of $Q(T_2)$ and label these vertices with x. Then the quiver $Q(T_2)$ is obtained from $Q(T_1)$ by the reflection at the vertex x (this vertex is either a sink or a source). If a diagonal that does not appear in T_1 or T_2 corresponds to the indecomposable representation V of $Q(T_1)$, then this diagonal corresponds to the representation C_xV of $Q(T_2)$.*

Proof. The only two arrows that change are the arrows corresponding to two triangle in the quadrangle $\square(i,j,k,l)$. It is clear that they were a sink or a source, as otherwise one of the triangles in either T_1 or T_2 would be internal. Let $[p,q]$ be a diagonal that does not appear in T_1 or T_2. The vectors $\underline{d}_{T_2,[p,q]}$ and $\underline{d}_{T_1,[p,q]}$ are the same outside the vertex x. One can easily verify case by case, depending which of the sides of $\square(i,j,k,l)$ are intersected by $[p,q]$, that the dimension vector $\underline{d}_{T_2,[p,q]}$ is obtained from $\underline{d}_{T_1,[p,q]}$ by reflection at vertex x. $\square$

Exercises.

Exercise 12.1.1. *Prove Lemma* 12.1.9.

Exercise 12.1.2. *Prove Lemma* 12.1.11.

Exercise 12.1.3. *Prove Proposition* 12.1.15

12.2. Cluster Combinatorics and Decorated Representations

Let Q be a Dynkin quiver. We define $Q_0^- = \{x^- \mid x \in Q_0\}$, $Q_1^- = \emptyset$ and $Q^- = (Q_0^-, Q_1^-)$. We also define $\widetilde{Q} = (\widetilde{Q}_0, \widetilde{Q}_1)$ where $\widetilde{Q}_0 = Q_0 \cup Q_0^-$ and $\widetilde{Q}_1 = Q_1 \cup Q_1^- = Q_1$.

For each vertex x of Q we associate the added vertex x^- of $\widetilde{Q}$ and there are no arrows incident to the added vertices. Let $S_x^- = S_{x^-}$ be the simple representation of $\widetilde{Q}$ at the vertex x^-. The category $\operatorname{Rep}(\widetilde{Q})$ is called the category of **decorated representations** of Q.

Every V from $\operatorname{Rep}(\widetilde{Q})$ can be written as sum $V = V^+ \oplus V^-$ such that $V^+(x^-) = 0$ and $V^-(x) = 0$ for $x \in Q_0$. We can view V^+ as a representation of Q, and V^- as a representation of Q^-. A homomorphism between $V = V^+ \oplus V^-$ and $W = W^+ \oplus W^-$ consists of a homomorphism $\phi^+ : V^+ \to W^+$ of representations of Q and a collection of linear maps $\phi(x^-) : V^-(x^-) \to V'^-(x^-)$ for $x \in Q_0$. We define the **signed dimension vector** $\underline{\mathrm{sdim}} V \in \mathbb{Z}^{Q_0}$ of the representation $V = V^+ \oplus V^-$ as

$$(\underline{\mathrm{sdim}} V)(x) = \underline{\dim}\, V(x) - \underline{\dim}\, V(x^-) \quad x \in Q_0. \tag{12.1}$$

Let us denote by Φ the root system associated to the graph of Q. Let Π denote the set of simple roots in Φ. We denote by Φ^+ the set of positive roots, and the set of **almost positive roots** is

$$\widetilde{\Phi}^+ = \Phi^+ \cup -\Pi.$$

For a positive root $\alpha \in \Phi^+$, let U_α be an indecomposable representation of dimension α of Q (and therefore also of $\widetilde{Q}$). This representation is unique up to isomorphism. If $\alpha = -\epsilon_x$ where ϵ_x is dimension vector of S_x, then we define $U_\alpha = S_x^-$.

We have the following obvious generalization of Gabriel's theorem

Proposition 12.2.1. *The map given by*

$$\alpha \mapsto [U_\alpha]$$

gives an isomorphism between almost positive roots and the isomorphism classes of indecomposable objects in $\mathrm{Rep}(\widetilde{Q})$. *The inverse is given by the function* $\underline{\mathrm{sdim}}(\cdot)$.

The support of a decorated represention V is

$$\mathrm{supp}(V) = \{x \in Q_0 \mid \dim V(x) \neq 0 \text{ or } \dim V(x^-) \neq 0\}.$$

Definition 12.2.2. Let $V = V^+ \oplus V^-, V' = V'^+ \oplus V'^-$ be two decorated representations. We define the **symmetrized extension space** of V and V' by

$$E_Q(V, V') = \mathrm{Ext}^1_Q(V^+, V'^+) \oplus \mathrm{Ext}^1_Q(V'^+, V^+) \oplus \bigoplus_{x \in Q_0} \mathrm{Hom}_{\mathbb{C}}(V^+(x), V'^-(x)) \oplus \bigoplus_{x \in Q_0} \mathrm{Hom}_{\mathbb{C}}(V^-(x), V'^+(x)). \tag{12.2}$$

The following proposition is clear from the definition.

Proposition 12.2.3. *A decorated representation* $V = V^+ \oplus V^-$ *satisfies* $E_Q(V, V) = 0$ *if and only if* $\mathrm{Ext}^1_Q(V^+, V^+) = 0$ *and* V^+ *and* V^- *have disjoint supports.*

Definition 12.2.4. The Q-**compatibility degree** of two roots $\alpha, \beta \in \widetilde{\Phi}^+$ is

$$(\alpha||\beta)_Q = \dim\ E_Q(U_\alpha, U_\beta). \tag{12.3}$$

Proposition 12.2.5.

(1) *For* $\alpha, \beta \in \widetilde{\Phi}^+$ *we have* $(\alpha||\beta)_Q = (\beta||\alpha)_Q$.

(2) *Let* $\bar{Q}$ *be a full subquiver of* Q *corresponding to a subset* $\bar{Q}_0$ *of* Q_0. *Let* $\alpha, \beta \in \widetilde{\Phi}^+$ *be two roots supported in* $\bar{Q}_0$. *Then we have*

$$(\alpha||\beta)_{\bar{Q}} = (\alpha||\beta)_Q.$$

(3) $E_{Q^{\mathrm{op}}}(V^*, V'^*) = E_Q(V, V')$ *and therefore*

$$(\alpha||\beta)_{Q^{\mathrm{op}}} = (\alpha||\beta)_Q.$$

Definition 12.2.6. A subset C of $\widetilde{\Phi}^+$ is Q-**compatible** if $(\alpha||\beta)_Q = 0$ for all $\alpha, \beta \in C$. The subset C is called a Q-cluster if it is a maximal (by inclusion) Q-compatible subset of $\widetilde{\Phi}^+$. A subset $C \subseteq \widetilde{\Phi}^+$ is called positive if $C \subseteq \Phi^+$.

For a subset $\bar{Q}_0 \subset Q_0$ we denote by $\Phi(\bar{Q}_0)$ the root system corresponding to the full subquiver of Q supported on $\bar{Q}_0$.

Proposition 12.2.7. *Let $\bar{Q}_0$ be a subset of Q_0. Then the map*

$$C \mapsto C \cup \{-\epsilon_x \mid x \in Q_0 \setminus \bar{Q}_0\}$$

is a bijection between all positive compatible subsets (resp. clusters) for $\Phi(\bar{Q}_0)$ and all compatible subsets (resp. clusters) for Φ with negative support $Q_0 \setminus \bar{Q}_0$.

Proposition 12.2.8. *Each cluster has $n := |Q_0|$ elements. The signed dimension vectors of each cluster form a $\mathbb{Z}$-basis of the space $\mathbb{Z}^{Q_0}$.*

Proof. In view of Proposition 12.2.7 it suffices to show this for positive clusters. But in such a case clusters correspond to tilting modules and compatible sets to partial tilting modules. The result now follows from Corollary 11.3.4 and Theorem 11.3.6 □

Proposition 12.2.9. *Each element of $\gamma \in \mathbb{Z}^{Q_0}$ has a unique Q-***cluster expansion***, i.e.,*

$$\gamma = \sum_{\alpha \in \widetilde{\Phi}^+} m_\alpha \alpha,$$

where $m_\alpha \in \mathbb{N}$ for all α and where $m_\alpha m_\beta = 0$ whenever $\alpha, \beta \in \widetilde{\Phi}^+$ are not Q-compatible.

Proof. It is enough to show that for each γ there exists a unique decorated representation $V = V^+ \oplus V^-$ such that $\underline{\mathrm{sdim}} V = \gamma$ and $E_Q(V, V) = 0$. The proposition follows from this statement because of the definition of Q-compatibility degree.

In view of Proposition 12.2.3 the condition $E_Q(V, V) = 0$ means that $\mathrm{Ext}^1_Q(V^+, V^+) = 0$ and that V^+ and V^- have disjoint supports in Q_0. Let δ be the dimension vector of V^+. Note that it is uniquely determined by γ, as $\delta = \sum_{x \in Q_0} \max(0, \gamma(x))\epsilon_x$. We choose V^+ to be the representation in $\mathrm{Rep}_\delta(Q)$ which lies in the open GL_δ-orbit. This is equivalent to $\mathrm{Ext}^1_Q(V^+, V^+) = 0$ by Corollary 11.3.4. Since the vector space V^- is uniquely determined by γ and δ the existence and uniqueness follow. □

The cluster expansion allows us to define a **generalized associahedron**. Notice that the cluster expansion is homogeneous, i.e., for $m \in \mathbb{N}$ we get the cluster expansion of $m\gamma$ by multiplying by m each summand in a cluster expansion of γ. This allows us to define the cluster expansion for the vectors in $\mathbb{Q}^{Q_0}$, and, by passing to the closure, for the vectors from $\mathbb{R}^{Q_0}$. We now look at the unit sphere in $\mathbb{R}^{Q_0}$ and we think of a point of this sphere as a ray of proportional (with positive ratio) dimension vectors.

Definition 12.2.10. Let Q be a Dynkin quiver. The simplicial complex Δ_Q is a triangulation of the unit sphere in $\mathbb{R}^{Q_0}$ whose cells of dimension k $(0 \le k \le n+1)$ correspond to the compatible subsets of cardinality $k+1$. The cell corresponding to a subset C consists of all dimension vectors whose cluster expansion contains exactly the elements from C with positive coefficients.

Example 12.2.11. For the quiver

$$\circ \longrightarrow \circ \longrightarrow \circ$$

we get the following simplicial complex:

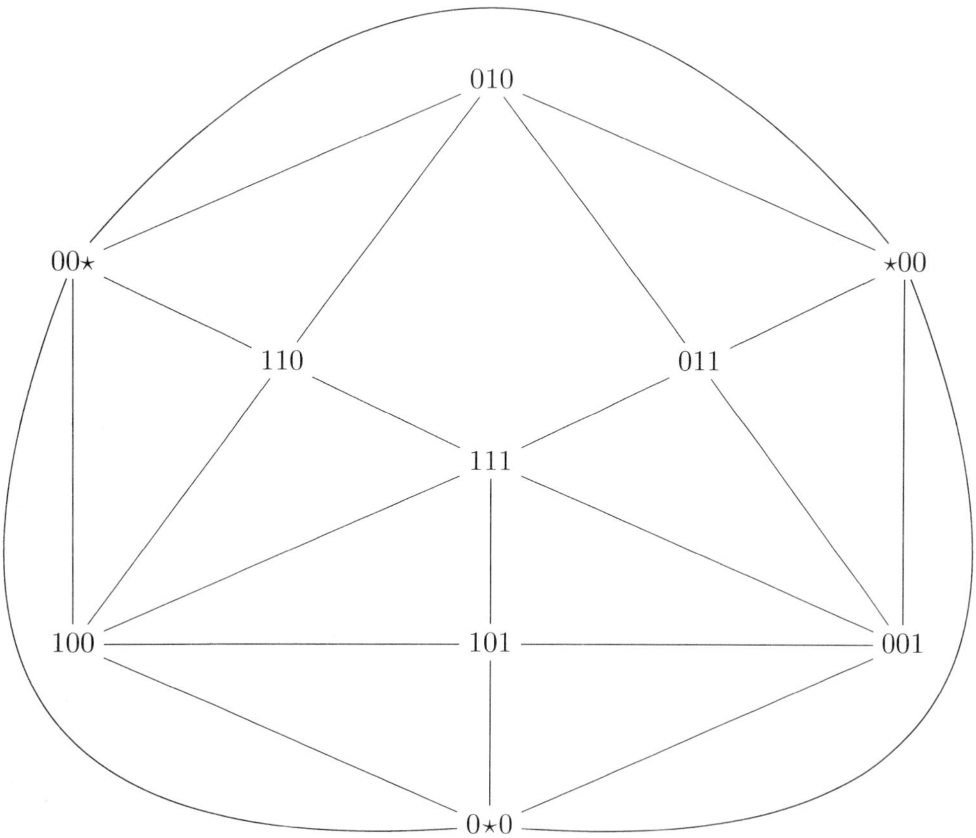

There are $9 = 6 + 3$ vertices, and 14 regions (including the outer region) corresponding to the 14 clusters.

The setup of decorated representations allows us to define the generalized reflection functors. We adopt the notation from Chapter 4.

Let Q be a Dynkin quiver and let $x \in Q_0$ be a source. Let $\{b_1, \dots, b_l\}$ be the arrows with $tb_j = x$. We extend the reflection functor C_x^- to decorated representations. Let V be a decorated representation. We have a canonical exact sequence

$$0 \longrightarrow S_x \otimes \operatorname{Hom}_Q(S_x, V^+) \longrightarrow V^+ \longrightarrow V' \longrightarrow 0. \tag{12.4}$$

By applying the long exact sequence of Ext's to the resolution from Exercise 2.2.2 of Chapter 2 we have the following exact sequence:

$$\begin{aligned} 0 \longrightarrow \operatorname{Hom}_Q(S_x, V^+) \longrightarrow V^+(x) \longrightarrow V^+(\mathrm{out}_x) \\ \longrightarrow \operatorname{Ext}^1_Q(S_x, V^+) \longrightarrow 0. \end{aligned} \tag{12.5}$$

In particular, $\operatorname{Hom}_Q(S_x, V^+)$ is isomorphic to the kernel of

$$V^+(b) : V^+ \to V(\mathrm{out}_x)$$

defined in Chapter 4. Now it is clear from the sequence (12.4) that

$$\operatorname{Hom}_Q(S_x, V') = 0,$$

i.e., V' does not have direct summands isomorphic to S_x. We define $C_x^-(V) = \bar{V}^+ \oplus \bar{V}^-$ by setting

$$\bar{V}^+ = C_x^-(V') \oplus (S_x \otimes V^-(x)), \tag{12.6}$$

$$\bar{V}^- = (S_x^- \otimes \operatorname{Hom}_Q(S_x, V^+)) \oplus \bigoplus_{x \neq y \in Q_0} (S_y^- \otimes V^-(y)). \tag{12.7}$$

More explicitly, C_x^- replaces the space $V(x)$ by the cokernel and $V(x^-)$ by the kernel of $V^+(b)$, leaving the other spaces unchanged.

To define C_x^- on morphisms, let $W = W^+ \oplus W^-$ be another decorated representation. Let $\phi : V \to W$ be a morphism of decorated representations. We write $\phi = \phi^+ \oplus \bigoplus_{y \in Q_0} \phi(y^-)$. We also define W', $\bar{W}^+$ and $\bar{W}^-$ in an analogous way to V', $\bar{V}^+$ and $\bar{V}^-$, respectively.

Since $\operatorname{Hom}_Q(S_x, W') = 0$, the map ϕ^+ induces the maps

$$\psi : \operatorname{Hom}_Q(S_x, V^+) \to \operatorname{Hom}_Q(S_x, W^+) \text{ and } \phi' : V' \to W'.$$

We define $C^-(\phi)$ as a direct sum of morphisms

$$C_x^-(\phi^+) \oplus \phi(x^-) : V^+ \to W^+$$

and

$$\psi \oplus \bigoplus\nolimits_{y \in Q_0 \setminus \{x\}} \phi(y^-) : \bar{V}^- \to \bar{W}^-.$$

Let Q be a Dynkin quiver and let $x \in Q_0$ be a sink. Let $\{a_1, \ldots, a_k\}$ will be the arrows with $ha_i = x$. We define the reflection functor C_x^+ to decorated representations. Let V be a decorated representation. We have a canonical exact sequence

$$0 \longrightarrow \mathrm{Hom}_Q(V^+, S_x) \longrightarrow V(x)^* \longrightarrow V(\mathrm{in}_x)^* \longrightarrow \mathrm{Ext}^1_Q(V^+, S_x) \longrightarrow 0 \tag{12.8}$$

by applying the long exact sequence of Ext's to the injective resolution of a simple S_x.

We also have a canonical exact sequence

$$0 \longrightarrow V' \longrightarrow V^+ \longrightarrow S_x \otimes \mathrm{Hom}_Q(V^+, S_x)^* \longrightarrow 0\,. \tag{12.9}$$

We define $C_x^+(V) = \bar{V}^+ \oplus \bar{V}^-$ by setting

$$\bar{V}^+ = C_x^+(V') \oplus (S_x^- \otimes V^-(x)), \tag{12.10}$$

$$\bar{V}^- = S_x^- \otimes \mathrm{Hom}_Q(V^+, S_x)^* \oplus \bigoplus_{y \in Q_0 \setminus \{x\}} S_y^- \otimes V^-(y). \tag{12.11}$$

The definition of morphisms $C_x^+(\phi)$ is left to the reader.

Let $\mathrm{Rep}^x(Q)$ denote the full subcategory of $\mathrm{Rep}(Q)$ consisting of representations which do not have S_x as a direct summand. We will write C_x for C_x^+ or C_x^-, whichever makes sense. We can summarize the properties of generalized reflection functors as follows.

Proposition 12.2.12. *For $x \in Q_0$ we have:*

1. *If $V \in \mathrm{Rep}^x(Q) \subset \mathrm{Rep}(\tilde{Q})$, then $C_x C_x V = V$.*
2. *$C_x S_x = S_x^-$ and $C_x S_x^- = S_x$.*
3. *$C_x(S_y^-) = S_y^-$ for $y \neq x$.*
4. *Suppose x is a source (resp. a sink) in Q, and $V \in \mathrm{Rep}^x(Q) \subset \mathrm{Rep}(\widetilde{Q})$. Let ϕ be any morphism of V to S_x (resp. from S_x to V). Then $C_x^-(\phi)$ (resp. $C_x^+(\phi)$) equals zero.*

Proposition 12.2.13. *For any vertex x which is a sink or a source in Q and for any decorated representation $V \in \mathrm{Rep}(\widetilde{Q})$ we have $C_x C_x(V) \cong V$.*

Proof. Assume that $x \in Q_0$ is a source (the case of the sink is similar). For a decorated representation $V = V^+ \oplus V^- \in \mathrm{Rep}(\widetilde{Q})$ denote by $W = W^+ \oplus W^-$ the representation $C_x^-(V)$. By definition we have

$$C_x^+ C_x^-(V) = \bar{W}^+ \oplus \bar{W}^-,$$

where

$$\bar{W}^+ = C_x^+(W') \oplus S_x \otimes W^-(x) \cong V' \oplus S_x \otimes \mathrm{Hom}_Q(S_x, V^+)$$

and

$$\bar{W}^- = S_x^- \otimes \mathrm{Hom}_{\sigma_x Q}(C_x^+(W') \oplus (S_x \otimes V^-(x)), S_x)^* \oplus \bigoplus_{y \neq x} S_y^- \otimes W^-(y^-)$$
$$\cong (S_x^- \otimes V^-(x)) \oplus \bigoplus_{y \neq x} (S_y^- \otimes V^-(y)) = V^-.$$

Note that $V' \oplus (S_x \otimes \mathrm{Hom}_Q(S_x, V^+))$ is (noncanonically) isomorphic to V^+ since the sequence (12.4) splits. This means that $C_x^+ C_x^-(V) \cong V$. □

Remark 12.2.14. The functor C_x^+ could be defined in terms of C_x^- by passing to the opposite quiver. More precisely, if x is a sink, then it is a source in the opposite quiver. We can set

$$C_x^+(V) = (C_x^-(V^\star))^\star.$$

One could also write more functorially, $C_x^+ = D \circ C_x^- \circ D$, where D is the duality functor. The reader can check that this is a functorial isomorphism.

Definition 12.2.15. For $x \in Q_0$ we define a piecewise-linear reflection $\tilde{\sigma}_x : \mathbb{Z}^{Q_0} \to \mathbb{Z}^{Q_0}$ by

$$\tilde{\sigma}_x(\alpha)(y) = \begin{cases} \alpha(y) & \text{if } y \neq x, \\ -\alpha(x) + \sum_{z-x} \max(\alpha(z), 0) & \text{if } y = x, \end{cases}$$

where we sum over all z that are adjacent to x in the graph $\Gamma(Q)$.

Lemma 12.2.16. *Suppose that $V = V^+ \oplus V^-$ is a decorated representation with $V^+(y) = 0$ or $V^-(y) = 0$ for all $y \in Q_0$. For any source or sink $x \in Q_0$ we have*

$$\underline{\mathrm{sdim}} C_x(V) = \tilde{\sigma}_x(\underline{\mathrm{sdim}} V).$$

Proof. Let x be a source (the case of the sink is similar). The functor C_x does not change any component $V^+(y)$, $V^-(y^-)$ for $y \neq x$, and this corresponds to the first part of the definition of $\tilde{\sigma}_x$. We have to deal with the x component. Using the definition of C_x we have

$$\underline{\mathrm{sdim}}(C_x(V))(x) = \dim C_x^-(V')(x) + \dim V^-(x) - \dim \mathrm{Hom}_Q(S_x, V^+).$$

In view of the exact sequence (12.4) we have

$$\dim C_x(V')(x) - \dim \mathrm{Hom}_Q(S_x, V^+) = -\dim V^+(x) + \dim V^+(\mathrm{out}_x).$$

Therefore,

$$\underline{\mathrm{sdim}}(C_x(V))(x) = -\dim V^+(x) + \dim V^+(\mathrm{out}_x) + \dim V^-(x)$$
$$= -\alpha(x) + \dim V^+(\mathrm{out}_x).$$

With the assumption $V^+(y) = 0$ or $V^-(y^-) = 0$ for all $y \in Q_0$, this agrees with the formula for $\tilde{\sigma}_x$. □

Corollary 12.2.17. *For all roots $\alpha \in \widetilde{\Phi}^+$ and all vertices in Q that are sinks or sources we have*

$$\underline{\mathrm{sdim}}(C_x U_\alpha) = \tilde{\sigma}_x(\underline{\mathrm{sdim}}(U_\alpha)).$$

Proof. Notice that $V^+(y) = 0$ or $V^-(y^-) = 0$ for all $y \in Q_0$ for every indecomposable object in $\mathrm{Rep}(\widetilde{Q})$. □

An easy consequence follows.

Corollary 12.2.18. *The transformation $\tilde{\sigma}_x$ transforms Δ_Q to $\Delta_{\sigma_x Q}$. These simplicial complexes are therefore isomorphic.*

Let us interpret these results in terms of symmetrized extension functors, but first we need a lemma.

Lemma 12.2.19. *Let $U \in \mathrm{Rep}(Q)$. Then, if $x \in Q_0$ is a source we get*

$$C_x^-(U)(x) \cong \mathrm{Ext}^1_Q(S_x, U).$$

If $x \in Q_0$ is a sink, then

$$C_x^+(U)(x) \cong \mathrm{Ext}^1_Q(U, S_x).$$

Proof. This follows from (12.5) and (12.8). □

Theorem 12.2.20. *Let $x \in Q_0$ be a source or a sink. We have*

$$E_{\sigma_x Q}(C_x V, C_x W) = E_Q(V, W)$$

for all decorated representations $V, W \in \mathrm{Rep}(\widetilde{Q})$.

Proof. Let us assume that $x \in Q_0$ is a source (the case of a sink is dual). We write $V = V^+ \oplus V^-$, $W = W^+ \oplus W^-$. We have (noncanonical) decompositions

$$V^+ = V' \oplus (S_x \otimes X) \text{ and } W^+ = W' \oplus (S_x \otimes Y),$$

where

$$X = \mathrm{Hom}_Q(S_x, V^+) \text{ and } Y = \mathrm{Hom}_Q(S_x, W^+).$$

The space $E_{\sigma_x Q}(C_x V, C_x W)$ decomposes as follows:

$$\begin{aligned}
&E_{\sigma_x Q}(C_x V, C_x W)\\
&\quad = \mathrm{Ext}^1_{C_x Q}(C_x(V'), C_x(W')) \oplus (\mathrm{Ext}^1_{C_x Q}(C_x(V'), S_x) \otimes W^-(x))\\
&\quad \oplus \mathrm{Ext}^1_{C_x Q}(C_x(W'), C_x(V'))^* \oplus (\mathrm{Ext}^1_{C_x Q}(C_x(W'), S_x) \otimes V^-(x))^*\\
&\oplus \mathrm{Hom}(C_x(V')(x), Y) \oplus \mathrm{Hom}(X, W^-(x)) \oplus \textstyle\bigoplus_{x \neq y} \mathrm{Hom}(V^-(y), W^+(y)).
\end{aligned}$$

Since the reflection functor induces induces the equivalence of categories $\mathrm{Rep}^x(Q)$ and $\mathrm{Rep}^x(\sigma_x Q)$ we have

$$\mathrm{Ext}^1_{\sigma_x Q}(C_x(V'), C_x(W') \cong \mathrm{Ext}^1_Q(V', W'), \tag{12.12}$$

$$\mathrm{Ext}^1_{\sigma_x Q}(C_x(W'), C_x(V') \cong \mathrm{Ext}^1_Q(W', V')^*. \tag{12.13}$$

Applying the second part of Lemma 12.2.19 to $U = C_x V'$ and using Proposition 12.2.13 we get

$$\mathrm{Ext}^1_{\sigma_x Q}(C_x V', S_x) \otimes W^-(x) \cong \mathrm{Hom}(V'(x), W^-(x)). \tag{12.14}$$

Similarly,

$$\mathrm{Ext}^1_{\sigma_x Q}(C_x W', S_x) \otimes V^-(x) \cong \mathrm{Hom}(V^-(x), W'(x)). \tag{12.15}$$

Now, substituting the obtained expressions into the decomposition of $\mathrm{Ext}^1_Q(C_x V, C_x W)$ and regrouping the terms, we can write it as a sum of the following four terms:

$$\mathrm{Ext}^1_Q(V', W') \oplus \mathrm{Hom}(X, C_x(W')(x)), \tag{12.16}$$

$$\mathrm{Ext}^1_Q(W', V')^* \oplus \mathrm{Hom}(C_x(V')(x), Y), \tag{12.17}$$

(12.18)
$$\mathrm{Hom}(V'(x), W^-(x)) \oplus \textstyle\bigoplus_{y \neq x} \mathrm{Hom}(V^+(y), W^-(y)) \oplus \mathrm{Hom}(X, W^-(x)),$$

$$\begin{aligned} \mathrm{Hom}(V^-(x), W'(x)) &\oplus \mathrm{Hom}(V^-(x), Y) \\ &\oplus \bigoplus_{y \neq x} \mathrm{Hom}(V^-(y), W^+(y)) \oplus \mathrm{Hom}(V^-(x), Y). \end{aligned} \tag{12.19}$$

To complete the proof we verify that these terms can be identified with the four terms in the decomposition (12.2), in the same order.

First we have

$$\begin{aligned} \mathrm{Ext}^1_Q(V^+, W^+) &= \mathrm{Ext}^1_Q(V' \oplus (S_x \otimes X), W' \oplus (S_x \otimes Y)) \\ &= \mathrm{Ext}^1_Q(V', W') \oplus (\mathrm{Ext}^1(S_x, W') \otimes X^*). \end{aligned}$$

This agrees with (12.16) by Lemma 12.2.19. The identification of the terms $\mathrm{Ext}^1_Q(W^+, V^+)^*$ with (12.17) is done by a dual argument. Finally the third and fourth terms are trivially identified with the last two terms of (12.2) . □

Remark 12.2.21. The isomorphism given in the theorem is not natural since C_x does not provide the equivalence of categories. But it can be shown (with the same proof as above) that there exists an isomorphism of functors $E_{\sigma_x Q}(C_x, C_x) \cong E_Q(C_x^2, C_x^2)$.

As an immediate consequence of Corollary 12.2.17 and Theorem 12.2.20 is the following result.

Proposition 12.2.22. *If $x \in Q_0$ is a source or a sink in Q, then*

$$(\tilde{\sigma}_x \alpha || \tilde{\sigma}_x \beta)_{\sigma_x Q} = (\alpha || \beta)_Q$$

for all $\alpha, \beta \in \widetilde{\Phi}^+$.

This result implies the following.

Corollary 12.2.23. *If Q and Q' are two quivers with the same underlying Dynkin graph, then the simplicial complexes Δ_Q and $\Delta_{Q'}$ are isomorphic to each other.*

Proof. If $Q' = \sigma_x Q$, then by Proposition 12.2.18 the piecewise linear function $\tilde{\sigma}_x$ transforms Δ_Q to $\Delta_{Q'}$. We know (Lemma 4.4.4) that by a sequence of reflections σ_x we can pass from any orientation of Q to another one. □

Exercises.

Exercise 12.2.1. *Let Q be the equioriented quiver of type $\mathbf{A}_n$. Consider the corresponding triangulation of the regular $(n+3)$-gon Π_{n+3} by choosing the diagonals $[1, i+2]$ for $1 \le i \le n$. We will think of the diagonal $[1, i+2]$ as the negative simple root $-\alpha_i$. Identify all the other diagonals with the positive roots of type $\mathbf{A}_n$ by counting their intersections with the diagonals $[1, i+2]$ $(1 \le i \le n+2)$. In this way we get a bijection between the diagonals in Π_{n+3} and almost positive roots $\widetilde{\Phi}^+$.*

(1) *Prove that for $\alpha, \beta \in \widetilde{\Phi}^+$ we have $(\alpha||\beta)_Q$ is equal to 1 if the corresponding diagonals intersect and to 0 otherwise.*

(2) *Prove that the clusters correspond to the triangulations of Π_{n+3} as defined in the Section 12.1.*

12.3. Triangulated Categories and Derived Categories

In this section we give the definition of a triangulated category and sketch the construction of the derived category of an abelian category. It is beyond the scope of this book to develop theory of triangulated and derived categories. We use the Gelfand-Manin book [**44**] as a principal reference for derived categories and Happel's book [**47**] as a principal reference to derived categories of quiver representations. We advise the reader not familiar with these theories to skip this section and study the case of derived category of quiver representations which can be understood more directly.

Let $\mathcal{C}$ be an additive category and $T : \mathcal{C} \to \mathcal{C}$ an automorphism (referred to as **a shift functor**). **A sextuple** (X, Y, Z, u, v, w) in $\mathcal{C}$ is a triple of morphisms

$$X \xrightarrow{u} Y \xrightarrow{v} Z \xrightarrow{w} TX .$$

Sometimes one uses for the sextuples the notation

$$\begin{array}{ccc} & Z & \\ {}^{w}\swarrow & & \nwarrow^{v} \\ X & \xrightarrow[u]{} & Y. \end{array}$$

A morphism of sextuples (X, Y, Z, u, v, w) and (X', Y', Z', u', v', w') is a triple (f, g, h) of morphisms making the following diagram commutative:

$$\begin{array}{ccccccc} X & \xrightarrow{u} & Y & \xrightarrow{v} & Z & \xrightarrow{w} & TX \\ \downarrow{\scriptstyle f} & & \downarrow{\scriptstyle g} & & \downarrow{\scriptstyle h} & & \downarrow{\scriptstyle Tf} \\ X' & \xrightarrow{u'} & Y' & \xrightarrow{v'} & Z' & \xrightarrow{w'} & TX'. \end{array}$$

If in this situation f, g, h are isomorphisms, then the triple (f, g, h) is called an isomorphism of sextuples.

Definition 12.3.1. A set $\mathcal{T}$ of sextuples (called triangles) is called a **triangulation** of $\mathcal{C}$ if the following axioms are satisfied:

(1) A sextuple isomorphic to a triangle is a triangle. Every morphism $u : X \to Y$ in $\mathcal{C}$ can be embedded into a triangle (X, Y, Z, u, v, w). The sextuple $(X, X, 0, 1_X, 0, 0)$ is a triangle.

(2) If (X, Y, Z, u, v, w) is a triangle, then $(Y, Z, TX, v, w, -Tu)$ is a triangle,

(3) Given two triangles (X, Y, Z, u, v, w) and (X', Y', Z', u', v', w') and morphisms $f : X \to X'$, $g : Y \to Y'$ such that $gu = u'f$ there exists $h : Z \to Z'$ such that (f, g, h) is a morphism of triangles,

(4) (The octahedral axiom). Consider triangles (X, Y, Z', u, i, i'), (Y, Z, X', v, j, j') and (X, Z, Y', vu, k, k'). Then there exist morphisms $f : Z' \to Y'$, $g : Y' \to X'$ such that the following diagram

commutes and its third row is a triangle:

$$\begin{array}{ccccccccc}
T^{-}Y' & \xrightarrow{Tk'} & X & \xrightarrow{1_X} & X & & & & \\
{\scriptstyle T^{-}g}\downarrow & & {\scriptstyle u}\downarrow & & {\scriptstyle vu}\downarrow & & & & \\
T^{-}X' & \xrightarrow{T^{-}j'} & Y & \xrightarrow{v} & Z & \xrightarrow{j} & X' & \xrightarrow{j'} & TY \\
 & & {\scriptstyle i}\downarrow & & {\scriptstyle k}\downarrow & & {\scriptstyle 1_{X'}}\downarrow & & {\scriptstyle Ti}\downarrow \\
 & & Z' & \xrightarrow{f} & Y' & \xrightarrow{g} & X' & \xrightarrow{Tij'} & TZ' \\
 & & {\scriptstyle i'}\downarrow & & {\scriptstyle k'}\downarrow & & & & \\
 & & TX & \xrightarrow{1_{TX}} & TX. & & & &
\end{array}$$

The additive category $\mathcal{C}$ with a set $\mathcal{T}$ of triangles satisfying these axioms is a *triangulated category*.

Let $(\mathcal{C}, T, \mathcal{T})$ and $(\mathcal{C}', T', \mathcal{T}')$ be triangulated categories. An additive functor $F : \mathcal{C} \to \mathcal{C}'$ is called **exact** if there exists an invertible natural transformation $\alpha : FT \to T'F$ such that (FX, FY, FZ, Fu, Fv, Fw) is in $\mathcal{T}'$ whenever (X, Y, Z, u, v, w) is in $\mathcal{T}$.

If an exact functor $F : \mathcal{C} \to \mathcal{C}'$ is an equivalence of categories we call it a **triangle-equivalence**. The categories $\mathcal{C}$ and $\mathcal{C}'$ are then called **triangle-equivalent**. The following result is straightforward.

Lemma 12.3.2. *Let $F : \mathcal{C} \to \mathcal{C}'$ be a triangle-equivalence and let G be a quasi-inverse of F. Then G is a triangle-equivalence.*

Definition 12.3.3.

(1) An additive functor $H : \mathcal{C} \to \mathcal{A}$ from a triangulated category $\mathcal{C}$ to an abelian category $\mathcal{A}$ is called a **(covariant) cohomological functor**, if for every triangle (X, Y, Z, u, v, w) we have a long exact sequence

$$\cdots \longrightarrow H(T^iX) \xrightarrow{H(T^iu)} H(T^iY) \xrightarrow{H(T^iv)} H(T^iZ) \xrightarrow{H(T^iw)} H(T^{i+1}X) \longrightarrow \cdots .$$

(2) An additive functor $H : \mathcal{C} \to \mathcal{A}$ from a triangulated category $\mathcal{C}$ to an abelian category $\mathcal{A}$ is called a **(contravariant) cohomological functor**, if for every triangle (X, Y, Z, u, v, w) we have a long exact sequence

$$\cdots \longrightarrow H(T^{i+1}X) \xrightarrow{H(T^iw)} H(T^iZ) \xrightarrow{H(T^iv)} H(T^iY) \xrightarrow{H(T^iu)} H(T^iX) \longrightarrow \cdots .$$

The principal example of a derived category is the derived category $D^b(\mathcal{A})$ of bounded complexes in an abelian category $\mathcal{A}$. It is defined as follows.

Let $\mathcal{A}$ be an abelian category. *A complex* $K^\bullet$ of objects in $\mathcal{A}$ is a collection of objects $\{K^n \mid n \in \mathbb{Z}\}$ and morphisms $d^n : K^n \to K^{n+1}$ such that $d^{n+1}d^n = 0$ for every $n \in \mathbb{Z}$. We represent a complex $K^\bullet$ as a diagram

$$\cdots \longrightarrow K^n \xrightarrow{d^n} K^{n+1} \longrightarrow \cdots$$

We define the *n-th homology object* of a complex $K^\bullet$

$$H^n(K^\bullet) = \ker d^{n+1} / \operatorname{im} d^n.$$

Let $K^\bullet, L^\bullet$ be two complexes of objects from $\mathcal{A}$. A morphism $f : K^\bullet \to L^\bullet$ is a collection of maps $f^n : K^n \to L^n$ in $\mathcal{A}$ such that for every $n \in \mathbb{Z}$ we have $d^n_K f^n = f^{n+1} d^n_L$. We can imagine the morphism of complexes as a diagram:

$$\begin{array}{ccccccc} \cdots \longrightarrow & K^n & \xrightarrow{d^n_K} & K^{n+1} & \longrightarrow \cdots \\ & \downarrow{\scriptstyle f^n} & & \downarrow{\scriptstyle f^{n+1}} & \\ \cdots \longrightarrow & L^n & \xrightarrow{d^n_L} & L^{n+1} & \longrightarrow \cdots . \end{array}$$

The complexes of object in $\mathcal{A}$ form an additive category denoted by $\operatorname{Com}(\mathcal{A})$. This is clearly an abelian category. The sequence $K^\bullet \to L^\bullet \to M^\bullet$ of complexes is exact if and only if for every n the sequence $K^n \to L^n \to M^n$ is exact. The short exact sequence

$$0 \to K^\bullet \to L^\bullet \to M^\bullet \to 0$$

induces a functorial long exact sequence of cohomology objects

$$\cdots \longrightarrow H^n(K^\bullet) \longrightarrow H^n(L^\bullet) \longrightarrow H^n(M^\bullet) \xrightarrow{\partial} H^{n+1}(K^\bullet) \longrightarrow \cdots .$$

The complex $K^\bullet$ is bounded if $K^n = 0$ for $n \ll 0$ and $n \gg 0$. We denote by $\operatorname{Com}^b(\mathcal{A})$ the category of bounded complexes.

The morphism $F^\bullet : K^\bullet \to L^\bullet$ is a **quasi-isomorphism** if all the induced maps $H^n(K^\bullet) \to H^n(L^\bullet)$ are isomorphisms in $\mathcal{A}$. We denote the class of quasi-isomorphisms by Qis. It is clear that the composition of quasi-isomorphisms is a quasi-isomorphism.

The derived category $D^b(\mathcal{A})$ we construct below is the localization of the category $K^b(\mathcal{A})$ with respect to the class of quasi-isomorphisms.

Two morphisms $f^\bullet, g^\bullet : K^\bullet \to L^\bullet$ are **homotopic** if there exists a collection of morphisms $h^n : K^n \to L^{n-1}$ $(n \in \mathbb{Z})$ such that

$$f^n - g^n = d^{n-1}_L h^n + h^{n+1} d^n_K$$

for all $n \in \mathbb{Z}$. The homotopy is an equivalence relation on the set of morphisms from $K^\bullet$ to $L^\bullet$. We denote it by $\sim$. One motivation for the definition of homotopy is the following theorem.

Lemma 12.3.4. *Let $f_1, f_2 : K^\bullet \to L^\bullet$ be two morphisms. Assume $f_1 \sim f_2$.*

(1) *For every morphism $g : L^\bullet \to M^\bullet$ we have $gf_1 \sim gf_2$,*

(2) *For every morphism $h : N^\bullet \to K^\bullet$ we have $f_1 h \sim f_2 h$.*

We define the **homotopic category** $K^b(\mathcal{A})$ to be the category whose objects are bounded complexes $K^\bullet$ of objects from $\mathcal{A}$ and whose morphisms are the equivalence classes of morphisms of complexes modulo homotopy.

For any $n \in \mathbb{Z}$ we define the shifted complex $K^\bullet[n]$ by setting $K^m[n] = K^{m+n}$, $d^m[n] = d^{m+n}$. For a morphism of complexes $f^\bullet : K^\bullet \to L^\bullet$ we define $f^\bullet[n] : K^\bullet[n] \to L^\bullet[n]$ by setting $f[n]^m = (-1)^n f^{m+n}$. The functor $T^n : \mathrm{Com}(\mathcal{A}) \to \mathrm{Com}(\mathcal{A})$ defined by $T^n(K^\bullet) = K^\bullet[n]$, $T^n(f^\bullet) = f^\bullet[n]$ is an automorphism of the category $\mathrm{Com}(\mathcal{A})$ which is called the **shift functor**.

Let $f^\bullet : K^\bullet \to L^\bullet$ be a morphism of complexes. The **cone** $C(f)$ of a morphism $f^\bullet$ is a complex with terms

$$C(f)^n = K^{n+1} \oplus L^n.$$

The differential is given in the matrix form

$$d^n_{C(f)} = \begin{pmatrix} d_K^{n+1} & 0 \\ f^{n+1} & d_L^n \end{pmatrix}.$$

The natural injections and projections define the exact sequence of complexes

$$0 \longrightarrow L^\bullet \xrightarrow{\ i\ } C(f)^\bullet \xrightarrow{\ p\ } K^\bullet[1] \longrightarrow 0\,.$$

The connecting homomorphism ∂^{i+1} in the long exact sequence

$$\cdots \longrightarrow H^i(L^\bullet) \longrightarrow H^i(C(f)^\bullet) \longrightarrow H^{i+1}(K^\bullet) \xrightarrow{\ \partial\ } H^{i+1}(L^\bullet) \longrightarrow \cdots$$

is equal to $H^{i+1}(f)$.

Let $K^\bullet, L^\bullet, f^\bullet$ be as above. The **cylinder** $\mathrm{Cyl}(f)$ is a complex with the terms

$$\mathrm{Cyl}(f)^n = K^n \oplus K^{n+1} \oplus L^n$$

with the differential given in the matrix form

$$d^n_{\mathrm{Cyl}(f)} = \begin{pmatrix} d_K^n & -1_{K^{n+1}} & 0 \\ 0 & -d_K^{n+1} & 0 \\ 0 & f^{n+1} & d_L^n \end{pmatrix}.$$

The natural injections and projections give the natural exact sequence of complexes

$$0 \longrightarrow K^\bullet \xrightarrow{\ j\ } \mathrm{Cyl}(f)^\bullet \xrightarrow{\ q\ } C(f)^\bullet \longrightarrow 0\,.$$

These sequences fit together in the following way.

Proposition 12.3.5. *For every morphism $f^\bullet : K^\bullet \to L^\bullet$ in* $\mathrm{Com}(\mathcal{A})$ *there exists a diagram with exact rows*

$$\begin{array}{ccccccccc}
 & & 0 & \longrightarrow & L^\bullet & \xrightarrow{i} & C(f)^\bullet & \xrightarrow{p} & K^\bullet[1] \longrightarrow 0 \\
 & & & & \downarrow{\scriptstyle s} & & \downarrow{\scriptstyle 1} & & \\
0 \longrightarrow & & K^\bullet & \xrightarrow{j} & \mathrm{Cyl}(f)^\bullet & \xrightarrow{q} & C(f)^\bullet & \longrightarrow & 0 \\
 & & \downarrow{\scriptstyle 1} & & \downarrow{\scriptstyle t} & & & & \\
 & & K^\bullet & \xrightarrow{f} & L^\bullet & & & &
\end{array}$$

which is functorial in $f^\bullet$. The morphisms s and t are quasi-isomorphisms, with $ts = 1_{L^\bullet}$. The composition st is homotopic to $1_{\mathrm{Cyl}(f)}$.

Proof. The morphisms s and t are defined by obvious projections and injections. The only nontrivial part of the proof is the definition of the homotopy between $1_{\mathrm{Cyl}(f)}$ and st. This is given in the matrix form by

$$h^n = \begin{pmatrix} 0 & 0 & 0 \\ 1_{K^n} & 0 & 0 \\ 0 & 0 & 0 \end{pmatrix}.$$

The reader can check all other statements without difficulty. □

We conclude this section by defining triangles in categories of complexes.

Definition 12.3.6. A triangle in the category $\mathrm{Com}^b(\mathcal{A})$ or $K^b(\mathcal{A})$ is a diagram

$$K^\bullet \xrightarrow{u} L^\bullet \xrightarrow{v} M^\bullet \xrightarrow{w} K^\bullet[1] .$$

A morphism of triangles is a diagram

$$\begin{array}{ccccccc}
K^\bullet & \xrightarrow{u} & L^\bullet & \xrightarrow{v} & M^\bullet & \xrightarrow{w} & K^\bullet[1] \\
\downarrow{\scriptstyle f^\bullet} & & \downarrow{\scriptstyle g^\bullet} & & \downarrow{\scriptstyle h^\bullet} & & \downarrow{\scriptstyle f^\bullet[1]} \\
K_1^\bullet & \xrightarrow{u_1} & L_1^\bullet & \xrightarrow{v_1} & M_1^\bullet & \xrightarrow{w_1} & K_1^\bullet[1] .
\end{array}$$

A **distinguished triangle** is a triangle which is quasi-isomorphic to the triangle defined by the cone and the cylinder of a morphism of complexes

$$K^\bullet \xrightarrow{j} \mathrm{Cyl}(f)^\bullet \xrightarrow{q} C(f)^\bullet \xrightarrow{p} K^\bullet[1]$$

with the maps taken from the diagram of Proposition 12.3.5.

Proposition 12.3.7. *Every short exact sequence of complexes in* $\mathrm{Com}^b(\mathcal{A})$ *is quasi-isomorphic to the middle row of the diagram from Prop.* 12.3.5.

Proof. Let

$$0 \longrightarrow K^\bullet \xrightarrow{f} L^\bullet \xrightarrow{g} M^\bullet \longrightarrow 0$$

be an exact sequence of complexes. We define the quasi-isomorphism of diagrams

$$\begin{array}{ccccccccc} 0 & \longrightarrow & K^\bullet & \xrightarrow{f} & L^\bullet & \xrightarrow{g} & M^\bullet & \longrightarrow & 0 \\ & & \uparrow{\scriptstyle 1_K} & & \uparrow{\scriptstyle t} & & \uparrow{\scriptstyle u} & & \\ 0 & \longrightarrow & K^\bullet & \xrightarrow{j} & \mathrm{Cyl}(f)^\bullet & \xrightarrow{q} & C(f)^\bullet & \longrightarrow & 0, \end{array}$$

where t is taken from the Proposition 12.3.5 and u is a composition of the projection on the second factor and g. The reader can check without difficulty that all diagrams commute. It remains to check that u is a quasi-isomorphism. The morphism g is an epimorphism of complexes, so u is also an epimorphism. We have

$$(\ker u)^n = (\ker t)^n = K^{n+1} \oplus K^n = K^n[1] \oplus K^n.$$

The differential of the complex $\ker u$ can be written in the matrix form

$$\begin{pmatrix} -d_{K[1]} & 0 \\ 1_{K[1]} & d_K \end{pmatrix}$$

so $\ker u$ can be identified with $C^\bullet(1_K)$. But this is an exact complex as can be seen from the long exact sequence of cohomology associated to the exact sequence which appears in the first row of Proposition 12.3.5 for $f^\bullet = 1_K$. This shows that u is a quasi-isomorphism. □

Definition 12.3.8. Let $\mathcal{C}$ be an arbitrary category and let S be a class of morphisms in $\mathcal{C}$. The localization of $\mathcal{C}$ with respect to S is a pair $(\mathcal{C}_S, q)$ where $\mathcal{C}_S$ is a category and $q : \mathcal{C} \to \mathcal{C}_S$ is a functor, satisfying the following universality property:

(1) for every $s \in S$ the morphism $q(s)$ is an isomorphism in $\mathcal{C}_S$,

(2) for every functor $f : \mathcal{C} \to \mathcal{D}$ such that $f(s)$ is an isomorphism in $\mathcal{D}$ for every $s \in S$ there exists a unique functor $f_S : \mathcal{C}_S \to \mathcal{D}$ such that $f = f_S q$.

Definition 12.3.9. Let $\mathcal{A}$ be an abelian category. The derived category $D^b(\mathcal{A})$ is the localization of the homotopic category $K^b(\mathcal{A})$ with respect to the class Qis of quasi-isomorphisms.

In [**44**], Chap. III it is proved that if $\mathcal{A}$ is an abelian category with enough of projective and injective objects, then the derived category $D^b(\mathcal{A})$ always exists.

12.4. The Derived Category of Quiver Representations

In this section we study the structure of the derived category of quiver representations. We show that every object in this category is represented by a stalk complex (i.e., a shifted module) and we make the objects, maps and distinguished triangles explicit. We prove that in this way we indeed get a triangulated category. The reader not acquainted with derived categories can take the results of this section as definitions.

We set $A = \mathbb{C}Q$ to be a path algebra of a quiver Q. We consider the abelian category $\mathcal{A} := A$–mod and the bounded derived category $D^b(\mathcal{A})$ of bounded complexes $K^\bullet$ with terms in $\mathcal{A}$. Recall that in $D^b(\mathcal{A})$ a quasi-isomorphism is an isomorphism.

We first analyze the structure of objects in the category $D^b(\mathcal{A})$. This structure is simple because the category A–mod is hereditary.

For any A-module V and for $n \in \mathbb{Z}$ we denote $V[n]$ the object in $D^b(\mathcal{A})$ represented by a complex $K^\bullet$ such that $K^n = V$ and $K^m = 0$ for $m \neq n$, with zero differential. We refer to the object $V[n]$ as to a **stalk complex**. The role of the objects $V[n]$ is explained by the following theorem.

Theorem 12.4.1. *Every indecomposable object in $D^b(\mathcal{A})$ is isomorphic to a stalk complex.*

Proof. In our case every $\mathbb{C}Q$-module has finite injective dimension, $D^b(\mathcal{A})$ is equivalent to the category $K^b(\mathcal{I})$ of complexes with injective terms (basically since every complex is quasi-isomorphic to its injective resolution); see [**44**], Chapter III.

Thus it is enough to show that an indecomposable bounded complex $I^\bullet$ of injective modules is quasi-isomorphic to a two term complex

$$\cdots \longrightarrow 0 \longrightarrow I^m \xrightarrow{d^m} I^{m+1} \longrightarrow 0 \longrightarrow \cdots$$

with the map d^m being surjective. Applying the shift we can assume that the complex $I^\bullet$ has a form

$$0 \longrightarrow I^0 \xrightarrow{d^0} I^1 \xrightarrow{d^1} \cdots \longrightarrow I^m \longrightarrow 0\,.$$

Consider the factorization

$$d^0 : I^0 \xrightarrow{g} X \xrightarrow{h} I^1\,,$$

where $X = \mathrm{im}(d^0)$, g is surjective and h is injective. Then the $\mathbb{C}Q$-module X is injective because our category is hereditary, and h is a section. This means we have an isomorphism of $\mathbb{C}Q$-modules

$$X \oplus C \xrightarrow{(h,u)} I^1 \qquad (12.20)$$

for some injective $\mathbb{C}Q$-module C. Since $hd^1 = 0$ we obtain an isomorphism of complexes $I^\bullet$ and the direct sum of complexes

$$\cdots 0 \longrightarrow 0 \oplus I^0 \xrightarrow{\begin{pmatrix} g & 0 \\ 0 & 0 \end{pmatrix}} X \oplus C \xrightarrow{\begin{pmatrix} 0 & 0 \\ 0 & ud^1 \end{pmatrix}} 0 \oplus I^2 \xrightarrow{\begin{pmatrix} 0 & 0 \\ 0 & d^2 \end{pmatrix}} 0 \oplus I^3 \longrightarrow \cdots$$

with the isomorphism given by identity maps and the map (12.20). Since $I^\bullet$ is indecomposable we conclude that one of the summands is zero. Since I^0 is nonzero by assumption, we see that the complex $I^\bullet$ is isomorphic to

$$0 \longrightarrow I^0 \xrightarrow{g} X \longrightarrow 0$$

and the theorem is proved. □

One also has a general fact about the morphisms in the derived category.

Proposition 12.4.2. *Let $\mathcal{A}$ be an abelian category with enough injective and enough projective objects. Let V, W be two objects in $\mathcal{A}$. Then*

$$\mathrm{Hom}_{D^b(\mathcal{A})}(V[m], W[n]) = \mathrm{Ext}^{m-n}_{\mathcal{A}}(V, W).$$

Proof. This is Remark III.5.4 b) in [**44**]. However, in our case of hereditary category this is very explicit. Consider the representations V and W with the injective resolution

$$0 \longrightarrow W \longrightarrow I^0 \longrightarrow I^1 \longrightarrow 0.$$

By applying the shift we can assume $n = 0$. The only meaningful cases are $m = 0$ and $m = 1$, as otherwise the maps from $V[0]$ to $I^\bullet$ are zero. In the case $m = 0$ we need to analyze the maps of stalk complexes which are just the maps of modules.

Consider the case $m = 1$. We need to analyze the maps of complexes

$$\begin{array}{ccccccc} 0 & \longrightarrow & 0 & \longrightarrow & V & \longrightarrow & 0 \\ & & \downarrow & & \downarrow f & & \\ 0 & \longrightarrow & I^0 & \xrightarrow{d^0} & I^1 & \longrightarrow & 0. \end{array}$$

To such a map we associate the extension

$$0 \longrightarrow W \longrightarrow E \longrightarrow V \longrightarrow 0,$$

where

$$E = I^0 \times_{I^1} V = \{(u, v) \in I^0 \times V \mid d^0(u) = f(v)\}.$$

It is routine to check that this gives the bijection

$$\mathrm{Hom}_{D^b(\mathcal{A})}(V[1], W) = \mathrm{Ext}^1_{\mathbb{C}Q}(V, W). \quad \square$$

The category $D^b(\mathbb{C}Q\text{–mod})$ is a triangulated category with the distinguished triangles coming from exact sequences

$$0 \longrightarrow U \xrightarrow{\ i\ } V \xrightarrow{\ p\ } W \longrightarrow 0$$

in $\mathbb{C}Q$–mod and the associated sequences

$$0 \longrightarrow V \longrightarrow W \longrightarrow U[1] \longrightarrow 0$$

represented by the sequence

$$0 \longrightarrow V \longrightarrow W \longrightarrow C(p) \longrightarrow 0,$$

and

$$0 \longrightarrow W \longrightarrow U[1] \longrightarrow V[1] \longrightarrow 0$$

represented by

$$0 \longrightarrow C(i) \longrightarrow U[1] \longrightarrow W[1] \longrightarrow 0$$

as well as all the shifts of these by $n \in \mathbb{Z}$.

Proposition 12.4.3. *The category $D^b(\mathbb{C}Q\text{–mod})$ with the distinguished triangles defined as above is a triangulated category.*

Proof. Checking the first three axioms is straightforward, and follows from the definitions. Let us look at the octahedral axiom. First let us notice that if any of the maps u, v, vu is zero, then one of the objects X, Y, Z has to be zero, because if a map in a distinguished triangle is zero, then either its domain or codomain is zero. Also, if Y is zero, then $vu = 0$ so X or Z is zero. So to cover these cases it is enough to consider $X = 0$ or $Z = 0$. If $X = 0$, then in the diagram of the octahedral axiom these maps from the second to third row are isomorphisms so the conclusion follows. If $Z = 0$, then we have $i' = f$, and the third row becomes the first column, so indeed it is a distinguished triangle.

Considering the cases when the maps u, v, vu are nonzero, we see that there are three possibilities:

(1) $X \xrightarrow{\ u\ } Y \xrightarrow{\ v\ } Z$ where $X, Y, Z \in \operatorname{Rep}(Q)$,

(2) $X \xrightarrow{\ u\ } Y \xrightarrow{\ v\ } Z[1]$ where $X, Y, Z \in \operatorname{Rep}(Q)$,

(3) $X \xrightarrow{\ u\ } Y[1] \xrightarrow{\ v\ } Z[1]$ where $X, Y, Z \in \operatorname{Rep}(Q)$.

In case (1) we have four possibilities. If both u and v are monomorphisms, then the octahedral axiom boils down to the obvious isomorphism $\operatorname{coker}(vu)/\operatorname{coker}(u) \cong \operatorname{coker}(v)$. Similarly, if both u and v are epimorphisms, the axiom reduces to $\ker(vu)/\ker(u) \cong \ker(v)$. The case of u being a proper epimorphism and v being a proper monomorphism is impossible because vu

is either a monomorphism or an epimorphism. In the remaining case of u being a monomorphism and v being an epimorphism, we have again two cases. Assume vu is a monomorphism. Then the diagram of octahedral axiom reduces to

$$\begin{array}{ccccc} X & \xrightarrow{1_X} & X & & \\ {\scriptstyle u}\downarrow & & \downarrow{\scriptstyle vu} & & \\ Y & \xrightarrow{v} & Z & \xrightarrow{j} & \ker(v)[1] \\ {\scriptstyle i}\downarrow & & \downarrow{\scriptstyle k} & & \downarrow{\scriptstyle \cong} \\ Y/X & \xrightarrow{f} & Z/X & \xrightarrow{g} & \ker(f)[1] \\ {\scriptstyle i'}\downarrow & & \downarrow{\scriptstyle k'} & & \\ X[1] & \xrightarrow{1_{X[1]}} & X[1] & & \end{array}$$

so the axiom is fulfilled. In the remaining case of vu being an epimorphism the diagram of octahedral axiom reduces to

$$\begin{array}{ccccc} X & \xrightarrow{1_X} & X & & \\ {\scriptstyle u}\downarrow & & \downarrow{\scriptstyle vu} & & \\ Y & \xrightarrow{v} & Z & \xrightarrow{j} & \ker(v)[1] \\ {\scriptstyle i}\downarrow & & \downarrow{\scriptstyle k} & & \downarrow{\scriptstyle \cong} \\ Y/X & \xrightarrow{f} & \ker(vu)[1] & \xrightarrow{g} & \ker(v)[1] \\ {\scriptstyle i'}\downarrow & & \downarrow{\scriptstyle k'} & & \\ X[1] & \xrightarrow{1_{X[1]}} & X[1] & & \end{array}$$

and again the axiom is fulfilled.

Let us look at case (2). We start with the morphisms

$$X \xrightarrow{u} Y \xrightarrow{v} Z[1] .$$

The second morphism is really an extension:

$$0 \longrightarrow Z \longrightarrow U \longrightarrow Y \longrightarrow 0 .$$

In the case when u is a monomorphism, it corresponds to an exact sequence:

$$0 \longrightarrow X \longrightarrow Y \longrightarrow W \longrightarrow 0 .$$

The diagram of octahedral axiom reduces to

$$\begin{array}{ccccc}
X & \xrightarrow{1_X} & X & & \\
\downarrow{\scriptstyle u} & & \downarrow{\scriptstyle vu} & & \\
Y & \xrightarrow{v} & (U \to Y) & \xrightarrow{j} & U[1] \\
\downarrow{\scriptstyle i} & & \downarrow{\scriptstyle k} & & \downarrow{\scriptstyle \cong} \\
W & \xrightarrow{f} & (U \to W) & \xrightarrow{g} & U[1] \\
\downarrow{\scriptstyle i'} & & \downarrow{\scriptstyle k'} & & \\
X[1] & \xrightarrow{1_{X[1]}} & X[1] & &
\end{array}$$

so the axiom is fulfilled.

If u is an epimorphism, so it corresponds to the exact sequence

$$0 \longrightarrow W \longrightarrow X \longrightarrow Y \longrightarrow 0\,,$$

the diagram of octahedral axiom corresponds to

$$\begin{array}{ccccc}
X & \xrightarrow{1_X} & X & & \\
\downarrow{\scriptstyle u} & & \downarrow{\scriptstyle vu} & & \\
Y & \xrightarrow{v} & (U \to Y) & \xrightarrow{j} & U[1] \\
\downarrow{\scriptstyle i} & & \downarrow{\scriptstyle k} & & \downarrow{\scriptstyle \cong} \\
(X \to Y) & \xrightarrow{f} & (X \oplus U \to Y) & \xrightarrow{g} & U[1] \\
\downarrow{\scriptstyle i'} & & \downarrow{\scriptstyle k'} & & \\
X[1] & \xrightarrow{1_{X[1]}} & X[1] & &
\end{array}$$

so again the axiom is fulfilled.

We leave case (3) to the reader. It is more convenient to shift by -1 and consider the case of the morphisms

$$X[-1] \xrightarrow{u} Y \xrightarrow{v} Z\,.$$

Then the proof is very similar to case (2).

Example 12.4.4. Let Q be a quiver:

$$1 \xrightarrow{a} 2 \xrightarrow{b} 3\,.$$

The category $D^b(\mathbb{C}Q)$ can be visualized like this.

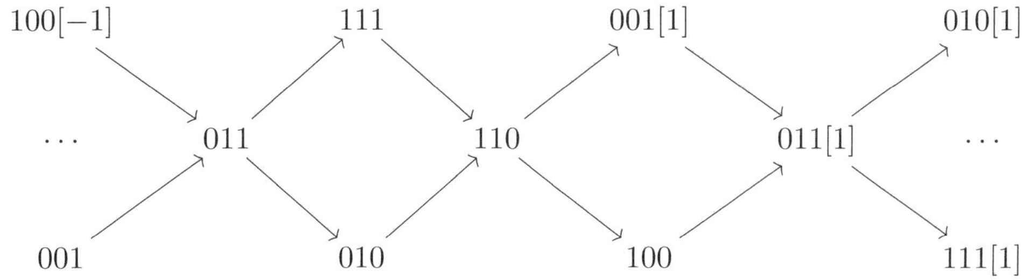

where $\alpha[i]$ denotes the stalk complex with the indecomposable of dimension vector α in the place i. The graph is **the Auslander-Reiten graph** of this derived category. Its vertices are the indecomposable objects and arrows are the irreducible morphisms. □

12.4.1. Orbit functors. In this section we assume that $\mathcal{A}$ is an abelian hereditary category with finite dimensional Hom and Ext^1 spaces. We can think of $\mathcal{A}$ as a subcategory of $D^b(\mathcal{A})$ by viewing a objects of $\mathcal{A}$ as a stalk complex in degree 0.

Definition 12.4.5. An auto-equivalence $G : D^b(\mathcal{A}) \to D^b(\mathcal{A})$ is an **orbit functor** if the following hold:

(1) G is an exact functor.

(2) For every indecomposable U in $\mathcal{A}$ there is only a finite number of $n \in \mathbb{Z}$ such that $G^n U \in \mathcal{A}$.

(3) There exists a positve integer N such that

$$\{U[n] \mid U \in \mathcal{A} \text{ indecomposable},\ n \in [-N, N]\}$$

represent the G-orbits in $D^b(\mathcal{A})$.

Definition 12.4.6. Let $G : D^b(\mathcal{A}) \to D^b(\mathcal{A})$ be an orbit functor. We define the **orbit category** $D^b(\mathcal{A})/G$ as follows:

(1) The objects in $D^b(\mathcal{A})/G$ are the G-orbits $\widetilde{X} = \{G^i X \mid i \in \mathbb{Z}\}$ of objects X from $D^b(\mathcal{A})$.

(2) We define

$$\mathrm{Hom}_{D^b(\mathcal{A})/G}(\widetilde{X}, \widetilde{Y}) = \bigoplus_{i \in \mathbb{Z}} \mathrm{Hom}_{D^b(\mathcal{A})}(G^i X, Y).$$

We have the following technical result which we do not prove in this book. The best reference is the paper [**56**].

Theorem 12.4.7 (Keller). *Let $\mathcal{A}$ be an Abelian hereditary category with finite dimensional* Hom *and* Ext^1 *spaces. Let G be an orbit functor on*

$D^b(\mathcal{A})$. The orbit category $D^b(\mathcal{A})/G$ is a triangulated category and the functor $\pi : D^b(\mathcal{A}) \to D^b(\mathcal{A})/G$ sending X to $\widetilde{X}$ is exact.

Remark 12.4.8. We have

$$\operatorname{Ext}^1_{D^b(\mathcal{A})/G}(\widetilde{X}, \widetilde{Y}) = \bigoplus_{i\in\mathbb{Z}} \operatorname{Ext}^1_{D^b(\mathcal{A})}(G^i X, Y).$$

We have the following easy consequences.

Proposition 12.4.9. *Let $\mathcal{D} = D^b(\mathcal{A})$, and let $G : \mathcal{D} \to \mathcal{D}$ be an orbit functor.*

(1) *The triangulated category $\mathcal{D}/G$ is a Krull-Schmidt category.*

(2) *The category $\mathcal{D}/G$ has Auslander-Reiten triangles induced by those in $\mathcal{D}$ and the Auslander-Reiten quiver of $\mathcal{D}/G$ is $\Gamma(\mathcal{D})/\phi(G)$ where $\Gamma(\mathcal{D})$ is an AR quiver for $\mathcal{D}$ and $\phi(G)$ is a graph isomorphism induced by G.*

(3) *We have the Serre duality formula*

$$\operatorname{Ext}^1_{\mathcal{D}/G}(\widetilde{X}, \widetilde{Y})^\star = \operatorname{Hom}_{\mathcal{D}/G}(\widetilde{Y}, \widetilde{\tau X})$$

which is functorial in $\widetilde{X}$ and $\widetilde{Y}$.

12.5. Cluster Categories

In this section Q is an acyclic quiver and $A = \mathbb{C}Q$. Let $\mathcal{A} = A$–mod. Consider the bounded derived category $D^b(\mathcal{A})$.

Definition 12.5.1. Consider the category $D^b(\mathcal{A})$ and its auto-equivalence $F := \tau^{-1}[1]$. The **cluster category** of the quiver Q is

$$\mathcal{C}Q := D^b(\mathcal{A})/F.$$

Proposition 12.5.2. *The category $\mathcal{C}Q$ has the following properties:*

(1) *$\mathcal{C}Q$ is a triangulated category.*

(2) *Each indecomposable object in $\mathcal{C}Q$ has a local ring of endomorphisms, so the Krull-Schmidt property holds in $\mathcal{C}Q$.*

(3) *The isomorphism classes of indecomposable objects are in bijection with the disjoint union of the set of isomorphic classes of indecomposable objects in $\mathcal{A}$ and the shifted indecomposable projective objects $P_x[1]$, $x \in Q_0$.*

Let $\mathcal{S}(Q)$ be a set of representatives of the indecomposable objects in $\mathcal{C}Q$.

We have

Proposition 12.5.3. *Let $X, Y \in \mathcal{S}(Q)$. Then:*

(1) $\mathrm{Hom}_{\mathcal{D}}(F^i X, Y) = 0$ *for all* $i \neq -1, 0$.

(2) *If X, Y does not lie on an oriented cycle in $\mathcal{D}$, then* $\mathrm{Hom}_{\mathcal{D}}(F^i X, Y) \neq 0$ *for at most one value of i.*

Example 12.5.4. Let Q be a quiver

$$1 \longrightarrow 2 \longrightarrow 3\ .$$

The category $\mathcal{C}(Q)$ can be visualized like this:

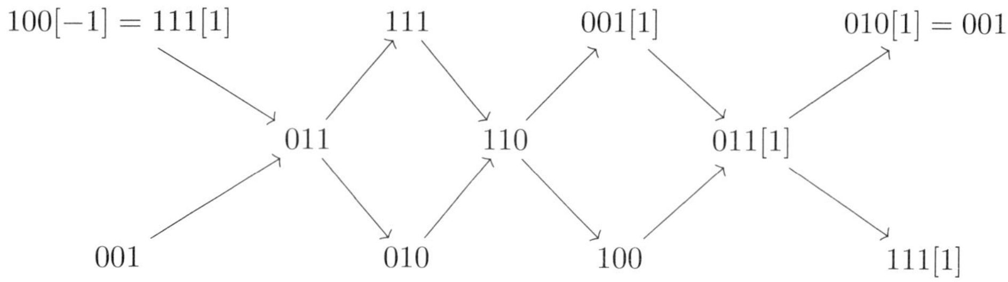

where $\alpha[i]$ denotes the stalk complex with the indecomposable of dimension vector α in the place i. The graph is the **Auslander-Reiten graph** of this cluster category. It has nine vertices which correspond to indecomposable objects (modules located in degree 0 and three shifted projectives), and the vertices are the irreducible maps.

We define **a tilting object in** $\mathcal{C}(Q)$ to be an object T with nonisomorphic indecomposable summands such that $\mathrm{Ext}^1_{\mathcal{C}(Q)}(T, T) = 0$. A tilting object is **a cluster** if it has n indecomposable direct summands. **A cluster complex** is a simplicial complex whose cells of dimension k consist of tilting objects in $\mathcal{C}(Q)$ with $k+1$ indecomposable summands.

Now we are ready to compare our approach with the approach via decorated representations.

Proposition 12.5.5. *Let Q be a Dynkin quiver. There is a natural bijection Θ between the indecomposable objects $\mathcal{S}(Q)$ in the cluster category and the set $\widetilde{\Phi}^+$ of almost positive roots. It takes a class of indecomposable module to its dimension vector, and a shifted projective (which is a minimal cover of a simple module) to the negative dimension of that simple module. Moreover, for two indecomposables $X, Y \in \mathcal{S}(Q)$ we have*

$$\dim \mathrm{Ext}^1_{\mathcal{C}(Q)}(X, Y) = (\Theta(X) || \Theta((Y))_Q.$$

Proof. This follows immediately from the description of Ext^1 in the cluster category and Definition 12.2.4. □

It follows that the compatible subsets of cardinality k correspond to tilting objects from $\mathcal{C}(Q)$. The cluster complex of $\mathcal{C}(Q)$ is isomorphic to the generalized associahedron Δ_Q.

Corollary 12.5.6. *Let Q be a Dynkin quiver with n vertices. Then the cluster complex $\mathcal{C}(Q)$ is a triangulation of the $(n-1)$-dimensional sphere.*

12.6. Cluster Tilted Algebras

Definition 12.6.1. Let Q be an acyclic quiver. For any tilting object X in the cluster category $\mathcal{C}(Q)$ we define the algebra

$$\mathcal{A}(X) := \mathrm{End}_{\mathcal{C}(Q)}(X).$$

The algebra $\mathcal{A}(X)$ is called a **cluster tilted algebra**.

We start with some examples.

Example 12.6.2. Take Q to be

$$1 \longrightarrow 2 \longrightarrow 3\ .$$

Its Auslander-Reiten quiver is

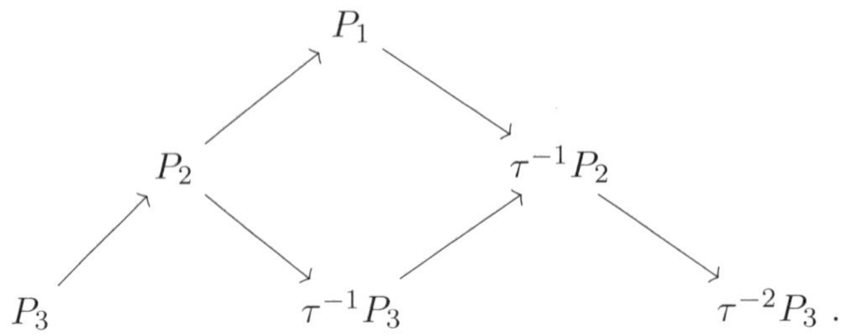

We define

$$X = X_1 \oplus X_2 \oplus X_3 := \tau^{-1}P_3 \oplus P_2 \oplus P_1.$$

Then one checks that $\mathrm{End}_{\mathcal{C}(Q)}(X) = \mathrm{End}_Q(X)$ and this algebra is isomorphic to the path algebra of the quiver

$$X_1 \longrightarrow X_2 \longleftarrow X_3\ .$$

This is the algebra $A(T)$ defined in Section 12.1.

Example 12.6.3. Let Q be as in the previous example, and

$$X = X_1 \oplus X_2 \oplus X_3 := P_3 \oplus P_1 \oplus \tau^{-2}P_3.$$

We have

$$\mathrm{Hom}_{\mathcal{C}(Q)}(X_1, X_2) = \mathrm{Hom}_Q(X_1, X_2) = K,$$
$$\mathrm{Hom}_{\mathcal{C}(Q)}(X_2, X_3) = \mathrm{Hom}_Q(X_2, X_3) = K.$$

We denote the generators of these spaces by a, b, respectively.

Let us look at $\mathrm{Hom}_{\mathcal{C}(Q)}(X_3, X_1)$. There is a nonzero homomorphism $F^{-1}X_3 = \tau[1](X_3) \to X_1$ in the derived category, it is a shift of the homomorphism $P_1 \to \tau^{-2}P_3$ in $\mathrm{AR}(Q)$. This gives a map $X_3 \to X_1$ in the cluster category and we denote it by c. So the algebra $\mathrm{End}_{\mathcal{C}(Q)}(X)$ is a 3-cycle quiver

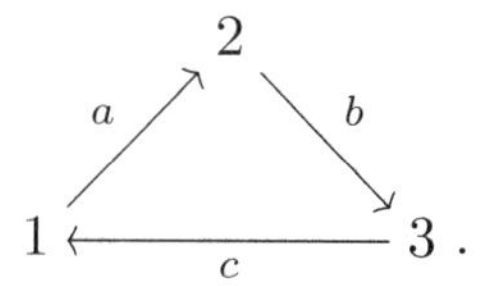

with the relations $ba = cb = ac = 0$. Again it is the algebra $A(T)$ defined in Section 12.1.

Example 12.6.4. Example 12.6.2 generalizes to the so-called sectional modules X. Their indecomposable direct summands are the connected subsets of $\mathrm{AR}(Q)$ which contain one indecomposable from each τ orbit. In such case the subquiver $\mathrm{AR}(Q)$ restricted to the vertices of X is a quiver Q' which is Q with a changed orientation. One checks easily that

$$\mathrm{End}_{\mathcal{C}(Q)}(X) = \mathrm{End}_Q(X) = \mathbb{C}(Q')^{\mathrm{op}}.$$

The examples above give an idea of what happens for the quiver of type $\mathbf{A}_n$. We use the combinatorial model from Section 12.1. For each triangulation T of the regular $(n+3)$-gon Π_{n+3} we defined the quivers $\mathrm{AR}(T)$ and finite dimensional algebras $A(T)$.

We have the following theorem of Caldero, Chapoton and Schiffler [**14**].

Theorem 12.6.5. *Let us consider the triangulation T of the regular $(n+3)$-gon Π_{n+3}. Let X be the corresponding tilting object in the cluster category $\mathcal{C}(\mathbf{A}_n)$. Then the algebra $\mathrm{End}_{\mathcal{C}(\mathbf{A}_n)}(X)$ is the path algebra of $A(T)$ defined in Section* 12.1.

Proof. Consider the cluster tilting object corresponding to the triangulation T of Π_{n+3}. Without loss of generality we can assume that we started with an equioriented quiver of type $\mathbf{A}_n$ whose vertices correspond to the diagonals $[1, i+2]$ $(1 \le i \le n)$, as in Exercise 12.2.1. The result follows from the combinatorial fact that for two diagonals $[i,j]$ and $[j,k]$ with a common vertex exactly one of the $\mathrm{Hom}_{\mathcal{C}(Q)}([i,j],[j,k])$ and $\mathrm{Hom}_{\mathcal{C}(Q)}([j,k],[i,j])$ is nonzero and one-dimensional, and this is the one for which the rotation from $[i,j]$ to $[j,k]$ is in the clockwise direction. This provides the generators of the Jacobson radical of the algebra $A(T)$ which are maps between pairs of sides in the triangles of our triangulation. The composition of two paths of length one inside of the internal triangle has to be zero, as the space of homomorphisms between these indecomposable objects is zero. □

The cluster tilted algebras can also described as a path algebras modulo relations that have a very special form. Suppose that Q is a quiver with oriented cycles. We define an equivalence relation $\equiv$ on the set of cycles which is generated by the relations

$$a_1 a_2 \cdots a_r \equiv a_2 a_3 \cdots a_r a_1$$

for all cyclic paths $a_1 a_2 \cdots a_r$. A **potential** for the quiver Q is a linear combination of equivalence classes of cyclic paths.

Example 12.6.6. An example of a potential for the quiver

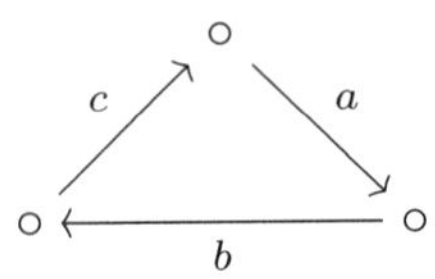

is $W = 2cba - 3bacbac = 2cba - 3cbacba$.

If $b \in Q_1$ is an arrow and $a_1 a_2 \cdots a_r$ is a cyclic path, then we define a partial derivative

$$\partial_b a_1 a_2 \cdots a_r = \sum_{\substack{i; \\ a_i = b}} a_{i+1} a_{i+2} \cdots a_r a_1 a_2 \cdots a_{i-1}$$

We can extend ∂_b to the set of potentials.

Example 12.6.7. In Example 12.6.6 we have

$$\begin{aligned} \partial_a W &= 2cb - 3cbacb - 3cbacb = 2cb - 6cbacb, \\ \partial_b W &= 2ac - 6acbac, \\ \partial_c W &= 2ba - 6bacba. \end{aligned}$$

Example 12.6.8. For the quiver

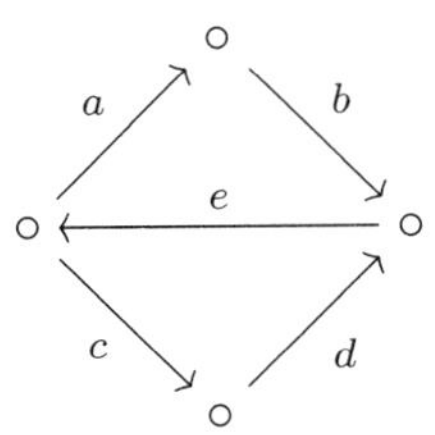

and the potential $W = edceba$ we get

$$\partial_e W = dceba + baedc.$$

For a potential W, we define the **Jacobian ideal** $J(W)$ as the (two-sided) ideal in the path algebra $\mathbb{C}Q$ generated by all partial derivatives $\partial_a W$ with $a \in Q_1$. The **Jacobian algebra** is the quotient $\mathbb{C}Q/J(W)$.

Theorem 12.6.5 allows to make a crucial observation about the structure of cluster tilting algebras. Let T again be a triangulation of the regular $(n+3)$-gon. Consider the quiver $\mathrm{AR}(T)$. Let S be the set of triangles, i.e., congruence classes of 3-cycles (with respect to the cyclic equivalence). Note that a given arrow can appear in at most one triangle in S. Consider the linear combination

$$W(T) := \sum_{cba \in S} cba.$$

The ideal $J(W(T))$ is exactly $J(T)$ and the Jacobian algebra $\mathbb{C}(Q(T))/J(W(T))$ is equal to the algebra $A(T)$.

We can restate Theorem 12.6.5 as follows.

Corollary 12.6.9. *The cluster tilting algebras* $\mathrm{End}_{\mathcal{C}(\mathbf{A}_n)}(X)$ *of the cluster tilting object* X *in the cluster category* $\mathcal{C}(\mathbf{A}_n)$ *is a Jacobian algebra of the quiver* $Q(T)$ *with the potential* $W(T)$.

It turns out that the cluster tilted algebras are Jacobian algebras for all cluster categories $\mathcal{C}(Q)$ where Q is any quiver without oriented cycles. This was proved in [**13**] by Buan, Marsh and Reiten. We recommend to the reader the paper [**2**] by Assem, Brüstle and Schiffler where they describe a relation of cluster tilted algebras to the tilted algebras, a class that was extensively studied in quiver representations.

We end this section with an example of a cluster tilted algebra of type D_n.

Example 12.6.10. Consider the quiver of type $\mathbf{D}_n$ with the following orientation of arrows:

$$\begin{array}{ccccccccccc} & & & & & & & & & & n-2 \\ & & & & & & & & & \swarrow & \\ 1 & \longrightarrow & 2 & \longrightarrow & \cdots & \longrightarrow & n-3 & \longrightarrow & n & & \\ & & & & & & & & & \nwarrow & \\ & & & & & & & & & & n-1\,. \end{array}$$

We consider the following tilting object. If $n = 2k$ is even, then

$$T(2k) = P_{n-2} \oplus \tau^{-1} P_{n-1} \oplus \tau^{-2} P_{n-2} \oplus \tau^{-3} P_{n-1} \oplus \dots \\ \oplus \tau^{-2(k-1)} P_{n-2} \oplus \tau^{-2k+1} P_{n-1}$$

and when $n = 2k+1$ is odd, we put

$$T(2k+1) = P_{n-2} \oplus \tau^{-1}P_{n-1} \oplus \tau^{-2}P_{n-2} \oplus \tau^{-3}P_{n-1} \oplus \dots \\ \oplus \tau^{-2(k-1)}P_{n-2} \oplus \tau^{-2k+1}P_{n-1} \oplus \tau^{-2k}P_{n-2}.$$

In both cases we denote the summands of $T(n)$ in the order listed above by $T_i(n)$ $(1 \leq i \leq n)$.

We claim that $T(n)$ is a tilting object and that its endomorphism algebra $\mathrm{End}_{\mathcal{C}(\mathbf{D}_n)}(T(n))$ is isomorphic to the following Jacobian algebra $\mathbb{C}Q/J(W)$ where Q is the quiver

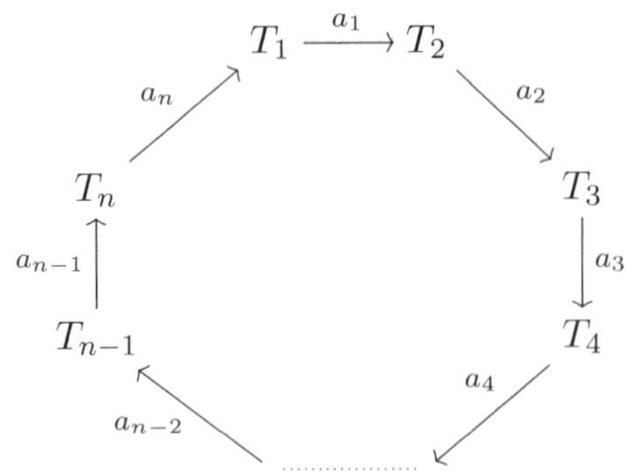

and W is the potential $a_n a_{n-1} \dots a_2 a_1$. In order to prove our statement we need to show that the homomorphism space from $T_i(n)$ to $T_j(n)$ has dimension 1 unless $j = i-1$ or $i = 1, j = n$ when it is equal to zero. Using the fact that cluster category does not depend on the orientation of the quiver we reduce to the case of the orientation we used, and we need to show that $\mathrm{Hom}_{\mathbb{C}Q}(T_1(n), T_i(n)) = \mathbb{C}$ unless $i = n$ when it is zero. This follows from the description of the Hom spaces for Dynkin quivers using the Auslander-Reiten quiver.

12.7. Bibliographical Remarks

The cluster categories became very important since their introduction in [**12**] when they became the first examples of categorification of cluster algebras which then served as models in similar situations arising from representation theory.

The combinatorial approach to clusters for type $\mathbf{A}_n$ is based on the paper [**14**] by Caldero, Chapoton and Schiffler. The second approach by almost positive roots is based on the paper [**64**] by Marsh, Reineke and Zelevinsky. These approaches seemed to be more suitable as an introduction instead of going right away to derived categories. We were trying to avoid the full discussion of derived categories, so we just included the proof of the fact that every object of derived category of quiver representations is represented by a stalk complex and that the cluster category is a triangulated category.

The reader interested in the subject should go through the famous paper [**12**] as well as the basic papers on cluster tilted algebras [**13**] and [**2**] where the connection to tilted algebras is clarified.

The cluster complexes can also be defined via the presentations of quiver representations; see [**51, 22**].

Another direction is related to cluster algebras themselves. From representations of Jacobian algebras one can construct cluster variables in the corresponding cluster algebra, using quiver Grassmannians (the fibers of the projection from Schofield incidence varieties to $\mathrm{Rep}_\alpha(Q)$ discussed in the chapter on semi-invariants).

This idea is due to Caldero and Chapoton and it was generalized to the acyclic cluster algebras, and the cluster algebras associated to triangulations of surfaces developed by Fomin, Shapiro and Thurston [**42**].

Notation

Index

Bibliography

[1] Abeasis, S., Del Fra, A., Kraft, H., The geometry of representations of A_m, Math. Ann. 256 (1981), 401–418.

[2] Assem, I., Brüstle, T., Schiffler, R., Cluster-tilted algebras as trivial extensions, Bull. London Math. Soc., 40 (2008), 151–162.

[3] Auslander, M., Reiten, I., Smalø, S., Representation Theory of Artin Algebras, Cambridge Studies in Advanced Mathematics, 36, Cambridge University Press, Cambridge, New York, 1995.

[4] Auslander, M., Reiten, I., Representation theory of Artin algebras III, Almost split sequences, Comm. in Algebra, 3 (1975), 269–310.

[5] Auslander, M., Reiten, I., Representation theory of Artin algebras IV, Invariants given by almost split sequences, Comm. in Algebra , 5 (1977), 443–518.

[6] Auslander, M., Reiten, I., Modules determined by their composition factors, Ill. Journal of Math., 29 (1985), 280–301.

[7] Belkale, P., Geometric proof of a conjecture by Fulton, Adv. in Math., 216 (2007), 346–357.

[8] Bernstein, I.N., Gelfand, I.M., Ponomarev, V.A., Coxeter functors and Gabriel's theorem, Russian Math. Surveys, 28 (1973), 17–32.

[9] Bobiński, G., Zwara, G., Normality of orbit closures for Dynkin quivers of type A_n, Manuscripta Math., 105 (2001), no. 1, 103–109.

[10] Bobiński, G., Zwara, G., Schubert varieties and representations of Dynkin quivers, Colloq. Math., 94 (2002), no. 2, 285–309.

[11] Brion, M., Multiplicity-free sub varieties of flag varieties, Commutative Algebra (Grenoble/Lyon, 2001), 13-23, Contemp. Math 331, Amer. Math. Soc., Providence, RI, 2003.

[12] Buan, A.B., Marsh, R., Reineke, M., Reiten, I., Todorov, G., Tilting theory and cluster combinatorics, Adv. in Math., 204 (2006), no. 2, 572–618.

[13] Buan, A.B., Marsh, R., Reiten, I., Cluster-tilted algebras, Trans. Amer. Math. Soc., 359 (2007), 323–332.

[14] Caldero, Ph., Chapoton, F., Schiffler, R., Quivers with relations arising from clusters (A_n case), Trans. Amer. Math.Soc., 358, no. 3, (2006), 359–376.

[15] Chindris, C., Derksen, H., Weyman, J., Counterexamples to Okounkov's log-concavity conjecture, Compos. Math. 143 (2007), no. 6, 1545–1557.

[16] Crawley-Boevey, W., Exceptional sequences of representations of quivers, Representations of algebras (Ottawa, ON 1992), CMS Conf. Proc., 14, Amer. Math. Soc., Providence, RI, 1993.

[17] Crawley-Boevey, W., Subrepresentations of general representations of quivers, Bull. London Math. Soc., 28 (1996), 363–366.

[18] Crawley-Boevey, W., Lectures on Representations of Quivers, preprint available from Crawley-Boevey's home page.

[19] Crawley-Boevey, W., Holland, M.P., Non-commutative deformations of Kleinian singularities, Duke Math. J., 92 (1998), 605–635.

[20] Crawley-Boevey, W., Van den Bergh, M., Absolutely indecomposable representations and Kac-Moody Lie algebras, Invent. Math. 155 (2004), no. 3, 537–559.

[21] Carroll, A., Weyman, J., Noncommutative Birational Geometry, Representations and Combinatorics, "Contemporary Mathematics" , Proceeding of the Special AMS session "Noncommutative birational geometry and cluster algebras" held during the Annual meeting of the AMS (Boston, 2012), edited by A. Berenstein, V. Retakh, 111–136, AMS, 2013.

[22] Derksen, H., Fei, J., General presentations of algebras, Adv. Math. 278 (2015), 210–237.

[23] Derksen, H., Makam, V., Polynomial degree bounds for matrix semi-invariants, Adv. Math. 310 (2017), 44-63.

[24] Derksen, H, Makam, V., Generating invariant rings of quivers in arbitrary characteristic, J. Algebra 489 (2017), 435–445.

[25] Derksen, H. Weyman, J. Semi-invariants of quivers and saturation for Littlewood-Richardson coefficients, J. Amer. Math. Soc. 13 (2000), no. 3, 467–479.

[26] Derksen, H., Weyman, J., On the canonical decomposition of quiver representations, Compositio Math. 133 (2002), no. 3, 245–256.

[27] Derksen, H. Weyman, J., On the Littlewood-Richardson polynomials, J. Algebra 255 (2002), no. 2, 247–257.

[28] Derksen, H., Weyman, J., Semi-invariants for quivers with relations, Special issue in celebration of Claudio Procesi's 60-th birthday, J. Algebra 258 (2002), no. 1, 216–227.

[29] Derksen, H., Weyman, J., Generalized quivers associated to reductive groups, Colloq. Math. 94 (2002), no. 2, 151–173.

[30] Derksen, H., Weyman, J., The combinatorics of quiver representations, Ann. Inst. Fourier (Grenoble) 61 (2011), no. 3, 1061–1131.

[31] Derksen, H. Schofield, A., Weyman, J., On the number of sub representations of a general quiver representation, J. London. Math. Sco. (2) 76 (2007), no. 1, 135–147.

[32] Derksen, H., Weyman, J., Zelevinsky, A., Quivers with potential and their representations I. Mutations, Selecta Math. (N.S.) 14 (2008), no. 1, 59–119.

[33] Derksen, H., Weyman, J., Zelevinsky, A., Quivers with potential and their representations II. applications to cluster algebras, J. Amer. Math. Soc. 23 (2010), no. 3, 749–790.

[34] Dlab, V., Ringel, M., Indecomposable Representations of Graphs and Algebras, Mem. Amer. Math. Soc., 173, 1976.

[35] Domokos, M., Relative invariants for representations of finite dimensional algebras, Manuscripta Math., 108 (2002), 123–133.

[36] Domokos, M., Zubkov, A.N., Semi-invariants of quivers as determinants, Transformation Groups 6 (2001), No. 1, 9-24.

[37] Donovan, P., Freislich, M.R., The Representation Theory of Finite Graphs and Associative Algebras, Carleton Lecture Notes, Ottawa 1973.

[38] Assem, I., Simson, D., Skowroński, A., Elements of the Representation Theory of Associative Algebras, London Mathematical Society, Student Texts 65, Cambridge University Press, 2006.

[39] Fei, J., Cluster algebras, invariant theory, and Kronecker coefficients I, Adv. Math. 310 (2017), 1064–1112.

[40] Fei, J. Cluster algebras and semi-invariant rings II: projections, Math. Z. 285 (2017), no. 3–4, 939–966.

[41] Fei, J., The upper cluster algebras of IART quivers I. Dynkin, arXiv 1603.02521.

[42] Fomin, S., Shapiro, M., Thurston, D., Cluster algebras and triangulated surfaces, I. Cluster complexes, Acta Math., 201 (2008), no 1, 83–146.

[43] Gabriel, P., Roiter, A.V., Representations of finite dimensional algebras, Encyclopaedia of Mathematical Sciences, no. 73, Springer-Verlag, 2nd edition, 1997.

[44] Gelfand, S.I., Manin, Y.I., Methods of Homological Algebra, Springer Monographs in Mathematics, Springer, 2003.

[45] Goodman, R., Wallach, N.R., Representations and Invariants of Classical Groups, Cambridge University Press, 1998,

[46] Gorodentsev, A.L., Rudakov, A.N., Exceptional vector bundles on projective spaces, Duke Math. J., 54 (1987), no. 1, 115–130.

[47] Happel, Triangulated categories in the Representation Theory of Finite Dimensional Algebras, London Mathematical Society Lecture Notes Series 119, Cambridge University Press, 1988.

[48] Hausel, T, Letellier, E., Rodrigues-Villegas, F., Positivity of Kac polynomials and DT-invariants of quivers, Annals of Mathematics (2) 177 (2013), no. 3, 1147–1168.

[49] Hille, L., On the volume of a tilting module, Abh. Math. Sem. Univ. Hamburg 76 (2006), 261–277.

[50] Humphreys, J., Linear Algebraic Groups, Graduate Texts in Mahematics, vol. 21, Springer, 1975.

[51] Igusa, K., Orr, K., Todorov, G., Weyman, J., Cluster complexes via semi-invariants, Compositio Mathematica 145, no. 4 (2009), 1001–1034.

[52] Ikenmeyer, Ch., Small Littlewood-Richardson coefficients, J. Algebraic Comb. 44 (2016), no. 1, 1–29; Erratum, J. Algebraic Comb. 44 (2016), no. 1, 31–32.

[53] Kac, V., Infinite root systems, representations of graphs and invariant theory, Invent. Math. 56 (1980), no. 1, 57–92.

[54] Kac, V., Infinite root systems, representations of graphs and invariant theory II, J. Algebra 78 (1982), no. 1, 141–162.

[55] Kac, V., Root systems, representations of quivers and invariant theory, in: Invariant Theory (Montecatini 1982), Lecture Notes in Mathematics, 996, Springer, Berlin, 1983).

[56] Keller, B., Triangulated orbit categories, Documenta Mathematica 10 (2005), 551-581, arXiv 0503240.

[57] King, A.D., Moduli of representations of finite-dimensional algebras, Quart. J. Math. Oxford Ser (2) 45 (1994), no, 180, 515–530.

[58] King, R., Tollu, Ch., Toumazet, F., Stretched Littlewood-Richardson polynomials and Kostka coefficients, CRM Proceedings and Lecture Notes 34 (2004), 99–112.

[59] Kleiner, M., The graded preprojective algebra of a quiver, Bull. London Math. Soc. 36 (2004), no. 1, 13–22.

[60] Kraft, H., Geometrische Methoden in der Invariantentheorie, Aspects of Mathematics, D1., Fried. Viehweg & Sohn, Braunschweig, 1984.

[61] Lakshmibai, V., Magyar, P., Degeneracy schemes, quiver schemes, and Schubert varieties, Internat. Math. Res. Notices 1998, no. 12, 627–640.

[62] Le Bruyn, L., Procesi, C., Semisimple representations of quivers, Trans. Amer. Mah.Soc. 317 (1990), no. 2, 585–598.

[63] MacDonald, I.G., Symmetric functions and Hall polynomials, 2nd edition, Oxford Mathematical Monographs, Oxford University Press, 1995.

[64] Marsh, R., Reineke, M., Zelevinsky, A., Generalized associahedra via quiver representations, Trans. Amer. Math. Soc, 355 (2003), 4171–4186.

[65] Mumford, D., Fogarty, J., Kirwan, F., Geometric Invariant Theory. Third edition. Results in Mathematics and Related res (2), 34, Springer-Verlag, Berlin, 1994.

[66] Nazarova, L.A., Representations of quivers of infinite type, Izv. Akad. Nauk SSSR, Ser. Mat. 37 (1973), 752–791.

[67] Obaid, M., Nauman, K., Al-Shammakh, W.S.M., Fakieh, W., Ringel, C.M., The number of complete exceptional sequences for a Dynkin algebra, Colloq. Math. 133 (2013), no. 2, 197–210.

[68] Ringel, C.M., The braid group action on the set of exceptional sequences of a hereditary Artin algebra, in: Contemporary Mathematics volume 171, 1994, 339–352.

[69] Ringel, C.M., Tame Algebras and Integral Quadratic Forms, Lecture Notes in Mathematics, 1099, 1984.

[70] Rotman, J., An Introduction to Homological Algebra, 2-nd edition, Universitext, No. 223, Springer, 2009.

[71] Sato, M., Kimura, T., A classification of irreducible prehomogeneous vector spaces and their relative invariants, Nagoya Math. J. 65 (1977), 1–155.

[72] Schofield, A., Semi-invariants of quivers, J. London Math. Soc. (2) (1991), no. 3, 385–395.

[73] Schofield, A., General representations of quivers, Proc. London Math. Soc. (3) 65 (1992), no. 1, 46–64.

[74] Schofield, A., The internal structure of real Schur representations, preprint, 1988.

[75] Schofield, A., Van den Bergh, M., Semi-invariants of quivers for arbitrary dimension vectors, Indag. Math. (N.S.) 12 (2001), no. 1, 125–138.

[76] Sherman, C., Quiver generalization of a conjecture of King, Tollu and Toumazet, J. Algebra 480 (2017), 487–504.

[77] Skowroński, A., Weyman, J., The algebras of semi-invariants of quivers, Transform. Groups 5 (2000), no. 4, 361–402.

[78] Zwara, G., Smooth morphisms of module schemes, Proc. London Math. Soc., (3) 84 (2002), no. 3, 539–558.